D1030007

The Fluoride Deception

CHRISTOPHER BRYSON

SEVEN STORIES PRESS
New York • London • Toronto • Melbourne

A Seven Stories Press First Edition

Seven Stories Press
140 Watts Street
New York, NY 10013
http://www.sevenstories.com

IN CANADA
Hushion House, 36 Northline Road, Toronto, Ontario M4B 3E2

IN AUSTRALIA
Palgrave Macmillan, 627 Chapel Street, South Yarra VIC 3141

LIBRARY OF CONGRESS CATALOGING-IN-PUBLICATION DATA
Bryson, Christopher, 1960–
The fluoride deception / Christopher Bryson.— 1st ed.
p. cm.
ISBN 1-58322-526-9 (alk. paper)
1. Fluorine—Toxicology. 2. Water—Fluoridation. 3. Fluorosis. I. Title.
RA1231.F55 B795 2004
615.9'25731—dc22
2003027046

College professors may order examination copies of Seven Stories Press titles for a free six-month trial period. To order, visit www.sevenstories.com/textbook, or fax on school letterhead to 212.226.1411.

Jacket design by Jess Morphew
Text design by India Amos

Printed in the USA

9 8 7 6 5 4 3 2 1

For my father

Toute entreprise humaine, fût-elle industrielle, est susceptible de perfectionnement!

Inscription on memorial to the sixty dead of the 1930 Meuse Valley disaster

It is not just a mistake for public health agencies to cooperate and collaborate with industries investigating and deciding whether public health is endangered—it is a direct abrogation of the duties and responsibilities of those public health organizations.

Scientist Clair Patterson to the U.S. Senate, 1966[1]

If you ain't thinking about Man, God and Law, you ain't thinking about nothin'.

Joe Strummer (1952–2000)

Contents

Foreword

THEO COLBORN

THE QUESTION OF whether fluoride is or is not an essential element is debatable. In other words, is the element, fluorine, required for normal growth and reproduction? On one hand there appears to be a narrow range of topical exposure in which it might prevent cavities. But if exposure is too high, it causes serious health problems. And could an individual who is totally deprived of fluoride from conception through adulthood survive? Definitive research to resolve these questions has never appeared in the public record or in peer-reviewed journals. It is important to keep this fact in mind as you read this book.

Chris Bryson informs us that fluorine is, indeed, an essential element in the production of the atom bomb, and there is good reason to believe that fluoridated drinking water and toothpaste—and the development of the atom bomb—are closely related. This claim sounded pretty far-fetched to me, and consequently I was extremely skeptical about the connection when I started reading the book. Bryson writes with the skill of a top-selling novelist, but it was *not* his convincing storytelling that made me finish the book. It was the haunting message that possibly here again was another therapeutic agent, fluoride, that had not been thoroughly studied before it was foisted on the public as a panacea to protect or improve health. Bryson reveals that the safety of fluoride became a firmly established paradigm based on incomplete knowledge. The correct questions were never asked (or never answered when they were asked), thus giving birth to false or bottomless assumptions that fluoride was therapeutic and safe. Certainly, the evidence Bryson unearthed in this book begs for immediate attention by those responsible for public health.

As the story unfolds, Bryson weaves pieces of what at first appears to be totally unrelated evidence into a tapestry of intrigue, greed, collusion, personal aggrandizement, corporate and government cover-up, and U.S. Public Health Service (USPHS) mistakes. While reading the book, I kept thinking back to 1950, three years after I got my BS degree in Pharmacy and the year I gave birth to my first child. Fluoride came on the market packaged in pediatric vitamin drops for infants. Mothers left the hospital with their new babies in their arms and prescriptions in their hands from their dismissing physicians for these fluoride-laced drops. About that time communities around the country began to add fluoride to their drinking water. The promised benefits of fluoride were so positive that my dentist friends began to wish that they had chosen dermatology instead of dentistry. At that same time pregnant women were being given a pharmaceutical, diethylstilbestrol (DES), to prevent miscarriages, as well as DES-laced prescription vitamins especially designed for pregnant women to produce big, fat, healthy babies. I felt good when I dispensed the fluoride and DES prescriptions—*they were products designed to prevent health problems rather than treat them.* Now I can only wonder how many children were harmed because I and others like me took the word of the National Institutes of Health (NIH), the USPHS, and the major pharmaceutical companies producing these products. We were caught up in the spin. We were blind to the corporate hubris and were swept along with the blissful enthusiasm that accompanies every new advance in modern technology and medicine.

The hazards posed by prenatal exposure to DES surfaced a lot sooner than those posed by fluoride. And although by 1958 it was discovered that DES caused a rare vaginal cancer that until that time had been found only in postmenopausal women, its use during pregnancy was not banned until 1971—thirteen years later. Even this year, 2003, new discoveries are being reported about the impact on health in the sons and daughters of the DES mothers, and now in their grandchildren. It is estimated that in the United States alone there are ten million daughters and sons. In comparison to DES, where exposure could be traced through prescription records, the extent of exposure to fluorides through drinking water, dental products, vitamins, and as Bryson points out, through Teflon, Scotchgard, Stainmaster, and other industrial and agricultural fluorinated products is practically unmeasurable.

Certainly the evidence Bryson presents in this book should cause those charged with protecting public health to demand answers about the developmental, reproductive, and functional role of fluorine in all living organisms. A lack of data on the safety of a product is not proof of safety. Evidence has only recently surfaced that prenatal exposure to certain fluorinated chemicals is dangerous, often fatal at high doses, and that—even at extremely low levels—such exposure can undermine the development of the brain, the thyroid, and the metabolic system. This evidence surfaced because industrial fluorine chemicals were suddenly being discovered in human and wildlife tissue everywhere they were looked for on earth. As a result, the U.S. Environmental Protection Agency (EPA) began to press the manufacturers of these products for data on their safety. It is no wonder that such chemicals never made it on the list of known endocrine disrupters, chemicals that undermine development and function. The studies were never done, or if they were, they were not available to the public. It is time that these chemicals, at the cumulative concentrations they are found in the environment, be tested thoroughly for their developmental, reproductive, and endocrine effects.

Whether or not Bryson's nuclear-bomb connection is ever confirmed without a doubt, this book demonstrates that there is still much that needs to be considered about the continued use of fluorine in future production and technology. The nuclear product that required the use of fluorine ultimately killed 65,000 people outright in one sortie over Japan. The actual number of others since then and in generations to come who will have had their health insidiously undermined by artificial exposure to fluorides and other fluorine chemicals with half-lives estimated in geologic time may well exceed that of the atom bomb victims millions and millions of times over.

Dr. Theo Colborn, coauthor of *Our Stolen Future: Are We Threatening Our Fertility, Intelligence, and Survival? A Scientific Detective Story* (1996)

Note on Terminology

THE TERMS *fluorine* and *fluoride* should not be confused in a book about chemical toxicity. Fluorine is an element, one of our planet's building blocks, an especially tiny atom that sits at the summit of the periodic table. Its lordly location denotes an unmatched chemical potency that is a consequence of its size and structure. The nine positively charged protons at the atom's core get little protection from a skimpy miniskirt of electrons. As a result, fluorine atoms are unbalanced and dangerous predators, snatching electrons from other elements to relieve their core tension. (A ravenous hunger for electrons explains why fluorine cuts through steel like butter, burns asbestos, and reacts violently with most organic material.)[1]

Mercifully, Mother Nature keeps fluorine under lock and key. Because of its extreme reactivity, fluorine is usually bound with other elements. These compounds are known as salts, or *fluorides*, the same stuff that they put in toothpaste. Yet the chemical potency of fluorides is also dramatic. Armed with a captured electron, the toxicity of the negatively charged fluoride ion now comes, in part, from its tiny size. (*Ionic* means having captured or surrendered an electron). Like a midget submarine in a harbor full of battleships, fluoride ions can get close to big molecules—like proteins or DNA—where their negative charge packs a mighty wallop that can wreak havoc, forming powerful bonds with hydrogen, and interfering with the normal fabric of such biological molecules.[2]

However—and please stay with me here, I *promise* it gets easier—somewhat confusingly, the words *fluorine* and *fluoride* are sometimes used interchangeably. A fluoride compound is often referred to, generically, as fluorine. (For example, the Fluorine Lawyers Committee was a group of corporate attorneys concerned about the medical and legal dangers from a great range of different industrial "fluorides" spilling from company smokestacks.)

In these pages I've tried to be clear when I'm referring to the element fluorine or to a compound, a fluoride. And because different fluoride compounds often have unique toxicities, where relevant or possible, I have also given the compound's specific name. Mostly, however, for simplicity's sake, I have followed convention and used the shorthand *fluoride* when referring to the element and its multiple manifestations, a procedure approved and used by the U.S. National Academy of Sciences.[3]

Acknowledgments

This book owes a debt of gratitude to many. First is my wife, Molly, whose love and encouragement pushed me to the starting line and carried me across the finish. My first encounter with fluoride came as a BBC radio journalist working in New York in 1993, when I was asked to find an "American angle" on water fluoridation. Ralph Nader put me in touch with scientists at the U.S. Environmental Protection Agency who opposed fluoridation.[1] As I followed that story, I met the medical writer Joel Griffiths, whose meticulous reporting became the foundation for several chapters in this book. His humor, poetry, and friendship made writing the book a simple pleasure. Librarians are foot soldiers of democracy, and a legion of them sacked archives for me from Tennessee to Washington State and from Denmark to London. Everywhere I was met with eager help digging out dusty files and courteous answers to the most foolish of questions. Special thanks to my favorite Metallica fan, Billie Broaddus, at the University of Cincinnati Medical Heritage Center, Marjorie Ciarlante at the National Archives in Washington, DC, and Donald Jerne at the Danish National Library of Science and Medicine. The book's spine is the authority of the many workers, scientists, and public officials who gave so freely of their time. Particular gratitude to Albert Burgstahler of the University of Kansas, the EPA's J. William Hirzy, Robert J. Carton, Phyllis J. Mullenix, Kathleen M. Thiessen of SENES Oak Ridge Inc., and Robert F. Phalen of the University of California at Irvine, who each spent long hours reviewing documents and medical studies for me.

I had the good fortune to serve an apprenticeship in the 1980s with the late Jonathan Kwitny, one of the nation's top investigative reporters. From his hospital bed, weak from radiation treatment, he encouraged me. "This is your book," he said. I was helped with financial support from the Fund for Investigative Journalism, Inc., and the Institute for Public Affairs. A bouquet to Dan Simon at Seven Stories Press, who clapped his hands in glee when told he'd be taking on the great industrial trusts of America. Special thanks to Lexy Bloom and Ruth Hein for their critical and conscientious editing; to George Mürer, Anna Lui, Chris Peterson, and India Amos for wrestling this octopus to the printer; and to the entire staff at Seven Stories Press for their passion and commitment.

Many helped in myriad other ways. This book is theirs, too. Gwen Jaworzyn, Janet Michel, Bette Hileman, *USA Today* and Peter Eisler, George Mavridis, Felicity Bryson and Vincent Gerin, Ruth Miller at the Donora Historical Society, Anne-Lise Gotzsche, Barbara Griffiths, Anthony and Nancy Thompson and family, Basil and Anne Henderson, Joan-Ellen and Alex Zucker, Nina and David Altschiller, Bill and Janey Murtha, Tom Webster, Naomi Flack, Ken Case, Bob Woffinden, Traude Sadtler, Gordon Thompson, Clifford and Russ Honicker, Jacqueline O. Kittrell, Ellie Rudolph, Robert Hall, Martha Bevis, John Marks, Chris Trepal, Carol Patton, Gar Smith at *Earth Island Journal*, Lennart Krook, Danny Moses at Sierra Club Books, Andreas Schuld, Erwin Rose and family, Roberta Baskin, the Connett family, Colin Beavan, Sam Roe, Karin and Hans Hendrik Roholm, Eleanor Krinsky, Allen Kline, Bill and Gladys Shempp (who put me up in their home in Donora one night), Elizabeth Ramsey, Lynne Page Snyder, and Peter Meiers, whom I never met nor spoke with but whose splendid research led me to the papers of Charles F. Kettering.

Thank you all.

Introduction

A Clear and Present Danger

Warning: Keep out of reach of children under 6 years of age. If you accidentally swallow more than used for brushing, get medical help or contact a Poison Control Center right away.

NEXT TIME YOU confront yourself in the bathroom mirror, mouth full of foam, take another look at that toothpaste tube. Most of us associate fluoride with the humdrum issue of better teeth and the promised fewer visits to the dentist. Yet the story of how fluoride was added to our toothpaste and drinking water is an extraordinary, almost fantastic tale. The plot includes some of the most spectacular events in human affairs—the explosion of the Hiroshima atomic bomb, for example. Many of the principal characters are larger than life, such as the "father of public relations" Edward L. Bernays, Sigmund Freud's nephew, who was until now more famous for his scheme to persuade women to smoke cigarettes.[1] And the twists and turns of the fluoride story are propelled by nothing less than the often grim requirements of accumulating power in the industrial era—the same raw power that is at the beating heart of the American Century.

Fluoride lies at the elemental core of some of the greatest fortunes that the world has ever seen, the almost unimaginable wealth of the Mellons of Pittsburgh and the Du Ponts of Delaware. And no wonder the warning on the toothpaste tube is so dramatic. The same potent chemical that is used to enrich uranium for nuclear weapons, to prepare Sarin nerve gas, and to wrestle molten steel and aluminum from the earth's ore is what we give to our children first thing in the morning and last thing at night, flavored with peppermint, strawberry, or bubble gum.

Fluoride is so muscular a chemical that it has become a lifeblood of modern industry, pumped hotly each day through innumerable factories, refineries, and mills. Fluoride is used to produce high-octane gasoline; to smelt such key metals as aluminum, steel, and beryllium; to enrich uranium; to make computer circuit boards, pesticides, ski wax, refrigerant gases, Teflon plastic, carpets, waterproof clothing, etched glass, bricks and ceramics, and numerous drugs, such as Prozac and Cipro.

Fluoride's use in dentistry is a sideshow by comparison. But its use in dentistry helps industry, too. How does it work? Call it elemental public relations. Fluoride is so potent a chemical that it's also a grave environmental hazard and a potential workplace poison. So, for the industry-sponsored scientists who first promoted fluoride's use in dentistry, linking the chemical to better teeth and stoutly insisting that, in low doses, it had no other health effect helped to change fluoride's image from poison to panacea, deflecting attention from the injury that factory fluoride pollution has long wreaked on workers, citizens, and nature.

Hard to swallow? Maybe not. The face-lift performed on fluoride more than fifty years ago has fooled a lot of people. Instead of conjuring up the image of a crippled worker or a poisoned forest, we see smiling children. Fluoride's ugly side has almost entirely escaped the public gaze. Historians have failed to record that fluoride pollution was the biggest single legal worry facing the atomic-bomb program following World War II. Environmentalists are often unaware that since World War II, fluoride has been the most damaging poison spilling from factory smokestacks and was, at one point during the cold war, blamed for more damage claims against industry than *all twenty other major air pollutants combined.*[2] And it was fluoride that may have been primarily responsible for the most notorious air pollution disaster in U.S. history—the 1948 Halloween nightmare that devastated the mill town of Donora, Pennsylvania—which jump-started the U.S. environmental movement.[3]

It's the same story today: more happy faces. Yet we are exposed to fluoride from more sources than ever. We consume the chemical from water and toothpaste, as well as from processed foods made with fluoridated water and fluoride-containing chemicals. We are exposed to fluorine chemicals from often-unrecognized sources, such as agricultural pesticides, stain-resistant carpets, fluorinated drugs, and such packaging as microwavable popcorn bags and hamburger wrappers, in addition to industrial air pollution and the fumes and dust inhaled by many workers inside their factories.

Fluoride's double-fisted trait of bringing out the worst in other chemicals makes it especially bad company. While a common air pollutant, hydrogen fluoride, is many times more toxic than better-known air pollution villains, such as sulfur dioxide or ozone, it "synergistically" boosts the toxicity of *these* pollutants as well.[4] Does fluoride added to our drinking water similarly increase the toxicity of the lead, arsenic, and other pollutants that are routinely found in our water supply? As we shall see, getting answers to such questions from the federal government, even after fifty years of endorsing water fluoridation, can prove impossible.

By the mid-1930s European scientists had already linked fluoride to a range of illnesses, including breathing problems, central-nervous-system disorders, and especially an array of arthritis-like musculoskeletal problems.[5] But during the cold war, in one of the greatest medical vanishing acts of the twentieth century, fluoride was systematically removed from public association with ill health by researchers funded by the U.S. military and big corporations. In Europe excess exposure to fluoride produced a medical condition described as "poker back" or "crippling skeletal fluorosis" among factory workers. But the chemical somehow behaved differently when it crossed the Atlantic, the industry-funded researchers implied, failing to produce such disability in the United States.[6] It was a deceit, as we shall see: scientific fraud on a grand and global scale; a lawyerly ruse to escape liability for widespread worker injury; a courtroom hustle made possible and perpetuated by the suppression of medical evidence and by occasional perjury.

Your history is all mixed up, say supporters of water fluoridation. The story of how fluoride was added to our toothpaste and water is a separate history, unrelated to fluoride's use in industry, they maintain. But there is only one story, not two. The tale of the dental "wonder chemical" and the mostly secret account of how industry and the U.S. military helped to create and polish that public image are braided too closely to distinguish between them. The stories merge completely in the conduct of two of the most senior American scientists who led the promotion of water fluoridation in the 1940s and 1950s, Dr. Harold Carpenter Hodge and Dr. Robert Arthur Kehoe.

Don't blame the dentists. They were taught that fluoride is good for teeth. Few realize that Dr. Hodge, the nation's leading fluoride researcher who trained a generation of dental school deans in the 1950s and 1960s, was the senior wartime toxicologist for the Manhattan Project. There he helped choreograph the notorious human radiation experiments in which hospital patients were injected with plutonium and uranium—without their knowledge or consent—in order to study the toxicity of those chemicals in humans. Hodge was similarly charged with studying fluoride toxicity. Building the world's first atomic bomb had required gargantuan amounts of fluoride. So, for example, on behalf of the bomb makers he covertly monitored one of the nation's first public water fluoridation experiments. While the citizens of Newburgh, New York, were told that fluoride would reduce cavities in their children, secretly blood and tissue samples from residents were sent to his atomic laboratory for study.[7]

Some dentists are unaware that much of the fluoride added to drinking water today in the United States is actually an industrial waste, "scrubbed" from the smokestacks of Florida phosphate fertilizer mills to prevent it from damaging livestock and crops in the surrounding countryside. In a sweetheart deal these phosphate companies are spared the expense of disposing of this "fluosilicic acid" in a toxic waste dump. Instead, the acid is sold to municipalities, shipped in rubber-lined tanker trucks to reservoirs across North America and injected into drinking water for the reduction of cavities in children. (So toxic are the contents of the fluoride trucks that in the aftermath of the September 11, 2001, terrorist attack, authorities were alerted to keep a watchful eye on road shipments of the children's tooth-decay reducer.)[8]

"I had no idea where the fluoride was coming from until the antifluoridationists pointed it out to me," Dr. Hardy Limeback, the head of Preventative Dentistry at the University of Toronto, Canada, and a former leading fluoridation supporter, told me. "I said, 'You have got to be wrong. That is not possible!'"

Those same phosphate manufacturers were members of an influential group of industries that sponsored Dr. Robert Kehoe's fluoride research at the University of Cincinnati during the 1940s and 1950s. Kehoe is better known today for his career-long defense of the safety of adding lead to gasoline (now discredited). But he was also a leading figure reassuring citizens and scientists of the safety of industrial fluoride and water fluoridation, while burying information about the chemical's toxic effects and privately sharing doubts with his corporate sponsors about the safety of even tiny amounts of the chemical.[9]

Not surprisingly, peering behind the fifty-year-old facade of smiling children with rows of picket-fence-white teeth is difficult. Industry is reluctant to have its monument to fluoride safety blackened or its role in dental mythmaking explored. Several of the archives I visited had gaping holes or missing documents, and some were closed entirely. And many scientists are reluctant to speak critically about fluoride—mindful of the fate of researchers who have questioned the government line. Scientists have been fired for their refusal to back down from their questions about the safety of fluoride, blackballed by industry, or smeared by propagandists hired by the U.S. Public Health Service and the American Dental Association.[10] "Bodies litter the field," one senior dental researcher told me when he learned that I was writing a book on fluoride.

Myths are powerful things. Mention of fluoride evokes a skeptically cocked eyebrow from liberals and conservatives alike and an almost reflexive mention of the 1964 Stanley Kubrick film *Dr. Strangelove*. The hilarious portrayal of General Jack D. Ripper as a berserk militarist obsessed with Communists adding fluoride to the nation's water became a cultural icon of the cold war—and perhaps the movie's most famous scene. (Today Nile Southern, the son of *Dr. Strangelove*'s screenwriter, Terry Southern, remarks that the news that U.S. military and industrial interests—not Communists—promoted water fluoridation is "just shocking. Terry and Stanley [Kubrick] would have been horrified by it.")[11]

The media caricature was largely false. The national grassroots struggle against water fluoridation was a precursor of today's environmental movement, with multicolored hues of political affiliation. It was led by veteran scientists with distinguished careers safeguarding public health, including the doctor who warned the nation about the dangers of cigarette smoking and the risk from allergic reaction to penicillin.[12] Yet instead of being seen as medical pioneers and minutemen, warning of the encroachment of industrial poisons, antifluoridationists are portrayed as unscientific and isolationist—the modern equivalent of believing that the earth is flat.

It is the U.S. medical establishment that is out on a limb, say critics. Adding to water a chemical so toxic that it was once used as rat poison was a uniquely American idea and is, increasingly, a lone American practice. Most European countries do not add fluoride to their water. Several nations have long since discontinued the practice, doubting its safety and worth.[13]

Fluoride may help teeth, but the evidence is not overwhelming. Although rates of dental decay have fallen significantly in the United States since the 1940s, similar improvements have been seen in countries where fluoride is not added to the water. Improved dental care, good nutrition, and the use of antibiotics may explain the parallel improvement. A largely sympathetic official review of fluoridation by the British government in 2000 found that most studies of the effectiveness of fluoridated water were of "moderate" quality and that water fluoridation may be responsible for 15 percent fewer cavities.[14] That's a far cry from the 65 percent reductions promised by the early promoters of fluoride. With revelations that such health problems as central nervous system effects, arthritis, and the risk of bone cancer were minimized or concealed entirely from the public by early promoters of fluoride, the possible benefit of a handful of better teeth might not be worth running the risk. "How many cavities would have to be saved to justify the death of one man from osteosarcoma?" asked the late Dr. John Colquhoun, the former chief dental officer of Auckland, New Zealand, and a fluoride promoter turned critic.[15]

"I did not realize the toxicity of fluoride," said Dr. Limeback, the Canadian. "I had taken the word of the public health dentists, the public health physicians, the USPHS, the USCDC, the ADA, the CDA [Canadian Dental Association] that fluoride was safe and effective without actually investigating it myself."

Even the theory of how fluoride works has changed. The CDC no longer argues that fluoride absorbed from the stomach via drinking water helps teeth. Instead, the argument goes, fluoride strikes at dental decay from *outside* the tooth, or "topically," where, among other effects, it attacks the enzymes in cavity-causing bacteria. Drinking fluoridated water is still important, according to the CDC, because it bathes the teeth in fluoride-enhanced saliva—a cost-effective way of reaching poorer families who may not have a balanced diet, access to a dentist, or the regular habit of brushing with fluoride toothpaste.[16]

But swallowing treated water allows fluoride into our bones and blood, where it may be harmful to other parts of the body, say critics. If fluoride can kill enzymes in tooth bacteria, its potentially crippling effects on other enzymes—the vital chemical catalysts that regulate much biological activity—must be considered.[17]

"When I investigated [such questions] I said, 'This is crazy. Let's take it out of the water because it is harming so many people—[not] simply the dental fluorosis [the white mottling on teeth caused by fluoride], but now we are seeing bone problems and possibly cancer and thyroid problems. If you are really targeting the poor people, let's give toothpaste out at the food banks. Do something other than fluoridate the water supply,'" said Dr. Limeback. "Then [the fluoride promoters] kept saying, 'Well, it is cost effective.' That is a load of crap–it is cost effective because they are using toxic waste, for crying out loud!"

History tells us that overturning myths is rarely easy. But we have been down this path before. The fluoride story is similar to the fables about lead, tobacco, and asbestos, in which medical accomplices helped industry to hide the truth about these substances for generations. Fluoride workers share a tragic fate with the souls who breathed beryllium, uranium, and silica in the workplace. Endless studies that assured workers that their factories and mines were safe concealed the simple truth that thousands of people were being poisoned and dying painful early deaths from these chemicals. So if this tale of how fluoride's public image was privately laundered sounds eerily familiar, maybe it's because the very same professionals and institutions who told us that fluoride was safe said much the same about lead, asbestos, and DDT or persuaded us to smoke more tobacco.

Lulled by half a century of reassurances from supporters of fluoride in the public health establishment, many doctors today have no idea of the symptoms of fluoride poisoning. A silent killer may stalk us in our ignorance. "There is a black hole out there, in terms of the public and scientific knowledge," says former industry toxicologist Dr. Phyllis Mullenix. "There is really no public health issue that could impact a bigger population. I don't think there is an element of this society that is not impacted by fluoride. It is very far-reaching and it is very disturbing."

Fifty years after the U.S. Public Health Service abruptly reversed course during the darkest days of the cold war—and endorsed artificial water fluoridation—it is time to recognize the folly, hubris, and secret agendas that have shackled us too long, poisoning our water, choking our air, and crippling workers. It is time, as the Quakers ask in life, to speak truth to power. Good science can sharpen the tools for change, but it will be public opinion and citizen action that strike those shackles free.

Major Figures in the Fluoride Story

EDWARD L. BERNAYS. A propagandist and the self-styled father of public relations, Bernays was Sigmund Freud's nephew. Among his clients were the U.S. military, Alcoa, Procter and Gamble, and Allied Signal. On behalf of big tobacco companies he persuaded American women to smoke cigarettes. He also promoted water fluoridation, consulting on strategy for the National Institute of Dental Research.

GERALD JUDY COX. A researcher at the Mellon Institute in the 1930s, where he held a fellowship from the Aluminum Company of America. Following Frary's (see below) suggestion, Cox reported that fluoride gave rats cavity-resistant teeth and in 1939 made the first public proposal to add fluoride to public water supplies.

HENRY TRENDLEY DEAN. The U.S. Public Health Service researcher who studied dental fluorosis in areas of the United States where fluoride occurred naturally in the water supply. His "fluorine-caries" hypothesis suggested that fluoride made teeth cavity-resistant but also caused unsightly dental mottling. Worried about toxicity, Dean opposed adding fluoride to water in Newburgh, New York, the site of the nation's first-planned water fluoridation experiment. In 1948 Dean became the first director of the National Institute of Dental Research (NIDR) and, in 1953, a top official of the American Dental Association.

OSCAR R. EWING. A top Wall Street lawyer for the Aluminum Company of America. As Federal Security Agency administrator for the Truman administration with jurisdiction over the Public Health Service, it was Ewing who, in 1950, endorsed public water fluoridation for the United States.

FRANCIS COWLES FRARY. As Director of Research at the Aluminum Company of America from 1918, Frary was one of the most powerful science bureaucrats in the United States and grappled with the issue of fluoride emissions from aluminum smelters. It was Frary who made early suggestions to Gerald Cox, a researcher at the Mellon Institute, that fluoride might make strong teeth.

GENERAL LESLIE R. GROVES. Head of the U.S. Army Corps of Engineers' Manhattan Project to build the world's first atomic bomb.

HAROLD CARPENTER HODGE. A biochemist and toxicologist at the University of Rochester who investigated fluoride for the U.S. Army's Manhattan Project, where he also supervised experiments in which unsuspecting hospital patients were injected with uranium and plutonium. After the war Hodge chaired the National Research Council's Committee on Toxicology and became the leading scientific promoter of water fluoridation in the United States during the cold war.

DUDLEY A. IRWIN. Alcoa's medical director who helped oversee Robert Kehoe's fluoride research at the Kettering Laboratory, and who met personally with top fluoride researchers at the National Institute of Dental Research (NIDR) following the verdict in the Martin air-pollution trial.

ROBERT A. KEHOE. As the Director of the Kettering Laboratory of Applied Physiology at the University of Cincinnati, Kehoe was the leading defender in the United States of the safety of leaded gasoline. Guided by a group of corporate attorneys known as the Fluorine Lawyers Committee, Kehoe similarly defended fluoride on behalf of a group of corporations that included DuPont, Alcoa, and U.S. Steel, all of which faced lawsuits for industrial fluoride pollution.

EDWARD J. LARGENT. A researcher at the Kettering Laboratory who defended corporations accused of fluoride pollution and spent a career negating the fluoride warnings of the Danish scientist Kaj

Roholm. Largent exposed his wife and son to hydrogen fluoride in a laboratory gas chamber.

NICHOLAS C. LEONE. The head of medical investigations at the federal government's NIDR who was in close communication with industry's Fluorine Lawyers and who, following the 1955 Martin verdict, met with Alcoa's Dudley Irwin and the Kettering Laboratory's Robert Kehoe to discuss how government water fluoridation safety studies could help industry.

WILLIAM J. MARCUS. A senior toxicologist in the EPA's Office of Drinking Water. In 1992, after he protested what he described as the systematic downgrading of the results of the government's study of cancer and fluoride, he was fired. A federal judge later ruled that he had been fired because of his scientific opinions on fluoride and ordered him reinstated.

PAUL AND VERLA MARTIN. Oregon farmers who were poisoned by fluoride from a Reynolds Metals aluminum plant. Their precedent-setting court victory in 1955 sparked emergency meetings between fluoride industry representatives and senior officials from the National Institute of Dental Research and launched a crash program of laboratory experiments at the Kettering Laboratory to prove industrial fluoride pollution "safe."

PHYLLIS J. MULLENIX. A leading neurotoxicologist hired by the Forsyth Dental Center in Boston to investigate the toxicity of materials used in dentistry. In 1994, after her research indicated that fluoride was neurotoxic, she was fired.

KAJ ELI ROHOLM. The Danish scientist who in 1937 published the book *Fluorine Intoxication*, an encyclopedic study of fluoride pollution and poisoning. He opposed giving fluoride to children.

PHILIP SADTLER. The third-generation son of a venerable Philadelphia family of chemists, Sadtler gave expert testimony during the 1940s and 1950s on behalf of farmers and citizens who claimed that they had been poisoned by industrial fluoride pollution. He blamed fluoride for the most notorious air pollution disaster in U.S. history, during which two dozen people were killed and several thousand were injured in Donora, Pennsylvania, over the Halloween weekend in 1948.

FRANK L. SEAMANS. A top lawyer for Alcoa, Seamans was also head of the group of senior attorneys known as the Fluorine Lawyers Committee, which represented big corporations in cases of alleged industrial fluoride pollution.

GEORGE L. WALDBOTT. A doctor and scientist and a leading expert on the health effects of environmental pollutants, Waldbott's research in the 1950s and 1960s on his own patients indicated that many people were uniquely sensitive to very small doses of fluoride. He founded the International Society for Fluoride Research and was a leader of the international and domestic opposition to water fluoridation.

COLONEL STAFFORD L. WARREN. Head of the Manhattan Project's Medical Section.

EDWARD RAY WEIDLEIN. Director of the Mellon Institute, where Cox carried out his studies.

1

Through the Looking Glass

At the children's entrance to the prestigious Forsyth Dental Center in Boston, there is a bronze mural from a scene in Alice in Wonderland. *The mural makes scientist Phyllis Mullenix laugh. One spring morning, when she was the head of the toxicology department at Forsyth, she walked into the ornate and marbled building and, like Alice, stepped through the looking glass. That same day in her Forsyth laboratory she made a startling discovery and tumbled into a bizarre wonderland where almost no one was who they had once appeared to be and nothing in the scientist's life would ever be the same again.*

AS SHE DROVE alongside the Charles River in the bright August sunshine of 1982 for her first day of work at the Forsyth Dental Center in Boston, toxicologist Phyllis Mullenix was smiling. She and her husband Rick had recently had their second daughter. Her new job promised career stability and with it, the realization of a professional dream.

Since her days as a graduate student Mullenix had been exploring new methods for studying the possible harmful effects of small doses of chemicals. By 1982 Dr. Mullenix was a national leader in the young science of neurotoxicology, measuring how such chemicals affected the brain and central nervous system. She and a team of researchers were developing a bold new technology to perform those difficult measurements more accurately and more quickly than ever before.

The system was called the Computer Pattern Recognition System.

It used cameras to record changes in the "pattern" of behavior of laboratory animals that had been given tiny amounts of toxic chemicals. Computers then rapidly analyzed the data. By detecting how the animals' behavior differed from that of similar "control" animals—that were not given the toxic agent—scientists were able to measure or "quantify" the extent to which a chemical affected the animals' central nervous system.

Previous such efforts had relied on subjective guesswork as to the severity of the chemical's toxic effect or on laborious and time-consuming efforts to quantify the changes the chemical made in behavior. The speed of the computers and the accuracy of the camera measurements in the Mullenix system, however, could potentially revolutionize the study of toxic chemicals.

As her car flew along the Charles River that summer morning in 1982, Mullenix knew that her new job and the support of the prestigious Forsyth Dental Center would finally allow her to complete the work on her new system.

Mullenix had caught the eye of Forsyth's director, John "Jack" Hein, some years earlier. He had attended one of her seminars at the Harvard Medical School, where she was a faculty member in the Department of Psychiatry. He had sat in the audience, dazzled, his mind racing. Hein remembers a "very bright" woman describing a revolutionary new technology, which he believed had the potential for transforming the science of neurotoxicology. "She had the world by the tail," said Hein. "There is nothing more exciting than a new methodology."[1]

Jack Hein wanted Mullenix to bring her new technology to Forsyth and to set up a modern toxicology laboratory. It would be the first such dental toxicology center in the country. Many powerful chemicals are routinely employed in a dentist's office, such as mercury, high-tensile plastics, anesthetics, and filling amalgams. Hein knew that an investigation of the toxicity of some of these materials was overdue.

The Forsyth director's boyish enthusiasm helped to sell Mullenix on the move. "I was very impressed with Dr. Hein," she said. "He was like a kid in a candy store. He couldn't wait for us to use the new methodology and apply it to some of the materials dentists work with."

Phyllis Mullenix's transfer to Forsyth was a move to one of Boston's most prestigious medical centers. The Forsyth Dental Infirmary for Children was established in 1910 to provide free dental care to Boston's poor children. By 1982, when Dr. Mullenix accepted Jack Hein's invitation, the renamed Forsyth Dental Center was affiliated with Harvard Medical School and had become one of the best-known centers for dental research in the world.

At the helm was Forsyth's director, Jack Hein, a well-known figure in American dental research. Hein had attended the University of Rochester in the 1950s, and there he had helped to develop the fluoride compound sodium monofluorophosphate (MFP). Colgate soon added MFP to its toothpaste, and Jack Hein became the company's dental director in 1955.[2] When he came to Forsyth in 1962, Hein was part of the new order in reshaping American dentistry—a changing of the guard then taking place in many dental schools and research centers.[3] Like Jack Hein, the new generation of leaders was uniform in its support of fluoride's use in dentistry.[4]

Forsyth had read the tea leaves well. While a previous Forsyth director, Veikko O. Hurme, had been an outspoken opponent of adding fluoride to public water supplies, Jack Hein's support came at the same time that Colgate poured cash into new facilities and fluoride research at Forsyth.[5] Additional funds came from research grants from other private corporations and from the federal National Institutes of Health (NIH). A sparkling new research annex, built in 1970, doubled the size of the Forsyth Center, with funds from the NIH and "major donors," such as Warner Lambert, Colgate Palmolive, and Lever Brothers.[6]

Jack Hein's track record as a fund-raiser for the Forsyth Center and his support for fluoride's use in dentistry owed much to his membership in an informal old boy's club of scientists who had also once done research at the University of Rochester. The University had been a leading center for fluoride research in the 1950s and 1960s, with many of its graduate students taking leading roles in dental schools and research centers around the United States.

In 1983, a year after Phyllis Mullenix arrived at Forsyth, director Hein introduced her to an elderly gentleman who had been Hein's professor and scientist mentor some thirty years earlier at the University of Rochester. The old man was a researcher with a distinguished national reputation—the first president of the Society of Toxicology, Mullenix learned, and the author of scores of academic papers and books. His name was Harold Carpenter Hodge, and his impeccable manners and formal dress left an indelible impression on Mullenix.

"I was impressed with Harold," she said. "He was very gentlemanly. He would never say an inappropriate word, and he always wore a white lab coat."

Hodge had recently retired from the University of San Francisco. Jack Hein had brought him to Forsyth for the prestige he would bring to Mullenix's new toxicology department, he said, and out of admiration for his former professor, who was then in his mid-seventies. "I thought it would be fun," Hein added.

Mullenix grew fond of Hodge. He seemed almost grandfatherly, ambling into her laboratory, chatting as her young children frolicked alongside. Hodge was especially fascinated by the new computer system for testing chemical toxicity. He would fire endless questions at Mullenix and her colleague, Bill Kernan from Iowa State University, Mullenix remembered. "He would quietly come up to my lab. And Harold would ask 'Why are you doing this?' and 'What are you doing?' and Bill [Kernan] would take great pains to explain every little scientific detail, showing him the rat pictures."

By the early 1980s Jack Hein's vision for the Forsyth Center included more than just dentistry. The canny fund-raiser believed that the new Mullenix technology could become another big money spinner for Forsyth—a winning weapon in the high-stakes field of toxic tort litigation, in which workers and communities allege they have been poisoned by chemicals. "It was an exciting new way of studying neurotoxicity," said Jack Hein, who would eventually assign Mullenix to spacious new offices and laboratories on the fourth floor of the Forsyth research annex.

Neurotoxicology was still a young science. If someone claimed to have been hurt by a chemical in the workplace or had been exposed in a pollution incident, finding the scientific truth was extraordinarily difficult. Big courtroom awards against industry often hinged on the subjective opinion of a paid expert witness and the unpredictable emotions of a jury, said Mullenix. "Industries did not like that. They felt that the answers were biased, and so the thought of taking investigator bias out of the system was very exciting to them. They thought this would help [industry] in court," she added.

The Computer Pattern Recognition System quickly attracted attention from other scientists, industry, and the media. The *Wall Street Journal* called the Mullenix technology "precise" and "objective."[7] Some of America's biggest corporations opened their wallets. The medical director of the American Petroleum Institute personally gave $70,000 to Mullenix. Monsanto gave $25,000. Amoco and Mobil chipped in thousands more, while Digital Equipment Corporation donated most of the powerful computer equipment.

"Several oil and chemical companies such as Monsanto Co. are supporting research on the system," the *Wall Street Journal* reported. "Questions are being raised more frequently about whether there are behavioral effects attributable to chemicals," a Monsanto toxicologist, George Levinskas, told the newspaper. The Forsyth system "has potential to give a better idea of the effects our chemicals might have," he added.[8]

In a letter of recommendation, Myron A. Mehlman, the former head of toxicology for the Mobil Oil Corporation, who was then working for the federal Agency for Toxic Substances and Disease Registry (ATSDR), called the Mullenix technology "a milestone for testing low levels of exposure of chemicals for neurotoxicity for the 21st Century. . . . The benefits of Professor Mullenix' discovery to Forsyth are enormous and immeasurable."[9]

Industry trusted Phyllis Mullenix. Since the 1970s the toxicologist had earned large fees consulting on pollution issues and the legal requirements of the Clean Air Act. Hired by the American Petroleum Institute, for example, she'd acted as scientific coordinator for that lobby group, advising it on proposed and restrictive new EPA standards for ozone. "Whenever it got technical they would dance me out," she said. "Every time EPA came out with another criteria document I would look for the errors."

Mullenix is not apologetic for waltzing with industry. Anybody could take her to the ball, she said, explaining, "I did not look at myself as a public health individual. I was amazed that the EPA did such shoddy work writing a criteria document. I thought that at the very least those documents should be factual."

At Harvard, Mullenix had been criticized by some academics

for her industry connections, a charge she calls "ridiculous." Said Mullenix, "No one group, be it government, academia or industry, can be right one hundred percent of the time. I don't see science as aligning yourself with one group. Industry can be right in one respect and they can be very wrong in another."

And Mullenix had other consulting work—for companies such as Exxon, Mobil, 3M, and Boise Cascade. Companies including DuPont, Procter and Gamble, NutraSweet, Chevron, Colgate-Palmolive, and Eastman Kodak all wrote checks supporting a 1987 conference she held titled "Screening Programs for Behavioral Toxicity."

Like many revolutionary ideas, the concept behind the Mullenix technology for studying central-nervous-system problems was simple. The spark of inspiration had come from Dr. Mullenix's graduate advisor at the University of Kansas Medical Center, Dr. Stata Norton. A slender and soft-spoken woman, Dr. Norton was one of the first prominent female toxicologists in the United States. She had won national recognition by demonstrating that there were "threshold" levels for the toxic effects of alcohol and low-level radiation on the fetus. Now retired to her summer cottage, surrounded by lush Kansas farmland, Dr. Norton's face opened in a smile as she remembered her former student. Normally, she said, graduate students rotated through the various laboratories at the Medical Center. But there was something different about Phyllis Mullenix.

"Phyllis came into my lab to do a short study—and she never left," Norton recalled, laughing.

Mullenix had a special willingness to grapple with complex new information, Norton said. When Norton was studying the effects of radiation on rats, Mullenix wanted to learn how the radiation had physically altered the rats' brains. She had never done that work before, Norton recalled, but her student stayed late at the lab, poring over medical journals, dissecting the rat's brains, and looking for tiny changes caused by the radiation. "I don't think she thought it was difficult," said Norton. "She was happy to jump on the project and get with it."

There was something else. Norton noticed her student had a fearless quality and a willingness to challenge conventional wisdom. The professor found it refreshing. "It takes a certain personality to stand up and do something different. Science is full of that, all the way from

Galileo," Norton said. "That doesn't mean you are right or you are wrong, but I can appreciate that in Phyllis because I am like that."

In the mid-1970s Stata Norton was a pioneer in the new field of behavioral toxicology, inventing new ways for measuring the ways chemicals affected behavior. At first Norton studied mice that had been trained or "conditioned" to behave in certain ways by receiving food rewards. Some scientists believed that by studying disruptions in this "conditioned" behavior, they could most accurately measure the toxic effects of different chemicals.

Norton was not so sure. One day, working with mice that had been trained to press a lever for food at precisely timed intervals, she suddenly wondered how the animals knew when to press the lever. "I looked in the box," she said. Inside she saw that each mouse seemed to measure the time between feeding by employing a "sequence" or pattern of simple activities such as sitting, scratching, or sniffing. "There was a rhythm," she explained. "They timed it by doing things."

Norton began her own experiments. She wondered if, by studying changes in this rhythm of "patterned" behavior during the time between feeding—as opposed to studying disruptions in the conditioned behavior exhibited for food rewards—she could get a more sensitive measurement of the toxicity of chemicals. Norton and Mullenix took thousands of photographs of rats that had been given a chemical poison and compared them with similar photographs of healthy "control" rats. They were able to detect changes in the sequences of the rats' behavior, even at very low levels of chemical poisoning. "We were all very excited," said Norton.

The spirit of independence and free inquiry in Stata Norton's laboratory inspired Phyllis Mullenix. It was the kind of environment she had grown up in. Her mother, Olive Mullenix, was a Missouri schoolteacher who'd ridden sixteen miles on horseback to her one-room schoolhouse each day and made her "own" money selling fireworks from a roadside stand. Her father, "Shockey" Mullenix (he had a shock of white hair), had left the farm with a dream to become a doctor. He settled for the workaholic life of a gas-station entrepreneur and trader in the small town of Kirksville, Missouri and the hope that his three children would realize his dreams. The son became a nuclear physicist for the Department of Energy; another daughter was a corporate Washington lawyer; and the youngest, Phyllis, the Harvard toxicologist.

In the late 1970s the Environmental Protection Agency grew interested in the Kansas research. The federal agency wanted a new way of measuring the human effects of low-level chemical contamination. The head of the EPA's neurotoxicology division, Lawrence Reiter, visited Stata Norton's laboratory. Phyllis Mullenix told him that the key to the success of the new technique was to speed up the time-consuming process of analyzing each frame of film. Mullenix thought that computers could do the job faster. The EPA agreed, and Mullenix became a consultant on a $4 million government grant awarded to Iowa State computer experts Bill Kernan and Dave Hopper. Kernan had worked previously for the Defense Department, writing some of its most elegant and sophisticated software.

"I was to train the physicist," said Mullenix. "The physicist would train the computer."

Developing the Computer Pattern Recognition System, as Mullenix's technology became known, took almost thirty years. Dr. Norton had begun studying her rats in the 1960s. When she passed the baton to Phyllis Mullenix in the 1970s, computers were barely powerful enough to handle the vast data-processing requirements for detecting subtle behavior changes and measuring chemical poisoning.

In Boston in the mid-1980s Mullenix grew incredibly busy. She now had two young daughters. She was consulting for industry. Her husband, Rick, was completing training as an air-traffic controller. And her father was seriously ill with emphysema 1500 miles away in Kirksville, Missouri.

Her Forsyth laboratory buzzed with activity. The new computers were hooked up by telephone to big data-processing units at Iowa State. By late 1987 the Computer Pattern Recognition System was almost ready. Forsyth printed brochures, touting a system that promised to "prevent needless exposure of the general public to the dangers of neurotoxicity, and industry to exaggerated litigation claims." Mullenix soon became a national pitchwoman for Forsyth, proclaiming a new day for corporations that feared lawsuits from workers and communities for chemical exposures. "I was hopped all over the country giving seminars on how this computerization was going to help the industrial situation," she said.

Director Jack Hein was anxious to illustrate the sensitivity of the new machine. He suggested that Mullenix start with fluoride, giving small doses to rats and testing them in the equipment. The longtime fluoride supporter wanted to test fluoride first, he said, in order to bolster the chemical's public image. "I was really interested in proving there were no negative effects," Hein said. "It seemed like a good way of negating the antifluoridationist arguments."

Mullenix shrugged. She didn't much care about fluoride. Secretly she thought that fluoride was a waste of her time and that Jack Hein was overreacting. "At Harvard the rule is publish or perish. And I didn't think that I would come up with anything that would be worth publishing," she said. "I'm used to studying hard-core neurotoxic substances, drugs like anticonvulsants, radiation, where it can totally distort the brain. I never heard anything about fluoride, except TV commercials that it is good for your teeth."

Hein introduced her to another young dental researcher, Pamela DenBesten, who had recently arrived at Forsyth. DenBesten was studying the white and yellow blotches, or mottling, on tooth enamel caused by fluoride known as dental fluorosis. Although Mullenix was lukewarm to the idea of using fluoride to test for central-nervous-system effects, DenBesten was more curious. She had noticed that when she gave fluoride to rats for her tooth-enamel studies, they did not behave "normally." While it was usually easy to pick up laboratory rats, the animals that had been fed fluoride would "practically jump out of the cage," DenBesten said.

The two women worked well together. Phyllis would often bring her two young daughters to work, and the Mullenix laboratory on the fourth floor became a sanctuary from the predominantly male atmosphere at Forsyth. DenBesten knew that Phyllis Mullenix had few friends at Forsyth. Many of the other researchers were hostile to the plainspoken toxicologist. DenBesten describes it as "gender-discrimination type stuff."[10]

Another Forsyth scientist, Dr. Karen Snapp, quickly made friends with Phyllis Mullenix. "I was always told that Phyllis was the batty woman up in the tower on the fourth floor," said Snapp. "I ran into her at lunch one day in the cafeteria. We started chatting, then we went out and had a coke together." Snapp found Mullenix refreshing, both for the quality of her science and her plainspoken manner. "She didn't bow down to the powers that be at Forsyth. A lot of people put up fronts and are very pious, and Phyllis was not that way at all—that is what I liked about her. She was very honest, very straightforward, you knew exactly where you stood," Snapp explained.

Snapp was also impressed with the rigor Mullenix brought to her scientific experiments. "She was very, very thorough. She at times had no idea what the outcome of an experiment was going to be. If she did an experiment and didn't get the result she thought she should get, she'd repeat it to make sure it was right, and [if the unexpected data held up] it's like, well—we change the hypothesis."

If Phyllis Mullenix was at first nonchalant about testing fluoride for central-nervous-system effects, that was not the attitude of perhaps the "oldest boy" at the Forsyth Center. She found that Dr. Harold Hodge, the affable old man in the freshly pressed lab coat, took what then seemed an almost obsessive interest in her fluoride work, firing endless questions about her methodology.

"He wanted to push me to do certain fluoride studies, and do this and do that, and how can I help?" said Mullenix.

2

Fireworks at Forsyth

The two white-coated scientists stared at each other, startled. High above Boston, surrounded by computer terminals and data print-outs and the bright lights of a modern toxicology laboratory, Phyllis Mullenix and Pamela DenBesten fell suddenly silent. Only the white rats in their cages scuttered and sniffed. The information slowly sank in. The scientists had repeated their experiment and, once again, the results were the same. They laughed, nervously.

"Oh shit," Dr. Phyllis Mullenix finally blurted out. "We are going to piss off every dentist in the country."

BY 1989 THE Mullenix team was getting its first results from the fluoride experiments. They had been gathering data for two years, giving the rats moderate amounts of fluoride, monitoring them in their cages, and then analyzing the data in the RAPID computer system, as her new technology was known. But something was wrong. The results seemed strange.

"Data was coming back that made me shake my head," said Mullenix. "It wasn't at all what we expected." Mullenix had expected that giving fluoride in drinking water would show no effect on the rats' behavior and central nervous system. Mullenix wondered if the problem was a bug in the new machinery. The team launched an exhaustive series of control experiments, which showed that the RAPID computers were working fine. All the results were "amazingly consistent," said Mullenix.

Fluoride added to their drinking water produced a variety of effects in the Forsyth rats. Pregnant rats gave birth to "hyperactive" babies. When the scientists gave fluoride to the baby rats following their birth, the animals had "cognitive deficits," and exhibited retarded behavior. There were sex differences, too. Males appeared more sensitive to fluoride in the womb; females were more affected when exposed as weanlings or young adults.

The two women told Jack Hein and Harold Hodge about the results. The men ordered them to repeat the experiments, this time on different rats. The team performed still more tests. Mullenix remembers that Harold Hodge kept asking her about the results, even though he was by now very ill. He had gone to his home in Maine but kept in contact by telephone. He asked every day.

By 1990 the data were crystal clear. The women had tested more than five hundred rats. "I finally said we have got enough animals here for statistical significance," said Mullenix. "There is a problem," she added.

The two women talked endlessly about what they had found. Mullenix was a newcomer to fluoride research, but Pamela DenBesten had spent her career studying the chemical. She suspected that they had made an explosive discovery and that dentists in particular would find the information important. "My initial gut reaction was that this is really big," said DenBesten. Although the Forsyth rats had been given fluoride at a higher concentration than people normally drink in their water—an equivalent of 5 parts per million as opposed to 1 part per million—DenBesten also knew that many Americans are routinely exposed to higher levels of fluoride every day. For example, people who drink large amounts of water, such as athletes or laborers in the hot sun; people who consume certain foods or juices with high fluoride levels; children who use fluoride supplements from their dentists; some factory workers, as the result of workplace exposure; or certain sick people, all can end up consuming higher cumulative levels of fluoride. Those levels of consumption begin to approach—or can even surpass, for some groups—the same fluoride levels seen in the Forsyth rats.

"If you have someone who has a medical condition, where they have diabetes insipidus where you drink lots of water, or kidney disease—anything that would alter how you process fluoride—then you could climb up to those levels," said DenBesten. She thought that the Forsyth research results would quickly be followed up by a whole series of additional experiments examining, for example, whether fluoride at even lower levels, 1 part per million, produced central-nervous-system effects. "I assumed it would take off on its own, that a lot of people would be very concerned," she added.

Jack Hein was excited as well, remembers Mullenix. (Harold Hodge had died before she could get the final results to him.)[1] "Hein said, 'I want you to go to Washington,'" Mullenix said. "'Go to the National Institute of Dental Research and give them a seminar. Tell them what you are finding.'"

Jack Hein knew that if more research on the toxicity of low-dose fluoride was to be done, the government's National Institutes of Health and the U. S. Public Health Service needed to be involved.

THE CAMPUS-STYLE GROUNDS of the federal National Institutes of Health (NIH), just north of Washington DC, have the leafy spaciousness of an Ivy League college. White-coated scientists and government bureaucrats in suits and ties stroll the tree-lined walkways that connect laboratories with office buildings. This is the headquarters of the U.S. government's efforts to coordinate health research around the country, with an annual budget of $23.4 billion forked out by US taxpayers.[2] The campus is the home of the different NIH divisions, such as the National Cancer Institute and the National Institute of Dental Research (NIDR), as it was then known. (Today it is known as the National Institute of Dental and Craniofacial Research.)

On October 10, 1990, Phyllis Mullenix and Jack Hein arrived at the NIH campus to tell senior government scientists and policy makers about her fluoride research. As director of the nation's leading private dental-research institute, Jack Hein was well-known and respected at NIH. He had helped to arrange the Mullenix lecture. Mullenix was no stranger to public-health officials either. One of the Institutes' biggest divisions, the National Cancer Institute, had awarded her a grant that same year totaling over $600,000. The money was for a study to investigate the neurotoxic effects of some of the drugs and therapies used in treating childhood leukemia. Many of those drugs and radiation therapies can slow the leukemia but are so powerful that they often produce central-nervous-system effects and can retard childhood intelligence. The government wanted Mullenix to use her new RAPID computer technology at Forsyth to measure the neurotoxicity of these drugs.

To present her fluoride data, Mullenix and Hein had flown from Boston, arriving a little early. Hein met up with some old friends from NIDR, while Mullenix strolled into the main hospital building on the Bethesda campus, killing time before her seminar. In the hallway, the scientist started to giggle. On the wall was a colorful posterboard display, recently mounted by NIH officials, titled "The Miracle of Fluoride."

"I thought how odd," remembered Mullenix. "It's 1990 and they are talking about the miracle of fluoride, and now I'm going to tell them that their fluoride is causing a neurotoxicity that is worse than that induced by some cases of amphetamines or radiation. I'm here to tell them that fluoride is neurotoxic."

She read on. Ironically, her trip to Washington fell on the historic fortieth anniversary of the Public Health Service's endorsement of community water fluoridation. Mullenix knew little about fluoride's history. The chemical had long been the great white hope of the NIDR, once promising to vanquish blackened teeth in much the same way that antibiotics had been a magic bullet for doctors in the second half of the twentieth century, beating back disease and infection.

Terrible teeth had stalked the developed world since the industrial revolution, when the whole-grain and fiber diet of an earlier agrarian era was often replaced by a poorer urban fare, including increased quantities of refined carbohydrates and sugars.[3] Cavities are produced when bacteria in the mouth ferment such sugars and carbohydrates, attacking tooth enamel, with the resulting acid penetrating into the tooth's core. Hope of a simple fix for bad teeth arrived in the 1930s, when a Public Health Service dental researcher named Dr. H. Trendley Dean reported finding fewer dental cavities in some parts of the United States, where there is natural fluoride in the water supply. Dean's studies became the scientific underpinning for artificial water fluoridation, which was begun in the 1940s and 1950s. Dean also became the first head of the NIDR. By the 1960s and 1970s, with rates of tooth decay in free fall across the United States, dental officials pointed a proud finger at the fluoride added to water and toothpaste. NIDR officials revered H. Trendley Dean as "the father of fluoridation."

"It was a major discovery by the Institute," said Jack Hein.

But opposition to fluoridation had been intense from the start. The postwar decline in rates of dental decay in developed nations had also occurred in communities where fluoride was *not* added to drinking water and had begun in some cases before the arrival of fluoride toothpaste.[4] Widespread use of antibiotics, better nutrition, improved oral hygiene, and increased access to dental care were also cited as reasons. And while medical and scientific resistance to fluoridation had been fierce and well-argued—the grassroots popular opposition was in many ways a precursor of today's environmental movement—Mullenix found the NIH's posterboard account of antifluoridation history to be oddly scornful. "They made a joke about antifluoridationists all being 'little old ladies in tennis shoes,'" she said. "That stuck in my mind."

Since Dean's day laboratory studies have forced a revolution in official thinking about how fluoride works.[5] While early researchers speculated that swallowed fluoride was incorporated "systemically" into tooth enamel even before the tooth erupted in a child's mouth—making it more resistant to decay—scientists now believe that fluoride acts almost exclusively from outside the tooth, or "topically" (such a "topical" effect has always been the explanation for how fluoride toothpaste functions, too). This new research says that fluoride defends teeth by slowing the harmful "demineralization" of calcium and phosphate from tooth enamel, which can leave teeth vulnerable to cavities. Fluoride also helps to "remineralize" enamel by laying down fresh crystal layers of calcium and a durable fluoride compound known as fluorapatite. And there is a third "killer" effect, in which the acid produced from fermenting food combines with fluoride, forming hydrogen fluoride (HF). This powerful chemical can then penetrate cell membranes, interfering with enzyme activity, and rendering bad bacteria impotent.[6]

"I still believe that fluoride works," says the Canadian dental researcher turned critic of water fluoridation, Dr. Hardy Limeback. "It works topically."

But these new ideas have not quenched the old debate. Dental officials now argue that water fluoridation produces a lifelong benefit not just for children; by bathing all teeth in water, officials argue, fluoride is continually repairing and protecting tooth enamel in teeth of all ages. Critics worry, however, that if hydrogen fluoride can inhibit bacteria enzymes in the mouth, then swallowing fluoride may unintentionally deliver similar "killer" blows to necessary bodily enzymes, thus also inhibiting the ones we need.[7]

Phyllis Mullenix, reading the NIH fluoride posters and preparing to give her speech on that fall day in 1990, knew almost nothing of the history of controversy surrounding fluoride. She was about to walk into the lion's den. She was stunned when she entered the lecture hall at the National Institutes of Health. It was packed. There were officials from the Food and Drug Administration. She spotted the head of the National Institute of Dental Research, Dr. Harald Löe, and she noticed men in uniform from the Public Health Service.

The lights dimmed. Mullenix told them about the new RAPID computer technology at Forsyth. At first the audience seemed excited. Then she outlined her fluoride experiment. She explained that the central-nervous-system effects seen in the rats resembled the injuries seen when rats were given powerful antileukemia drugs and radiation therapies. The pattern of central-nervous-system effects on the rats from fluoride "matched perfectly," she said.

The room fell suddenly quiet. She attempted a joke. "I said, 'I may be a little old lady, but I'm not wearing tennis shoes,'" she remembers. "Nobody was laughing. In fact, they were really kind of nasty."

The big guns from the NIH opened up. Hands shot into the air. "They started firing question after question, attacking me with respect to the methodology," remembered Mullenix. She answered their questions patiently, and finally, when there were no more hands in the air, she and Jack Hein climbed into a cab and headed for the airport.

Jack Hein is reluctant to discuss these long-ago events. It was a messy ending to his career. He retired from Forsyth the following year, in 1991. He agrees that the Mullenix fluoride results were unpopular but adds that data showing fluoride damage to the central nervous system should have been "vigorously" followed up. "That perspective had never been looked at before," he remarks. "It turned out there was something there." Hein believes that getting the NIDR and the government to change their position on fluoride, however, is a difficult task. Many senior public-health officials have devoted their professional careers to promoting fluoride. "NIDR really fought hard showing that fluoride was effective," Hein says.

"It was a major discovery by the Institute. "They did everything they could to promote it."[8]

Hein made a final effort to sound a warning on fluoride. He told Mullenix that he was going to call a meeting of industry officials whose products contained fluoride. Like Mullenix, Hein had spent a career cultivating ties with various large-scale industries. He sent her a note listing the "people who are coming for a private 'Fluoride Toxicity'" conference that would be held in his Forsyth office. "He said, 'NIDR were being stupid, the industries will respond better,'" Mullenix recalls.

Several months after the Washington seminar, Phyllis Mullenix sat at the table in Jack Hein's office with representatives from three of the world's most powerful drug companies: Unilever, Colgate-Palmolive, and SmithKline Beecham. Anthony Volpe, Colgate-Palmolive's Worldwide Director of Clinical Dental Research, was there, and so was Sal Mazzanobile, Director of Oral Health Research for Beecham. The senior scientist Joe Kanapka was sent by the big transnational company Unilever.

Mullenix outlined her fluoride findings. The men took notes. Suddenly Joe Kanapka of Unilever leaned back in his chair with an exasperated look. "He said, 'Do you realize what you are saying to us, that our fluoride products are lowering the IQ of children?'" remembers Mullenix. "And I said, 'Well yes, that is what I am saying to you.'" As they left, the men "slapped me on the back," Mullenix said, telling her, "We will be in touch, we need to pursue this."

The next day a note from Jack Hein's office arrived with the telephone numbers of the industry men, so that she could follow up. "I did call them," says Mullenix. "And I called. And the weeks went by and the months went by." Eventually Joe Kanapka from Unilever called back, she remembers. "He says, 'I gave it to my superiors and they haven't gotten back to me.'"

Contacted recently, Joe Kanapka said that he had visited Forsyth "many times" but had no memory of the fluoride conference. When asked if he had once worried that his products might be hurting children's intelligence, he replied, "Oh God, I don't remember anything like that, I'm sorry." He explained that open-heart surgery had temporarily impaired his memory. "I don't remember who Mullenix is," he added.

Beecham's Sal Mazzanobile remembers the meeting. The fluoride data presented that day were "preliminary," he recalled. Mullenix never called him again, he claims, and he therefore presumed her data were inaccurate. "I can't see why, if somebody had data like that, they would not follow up with another study in a larger animal model, maybe then go into humans," he said. "It could be a major health problem."

Did the director of consumer brands at Beecham—makers of several fluoride products—call Mullenix himself or find out if her data were ever published? "I wasn't the person responsible to follow up, if there was a follow-up," Mazzanobile answered. He did not remember who at Beecham, if anybody, might have had responsibility for keeping apprised of the Mullenix research.

Procter and Gamble followed up on Mullenix's warning. They flew her out to their Miami Valley laboratories in Cincinnati. Mullenix flew home with a contract and some seed money to begin a study to look at the effects of fluoride on children's intelligence. Shortly afterward, however, "they pulled out and I never heard from them again," recalls Mullenix.

In 1995 Mullenix and her team published their data in the scientific journal *Neurotoxicology and Teratology.* Their paper explained that, while a great deal of research had already been done on fluoride, almost none had looked at fluoride's effects on the brain. And while earlier research had suggested that fluoride did not cross the crucial blood brain barrier, thus protecting the central nervous system, Mullenix's findings now revealed that "such impermeability does not apply to chronic exposure situations."[9]

When the baby rats drank water with added fluoride, the scientists had measured increased fluoride levels in the brain. And more fluoride in the brain was associated with "significant behavioral changes" in the young rats, which resembled "cognitive deficits," the scientists reported. The paper also suggested that when the fluoride was given to pregnant rats, it reached the brain of the fetus, thus producing an effect resembling "hyperactivity" in the male newborns.

The Mullenix research eventually caught the attention of another team of Boston scientists studying central-nervous-system problems. They produced a report in 2000 reviewing whether toxic chemicals had a role in producing what they described as "an epidemic of developmental, learning and behavioral disabilities" in children. Their report considered the role of fluoride, and focused on the Mullenix research in particular. "In Harm's Way—Toxic Threats to Child Development" by the Greater Boston chapter of Physicians for Social Responsibility described how 12 million children (17 percent) in the United States "suffer from one or more learning, developmental, or behavioral disabilities." Attention deficit and hyperactivity disorder (ADHD) affects 3 to 6 percent of all schoolchildren, although recent evidence suggests the prevalence may be much higher, the scientists noted. Not enough is known about fluoride to link it directly to ADHD or other health effects, the report pointed out. Nevertheless, the existing research on fluoride and its central-nervous-system effects were "provocative and of significant public health concern," the team concluded.

The Mullenix research surprised one of the authors of the report, Dr. Ted Schettler. He had previously known almost nothing about fluoride. "It hadn't been on my radar screen," he said. Most startling was how few studies had been done on fluoride's central-nervous-system effects. Schettler turned up just two other reports, both from China, suggesting that fluoride in water supplies had reduced IQ in some villages."That just strikes me as unbelievable quite frankly," he said. "How this has come to pass is extraordinary. That for forty years we have been putting fluoride into the nation's water supplies—and how little we know about [what] its neurological developmental impacts are. . . . We damn well ought to know more about it than we do."

Does Mullenix's work have any relevance to children? Schettler does not know. Comparing animal studies to humans is an uncertain science, he explained. Nor was Schettler familiar with Mullenix's computer testing system. But the toxic characteristics and behavior of other chemicals and metals, such as lead and mercury, concern him. For those pollutants, at least, human sensitivity is much greater than in animal experiments; among humans, it is greater in children than in adults. The impact of other toxic chemicals on the developing brain is often serious and irreversible.

So is the Mullenix work worth anything? "I don't know the answer to that," Schettler said. "But what I do draw from it is that it is quite plausible from her work and others that fluoride interferes with normal brain development, and that we better go out to get the answers to this in human populations."

The burden of testing for neurological effects falls on the Public Health Service, which has promoted water fluoridation's role in dental health for half a century. "Whenever anybody or any organization attempts a public health intervention, there is an obligation to monitor emerging science on the issue—and also continue to monitor impacts in the communities where the intervention is instituted. So that when new data comes along that says, 'Whoa, this is interesting, here is a health effect that we hadn't thought about,' we better have a look at this to make sure our decision is still a good one," Schettler said.

Phyllis Mullenix says that she carried the ball just about as far as she could. Following the seminar at NIH, Harald Löe, the director of the National Institute of Dental Research, had written to Forsyth's director Jack Hein on October 23, 1990, thanking him and Mullenix for their visit and confirming "the potential significance of work in this area." He asked Mullenix to submit additional requests for funding. "NIDR would be pleased to support development of such an innovative methodology which could have broad significance for protecting health," Löe wrote.[10]

"I was very excited about that," said Mullenix. "I took their suggestions in the letter. [However] every one of them ended up in a dead end."[11] Mullenix now believes that the 1990 letter was a cruel ruse—to cover up the fact that the NIH had no interest in learning about fluoride's potential central-nervous-system effects. "What they put in writing they had no intentions [of funding]. It took years to figure that out," she says.

Dr. Antonio Noronha, an NIH scientific-review adviser familiar with Dr. Mullenix's grant request, says a scientific peer-review group rejected her proposal. He terms her claim of institutional bias against fluoride central-nervous-system research "farfetched." He adds, "We strive very hard at NIH to make sure politics does not enter the picture."[12]

But fourteen years after Mullenix's Washington seminar the NIH still has not funded any examination of fluoride's central-nervous-system effects and, according to one senior official, does not currently regard fluoride and central-nervous-system effects as a research priority. "No, it certainly isn't," said Annette Kirshner, a neurotoxicology specialist with the National Institute of Environmental Health Studies (NIEHS). Dr. Kirshner confirmed that although "our mission is to look into the effects of toxins [and] adverse environmental exposures on human health," she could recall no grants being given to study the central-nervous-system effects of fluoride. "We'd had one or two grants in the past on sodium fluoride, but in my time they've not been 'neuro' grants, and I've been at this institute about thirteen and a half years." Does NIEHS have plans to conduct such research? "We do not and I doubt if the other Institutes intend to," said Dr. Kirshner by e-mail.

Nor do the government's dental experts plan on studying fluoride's central-nervous-system effects any time soon. In an e-mail sent to me on July 19, 2002, Dr. Robert H. Selwitz of the same agency wrote that he was "not aware of any follow-up studies" nor were the potential CNS effects of fluoride "a topic of primary focus" for government grant givers. Dr. Selwitz is the Senior Dental Epidemiologist and Director of the Residency Program in Dental Public Health, National Institute of Dental and Craniofacial Research, NIH. At first he appeared to suggest that the Mullenix study had little relevance for human beings, telling me that her rats were "fed fluoride at levels as high as 175 times the concentration found in fluoridated drinking water."

But his statement was subtly misleading. Rats and humans have very different metabolisms, and in laboratory experiments these differences must be compensated for. The critical measurement in studying effects on the central nervous system is not how much fluoride is given to the laboratory animals but how much of the chemical, after they drink it, subsequently appears in the animals' blood. The amount of fluoride in the blood of the Mullenix rats—a measurement known as the blood serum level—had been the equivalent of what would appear in the blood of a human drinking about 5 parts per million of fluoride in water. This, of course, is just five times the level the government suggests is "optimal" for fluoridated water—1 ppm. I asked Dr. Selwitz, therefore, if it was fair to portray the Mullenix rats as having drunk "175 times" the amount of fluoride that citizens normally consume from fluoridated water.

Wasn't the "blood serum" measurement and comparison more relevant? Wasn't his statement, inadvertently at least, misleading?

Dr. Selwitz, who had just been ready to dispense medical arguments and implied reassurances as to why Mullenix's research was not relevant to human beings, now explained that he could not answer my question. "The questions you are asking in your recent e-mail message involve the field of fluoride physiology," wrote the senior dental epidemiologist at NIDCR. "This subject is not my area of expertise."

FAR FROM USHERING in new opportunities for scientific research, Mullenix's fluoride studies appear to have spelled the death knell for her once-promising academic career. When Jack Hein retired from Forsyth on June 30, 1991, the date marked the beginning of a very different work environment for Phyllis Mullenix. She gave a seminar at Forsyth on February 20, 1992, outlining what she had discovered and explaining that she hoped to publish a major paper about fluoride toxicity with Pamela DenBesten.

"That's when my troubles started," said Mullenix.

Pam DenBesten had been worried about the Boston seminar. Senior researchers at Forsyth, such as Paul DePaola, had published favorable research on fluoride since the 1960s. The seminar was "ugly," says Mullenix. DenBesten describes the scientists' response as "angry" and "sarcastic." "She was risking their reputation with NIH," DenBesten explains.

Karen Snapp remembers "hostile" questioning of Mullenix by the audience. "They looked upon Phyllis's research as a threat. The dental business in this country is focused on fluoride. They felt that funding would dry up. We are supposed to be saying that fluoride is good for you, whereas somebody is saying maybe it is not good for you. . . . In their own little minds, they were worried about that."

The following day Forsyth's associate director, Don Hay, approached Mullenix. "He said, 'You are going against what the dentists and everybody have been publishing for fifty years, that this is safe and effective. You must be wrong,'" Mullenix recalled. "He told me, 'You are jeopardizing the financial support of this entire institution. If you publish these studies, NIDR is not going to fund any more research at Forsyth.'"

Karen Snapp also remembers Don Hay as opposing publication of the paper. "He didn't believe the science. He didn't believe the results—and he did not think the paper should go out." Both Snapp and Mullenix were concerned that somehow Don Hay would prevent the paper from being published. "I think we were even laughing about it, saying 'I think in America we have something called freedom of the press, freedom of speech?'" Snapp recalls.

Don Hay calls allegations that he considered suppressing the Mullenix research "false." He told *Salon.com*: "My concern was that Dr. Mullenix, who had no published record in fluoride research, was reaching conclusions that seemed to differ from a large body of research reported over the last fifty years. We had no knowledge of the acceptance of her paper prior to the time she left [Forsyth]."[13]

Editor Donald E. Hutchings of *Neurotoxicology and Teratology*, where the Mullenix paper was published, says that there was no effort to censor or pressure him in any way. Her study was first "peer-reviewed" by other scientists, revised, and then accepted. "Was I called and told that 'If you publish this we are going to review your income taxes, [or] send you a picture of J. Edgar Hoover in a dress?' No," he said. Hutchings was a little bemused, however, to get such a critical paper on fluoride from a Forsyth researcher. He knew that Forsyth had long been a leading supporter of a role for fluoride in dentistry. "It almost strikes me like you are working in a distillery and you are doing work studying fetal alcohol syndrome. That is not work that they are going to be eager to be sponsoring. I didn't care—it wasn't my career. I thought it was really courageous of her to be doing that."

On May 18, 1994—just days after the paper had been accepted—Forsyth fired Mullenix. The termination letter merely stated that her contract would not be renewed. There was no mention of fluoride. A new regime was now installed at the Center. The toxicology department was closed, and a new Board of Overseers had been established, with the mission "to advise the Director in matters dealing with industrial relationships."[14]

Mullenix describes the final couple of months at Forsyth as the lowest ebb in her career. The big grant from the National Cancer Institute had dried up and her laboratory conditions were horrible, she said. "The roof leaked, they destroyed the equipment, they destroyed the animals. That was the lowest point, right before I physically moved out in July 1994. Nobody would even talk to me."

Her mother remembers Phyllis calling frequently that summer. "She was very upset about it," said Olive Mullenix. At first she wondered if her daughter had done something wrong. Phyllis explained that her fluoride research had been unpopular. "There was no use to get angry," said Olive Mullenix. "She was honest about what she found and they didn't like it."

Stata Norton got calls too from her former student. Norton was not surprised at the hostile response from Forsyth. She knew that clean data can attract dirty politics. "There are situations in which people don't want data challenged, they don't want arguments," said Norton.

The implications of Mullenix's work have been buried, according to her former colleague, the scientist Karen Snapp. "Is it fair to say that we don't know the answer to the central-nervous-system effects of the fluoride we currently ingest? I think that Phyllis got just the tip of the iceberg. There needs to be more work in that area," Snapp said.

Jack Hein wishes that he had approached things differently. He knew that the scientific landscape of the last fifty years was "littered with the bodies of a lot of people" who, like Phyllis Mullenix, "got tangled up in the fluoride controversy." His team should have tested other dental materials before tackling fluoride, said Hein. "It would have been better if we had done mercury and then fluoride," he said. "Less controversial."

It would have made no difference, believes Mullenix. Nor does she believe another scientist would have been treated differently. She had stellar academic credentials, powerful industry contacts, and hard scientific data about a common chemical. "That is the sad part of it," she said. "I thought I had the people back then. I thought you could reason one scientist to another. I don't know that there is anything I could have done differently, without just burying the information."

Mullenix no longer works as a research scientist. Since her fluoride discovery at Forsyth a decade ago, she has received no funding or research grants. "I liked studying rats," she said. "I probably would have continued working with the animals my entire life. Now," she added, "I don't think I will ever get to work in a laboratory again."

Jack Hein and Pamela DenBesten knew about fluoride's bizarre undertow, one that could pull and snatch at even the most established scientist, and they were able to swim free from the Forsyth shipwreck. But Mullenix was dragged down by a tide that no one warned her about. "I didn't understand the depth," she said. "And to me, in my training, you pay no attention to that. The data are the data and you report them and you publish and you go from there."

Mullenix is disappointed at the response of her fellow scientists. Jack Hein walked off into the sunset of retirement. Most of her former colleagues were reluctant to support her call for more research on fluoride, she said. Instead of saying "maybe scientifically we should take another look, everybody took cover, they all dove into the bushes and wouldn't have anything to do with me."

Olive Mullenix did not raise her daughter that way. "You can't just walk away from something like this," Phyllis Mullenix said. "I mean, they had to find out that thalidomide was wrong and change. Why should fluoride be any different?"

"A Spooky Feeling"

ONE HOT JULY evening in 1995 the phone rang. Dr. Phyllis Mullenix was in her office, upstairs in her Andover, Massachusetts, home. Scientific papers were strewn on the floor. She had been depressed. Her firing from Forsyth the previous summer had hit the family hard. Her daughters were applying to college; she and her husband, Rick, were quarreling about money.

She lifted the receiver. A big bass voice boomed an apology from New York City for calling so late. Mullenix did not recognize the speaker. She settled back into her favorite white leather armchair. Joel Griffiths explained that he was a medical writer in Manhattan. He had a request. Would Mullenix look at some old documents he had discovered in a U.S. government archive? The papers were from the files of the Medical Section of the Manhattan Project, the once supersecret scientific organization that had built the world's first atomic bomb.

Mullenix rolled her eyes. It was late. Rick, now an air traffic controller, was trying to sleep in the next room. The atom bomb, Mullenix thought! What on earth did that have to do with fluoride?

Mullenix's own patience was growing thin. Since her research had become public, she had been bombarded with phone calls and letters from antifluoride activists. Some of the callers had been battling water fluoridation since the 1950s. Late-night radio talk shows were especially hungry to speak with the Harvard scientist who thought that fluoride was dangerous. They called her at three or four in the morning from across the country and overseas. Usually "there was no thank you note, and you never heard from them again," Mullenix said.

The New York reporter dropped a bombshell. Dr. Harold Hodge, Mullenix's old laboratory colleague, was described in the documents as the Manhattan Project's chief medical expert on fluoride, Griffiths told her. Workers and families living near atomic-bomb factories during the war had been poisoned by fluoride, according to the documents, and Harold Hodge had investigated.

Mullenix felt a sudden "spooky" feeling. She shifted in her chair. Harold Hodge was now dead, but as the journalist continued, Mullenix cast her mind back to the days in her Forsyth laboratory with the kind old gentleman, the grandfatherly figure who had sometimes played with her children.

"All he did was ask questions," she told Griffiths. "He would sit there and he would nod his head, and he would say, 'You don't say, you don't say.'" Once, Mullenix recalled, as Hodge watched her experiments, he had briefly mentioned working for the Manhattan Project. But he had never said that fluoride had anything to do with nuclear weapons—or that he had once measured the toxic effects of fluoride on atomic-bomb workers. Yes, Mullenix told the journalist, she wanted to see the documents.

Some days later a colleague of Griffiths's arrived at the Mullenix home. Clifford Honicker handed her a thick folder of documents. Honicker was part of a small group of researchers and reporters who had unearthed many of the ghoulish medical secrets of the Manhattan Project and the Atomic Energy Commission. Those secrets had included details about scores of shocking cold-war human radiation experiments on hospital patients, prisoners, pregnant women, and retarded children.

For years the media had ignored the information about human experimentation that Honicker and others were discovering. Finally, in 1995, an investigative journalist named Eileen Welsome had won a Pulitzer Prize for revealing how atomic-bomb-program doctors had injected plutonium into hospital patients in Tennessee and New York. She uncovered the names of the long-ago victims. Harold Hodge had planned and supervised many of those experiments, the documents showed. President Bill Clinton ordered an investigation. His energy secretary, Hazel O'Leary, began a new policy of openness. And Honicker and others had gained access to newly declassified cold-war documents—including much of the new information on fluoride.

That night, after Honicker left, Mullenix settled in her chair and began to read. Her face drained as she read one memo in particular. The fifty-year-old document mentioned Harold Hodge—and discussed fluoride's effects on the brain and central nervous system. It was the same work she had done at the Forsyth Dental Center.

"I went white. I was outraged," said Mullenix. "I was hollering and pacing the floor. He wrote this memo saying that he knew fluoride would affect the central nervous system!"

The central-nervous-system memo—stamped "secret"—is addressed to the head of the Manhattan Project's Medical Section, Colonel Stafford Warren, and dated April 29, 1944. It is a request to conduct animal experiments to measure the central-nervous-system effects of fluoride. Dr. Harold Hodge wrote the research proposal.

"Clinical evidence suggests that uranium hexafluoride may have a rather marked central nervous system effect. . . . It seems most likely that the F [code for fluoride] component rather than the T [code for uranium] is the causative factor," states the memo.[15]

A light flashed on for Mullenix. At the time, in 1996, she was still sending grant requests to the National Institutes of Health in Washington, DC, asking to continue her studies on fluoride's central-nervous-system effects. A panel of NIH scientists had turned down the application, flatly telling her, "Fluoride does not have central nervous system effects." Mullenix realized the absurdity of what she had been doing. Harold Hodge and the government had suspected fluoride's toxic effects on the human central nervous system for half a century.

She read on. The 1944 memo explained why research on fluoride's central-nervous-system effects was vital to the United States' war effort. "Since work with these compounds is essential, it will be necessary to know in advance what mental effects may occur after exposure. . . . This is important not only to protect a given individual, but also to prevent a confused workman from injuring others by improperly performing his duties."

"All of a sudden it dawned on me," said Mullenix. "Harold Hodge, back in the 1940s, had asked the military to do a study that I had done at Forsyth. . . . Hodge knew this fifty years ago. Why didn't he tell me what he was interested in? Why didn't he say to me, 'This stuff, I know, is a neurotoxin?'" All he did was ask questions, and he would sit there and he would nod his head and he would say, 'You don't say, you don't say.' He never once said, 'I know it is a neurotoxin, I know it causes confusion, lassitude, and drowsiness.'"

Today Mullenix calls Harold Hodge a "monster" for his human-radiation experiments. In retrospect she compares sharing a laboratory with him with "being in a movie theater, sharing popcorn with the Boston Strangler."

Had the two Rochester alumni—Jack Hein and Harold Hodge—manipulated the toxicologist to perform the fluoride studies that Hodge had proposed fifty years earlier, she wondered. Did they let Mullenix take the fall when her experiments proved what Hodge had already suspected? At first, Mullenix had shown no interest in studying fluoride, she remembered. "It seems strange that a neurotoxicology person was brought into a dental institution to look at fluoride," Mullenix said. "I felt that I had really been lied to, or led along," she added, "used like a little puppet."

Mullenix called up Jack Hein. He denied knowing anything about Harold Hodge's long-ago Manhattan Project fears that fluoride was a neurotoxin, she said. And instead, he offered to pass the explosive information on to the government, telling Mullenix, "Shouldn't you tell the NIDR—do you want me to help you take it to the NIDR?" (Hein may have known far more than he told Mullenix, however. In a 1997 interview with the United Kingdom's Channel Four television, he disclosed that one of the primary concerns of Manhattan Project toxicologists *had* been fluoride's effects on the central nervous system.)[16]

The next day Dr. Mullenix called the head of the National Institute of Dental Research, Dr. Harold Slavkin. She hoped the nation's top dental officer would be concerned about the wartime memo. Instead, she remembers, "He got very nasty about it. He basically pushed me off, like I was some kind of a crackpot." She thought that NIDR would be interested in the memos, that the institute would want to read them. But he treated her as if she were "some kind of a whacko," she recalls. She put the telephone down and a terrible truth dawned on her. The public guardians at the National Institutes of Health, like Harold Hodge, also had a double identity. It seemed they, too, were keepers of cold war national-security secrets—bureaucratic sentries at the portcullis of the nuclear-industrial state.

3

Opposite Sides of the Atlantic

Copenhagen: Crucible of Discovery

KAJ ELI ROHOLM had a passion for life and medicine. The son of a Danish sea captain and an immigrant Polish Jew, Roholm shone briefly as one of Europe's brightest stars. During the 1920s and 1930s, when Copenhagen glowed as a crucible of scientific discovery and Nils Bohr and a cadre of physicist disciples laid the theoretical foundation for nuclear fission, Kaj Roholm had advanced the healing arts.[1]

"He was a very vital and lively person," remembered the ninety-five-year-old Georg Brun, who met Roholm almost a lifetime ago, when both were young doctors training in a Danish hospital. They had talked eagerly about politics, history, and medicine.[2] Although a handful of specialists around the world today remember Roholm for his "great and lasting" study of fluoride toxicity, he was also a pioneer in the use of biopsy samples to study the human liver, an expert in infectious and occupational diseases, and a tireless advocate for public health.[3] "He was interested in everything," said Brun.

As Copenhagen's Deputy Health Commissioner in the late 1930s, the thirty-eight-year-old led his fellow doctors in campaigns against diphtheria and venereal diseases and in campaigns to improve the health of newborn children. He harnessed modern media to his public-health agenda, producing films, radio advertisements, posters, and brochures; and he arranged for wartime distribution of a hundred thousand copies of his pamphlet, "What

Everyone Wants to Know about Infectious Diseases."[4] When the Nazis marched into Denmark in April 1940, the doctor remained at his post. Although Copenhagen won the wartime reputation of a humane city—where Jews escaped much of the violence occurring in other occupied European cities—Roholm described occupation conditions as "awful."[5]

A quirk in the Earth's geology drew Roholm to fluoride. Virtually the entire world's supply of the fluoride-containing mineral known as cryolite was found, at the time, in a single deposit beneath the Danish colony of Greenland. Cryolite is an Eskimo word meaning *ice stone*. Trade in the brilliant white rock had grown rapidly in the early twentieth century, after researchers learned that aluminum could be made more cheaply by using electricity to melt the ice stone in a glowing-hot "pot," along with refined bauxite ore. A great river of this aluminum had armed soldiers with munitions and lightweight equipment during World War I.[6]

As the cryolite ships arrived in Denmark, the ice stones were hauled to the Øresund Chemical Works in Copenhagen, where a heavy cloud of cryolite dust filled the factory air and where a medical mystery preoccupied doctors. Inside the plant the Danish workers were stricken with multiple ailments, including a bizarre crippling of their skeletons known as poker back. Professor P. Flemming Møller of the Rigshospital suspected that fluoride was responsible; cryolite contains more than 50 percent fluoride. In 1932 Møller labeled the disease "cryolite intoxication" and suggested that a young doctoral candidate, Kaj Roholm, study the newly discovered condition.[7]

Roholm seized the challenge with the passion of youth. He listened carefully to the complaints of the Copenhagen cryolite workers, examining them with the use of X-rays. He conducted his own laboratory experiments, feeding fluoride to pigs, rats, and dogs in order to study its biological effects. A shocking picture emerged of a chemical with a venomous and hydra-headed capacity for harm. Silently and insidiously fluoride stole into the workers' blood—from swallowed dust, Roholm reported, with the poison accumulating in teeth, bones, and quite possibly the workers' kidneys and lungs.[8] Eighty-four percent of the workers at the cryolite plant had signs of osteosclerosis. Their bones sopped up fluoride like sponges, wreaking havoc on their skeletons, immobilizing spinal columns, malforming knees and hips, and even thickening some men's skulls. Half the employees had a lung condition known as pulmonary fibrosis and many suffered from an emphysema-like affliction.[9] And in a disease process that resembled the effects of aging, the workers' ligaments grew hard and sprouted bony spines, while their bones became lumpy and irregular in shape.[10] "Arthritic and rheumatic afflictions have a marked frequency" among the employees, Roholm stated, and serious stomach problems were commonplace; several cryolite workers also had chronic skin rashes and pussy sores on their chest and back, especially in the summer.

Fluoride probably poisoned the central nervous system as well. "The marked frequency of nervous disorders after employment has ceased might indicate that cryolite has a particularly harmful effect on the central nervous system," Roholm noted.[11] He called the disease "fluorine intoxication" and suspected that it was fluorine's ability to poison enzymes—the chemical messengers that regulate much bodily activity—that made it a threat on so many biological fronts. "We must assume that the effect of fluorine on protoplasm and on enzymatic processes is capable of causing profound changes in the metabolism of the organism," Roholm added.[12]

The scientist also examined fluoride's effects on teeth. There had been scientific speculation since the nineteenth century that because ingested fluoride was deposited in teeth and bone, it was therefore necessary for healthy teeth.[13] A team at Johns Hopkins University tested that theory in 1925, feeding rats fluoride, but found that it made their teeth weaker.[14] Roholm found the same thing. The workers' teeth he studied were bad, and the worst teeth had the most fluoride in them. Lactating mothers in the Copenhagen factory had even poisoned their own children; since fluoride passed though their breast milk, children who had never been inside the plant developed mottled teeth—evidence that mother and child had been exposed to an industrial chemical.[15]

Roholm's conclusions on fluoride and teeth were blunt. "*The once general assumption that fluorine is necessary to the quality of the enamel rests upon an insufficient foundation. Our present knowledge most decidedly indicates that fluorine is not necessary to the quality of that tissue, but that on the contrary the enamel organ is electively sensitive to the deleterious effects of fluorine*," he wrote (emphasis in original).[16] His medical recommendation: "Cessation of the therapeutic use of fluorine compounds for children."[17] In other words, more than sixty years ago the world's leading fluoride scientist rejected the notion that fluoride was needed for stronger teeth, agreeing with earlier studies that found that fluoride weakened the enamel—and explicitly warning against giving fluoride to children.

Roholm continued his investigation. He traveled to places where he suspected that similar such fluoride intoxication had occurred, and he read widely in the great libraries of Berlin and London. A clear picture emerged: the scientist saw how fluoride's chemical potency had long caused problems in the natural world and that its usefulness to modern industry was increasingly causing problems in human affairs.[18] In Iceland he saw grazing sheep that were emaciated and crippled, their teeth weakened, with a disease called *gaddur.* Their forage had become contaminated with fluoride spewed into the biosphere from deep inside the earth during volcanic eruptions. The disease especially injured young animals.[19] In the United States, such natural fluoride had plagued the westward-sweeping migrants in Texas, South Dakota, Arizona, and Colorado. These thirsty pioneers had sunk wells deep into the desert but drew water that was contaminated with fluoride. The poison produced an ugly tooth deformity known as Colorado Brown Stain or Texas Teeth. (Today that deformity is known by the medical term dental fluorosis and is an early indicator of systemic fluoride poisoning. A more severe form of poisoning, produced by earth-bound natural fluoride, known as crippling skeletal fluorosis, is also widespread in much of the Third World, where lack of nutrition often worsens the fluoride's effects.)[20]

Roholm saw that in the industrial world fluoride had become a bedrock for key manufacturing processes; 80 percent of the world's supply of fluorspar, the most commonly used fluoride mineral, was used in metal smelting; steel, iron, beryllium, magnesium, lead, aluminum, copper, gold, silver, and nickel all used it in production.[21] (The word fluoride comes from the Latin root fluor meaning "to flux" or "to flow." Fluoride has the essential property of reducing the temperature at which molten metal is "fluxed" from superheated ore.) Brickworks, glass and enamel makers, and superphosphate fertilizer manufacturers each used raw materials that included enormous volumes of fluoride. And at DuPont's Kinetic Chemicals in New Jersey, scientists were giving birth to a new global industry of "organic" or carbon-based fluoride products, engineering man-made fluoride and carbon molecules to mass-produce a popular new refrigerant known as Freon.

Roholm saw that what had long befallen the natural world was now increasingly happening to human beings, and by their own hand. Industry's growing appetite for fluoride presented a special threat to workers and surrounding communities. The Dane studied case after case in which factory fluoride hurt workers and contaminated surrounding areas—and where angry lawsuits had been launched for compensation. In Freiburg, Germany, for example, smelters had been compensating their neighbors for smoke-damaged vegetation since 1855. In 1907 it was finally confirmed that fluoride smoke from those smelters had poisoned nearby cattle.[22] Similar damage to plants and cattle was seen elsewhere in Europe, near superphosphate fertilizer plants, brickworks, iron foundries, chemical factories, and copper smelters.[23] But although the damage was widespread, information about its chemical cause was less available. "The toxicity of fluorine compounds is considerable and little known in industry," Roholm wrote.

Science was partly to blame, he suggested. The industrial revolution, for example, had been fueled with coal, which had darkened the skies over cities such as Pittsburgh, Glasgow, Manchester, and London. But air pollution investigators had focused the blame for subsequent environmental damage and human injury on sulfur compounds rather than on the large quantities of fluoride frequently found in coal.[24]

Roholm suggested that even the century's worst industrial air pollution disaster to date, in Belgium's Meuse Valley—which killed sixty people and injured several thousand in December 1930—had been caused by fluoride, not sulfur. During the Meuse Valley incident thousands of panicked local citizens had scrambled up hillsides to flee choking gases during three days of horror. Roholm proposed that fluoride from the nearby factories had been trapped by a temperature inversion, then dissolved in moisture and carried by particles of soot deep into the victims' lungs.[25] Roholm thought that disaster investigators had overlooked both the toxicity and the prevalence of fluoride pollution from nearby zinc, steel, and phosphate plants. He calculated that tens of thousands of pounds of the chemical were spilled each day from the local factories, etching windows, crippling cattle, damaging vegetation, and making citizen lawsuits in the Meuse Valley "a well known phenomenon."[26]

Roholm singled out the new global aluminum industry. He studied a lawsuit against a Swiss manufacturer in which it was alleged that fluoride fallout during World War I had hurt cattle and vegetation. Animal injury was again found near an Italian aluminum plant in 1935; the following year scientists found health problems inside a Norwegian aluminum smelter, where workers suffered sudden gastric pains and vomiting, bone changes, and symptoms resembling bronchial asthma.[27] "A special position is occupied by aluminum works," Roholm wrote, "inasmuch as the damaged vegetation especially has caused secondary animal diseases."[28] He advocated government action: "Factories giving off gaseous fluorine compounds should be required to take measures for their effective removal from chimney smoke."[29]

Roholm's monumental 364-page study, *Fluorine Intoxication*, was published in 1937 and was quickly translated into English. It contained references to 893 scientific articles on fluoride. The trust and cooperation of the Danish cryolite industry was necessary to make his study. Nevertheless, the book was a warning to corporations: they must pay attention to their factory conditions and to the insidious—often misdiagnosed—effects of fluoride on workers. Roholm had several clear recommendations for employers and doctors, among them:

- Recognition of chronic fluorine intoxication as an occupation disease rating for compensation.
- Prohibition against employment of females and young people on work with fluorine compounds developing dust or vapor.
- Demand that industrial establishments should neutralize waste products containing fluorine.[30]
- A prohibition against the presence of fluorine in patent medicine may be necessary.[31]

Pittsburgh 1935

IT WAS A May morning in Pittsburgh, and a watery spring sun struggled through the smoky haze. Inside his office at the Mellon Institute, the director, Ray Weidlein, put down his newspaper in satisfaction. Several dailies had picked up a press release he had recently issued:

"New attack on Tooth Decay . . . to be carried on at the Mellon Institute" headlined a May 1, 1935, example in the *Youngstown (OH) Telegram*. Mellon researchers had "found evidence that the presence of a factor in the diet at a crucial period of tooth formation leads to the development of teeth resistant to decay," the newspaper proclaimed. A Mellon scientist, Gerald J. Cox, was to lead the hunt for the mysterious "factor" improving teeth, and Pittsburgh's well-known Buhl Foundation would fund the research on rodents.[32]

Since tooth decay was a major problem in the industrialized United States, the story must have seemed liked good news to most readers, and especially to dentists. But the headlines were certainly welcome good press for Ray Weidlein. Several of the big industrial corporations who funded the Mellon Institute's work had recently been dragged through the pages of the nation's media with some very unflattering stories—and were increasingly under attack from Congress and the courts. That spring *Time* magazine was one of several papers and magazines that had carried accounts of the horrific events at Gauley Bridge in West Virginia, where several hundred mostly black migrant miners had died from silicosis contracted while drilling a tunnel for the Union Carbide Company during 1931–1932. News of what would be America's worst industrial disaster to date had filtered out from Appalachia slowly, but by 1935 the West Virginia deaths had become a full-blown national scandal. Hundreds of lawsuits had been filed against Union Carbide and its contractors. Reporters were daily scrutinizing the often appalling rates of occupational illness in other industries. And sympathetic citizen juries were regularly awarding millions of dollars to injured workers, provoking a full-blown financial emergency for several leading industrial corporations—and panic among their insurers. In January Congress would hold hearings, and Gauley Bridge would, for many Americans, come to symbolize a callous disregard by powerful corporations for workers' health.[33]

Ray Weidlein and the Mellon Institute were in full crisis mode that spring of 1935, helping Union Carbide and other top corporations contain public outrage over the workplace carnage—and head off draconian legislation for better pollution control inside factories. The corporate strategy was clear: get dominion over basic science, wrestle control of health information from labor groups, and in turn, reinvest that medical expertise in the hands of industry-anointed specialists. These steps were seen as the "anti-toxin for the agitation against private enterprise," according to one of Weidlein's correspondents.[34] The besieged corporations organized a lobbying group known as the Air Hygiene Foundation because, as the group noted, "sound laws must be based on sound facts"; and, perhaps more importantly, because "half a billion dollars in damage suits have been filed against employers in occupational disease claims."[35]

Headquartered at the Mellon Institute, in 1937 the Air Hygiene Foundation had a membership list sporting many of the best-known names in industry, including Johns-Manville, Westinghouse, Monsanto, U.S. Steel, Union Carbide, Alcoa, and DuPont. And for the better part of the next thirty years the organization—later renamed the Industrial Hygiene Foundation—would profoundly shape the public debate over air pollution, goading members to voluntarily improve work conditions inside their factories, thus avoiding legal mandates, and sponsoring medical research that bolstered industry's medicolegal position in the courtroom. Such research, much of it done at the Mellon Institute, was "important from both medical and legal standpoints in the preparation of court cases," Ray Weidlein stated.[36]

An example of the Foundation's success in influencing the contest over air pollution and occupational hazards was the effort to "investigate" asbestos. One of the Foundation's members, Johns-Manville, was a top asbestos producer. The tiny fibers had been linked to ill health in workers since 1918. But as late as 1967 Dr. Paul Gross was using the Industrial Hygiene Foundation's laboratory to conduct influential medical research, permitting Foundation members to dispute the claim that asbestos fibers were uniquely dangerous.[37] His conclusions were erroneous—reportedly suspected as such even by his fellow Mellon scientists—yet corporate profits and worker pain were prolonged for a generation while the Mellon Institute continued grinding out its industry-backed "research."[38] We can blame today's flood of death and disease in asbestos workers—and the $54 billion in court awards against industry—at least partly on the Air Hygiene Foundation and the long-ago diligence of the Mellon Institute and its director, Dr. E. R. Weidlein.[39]

If Ray Weidlein smiled over the press release heralding Cox's dental studies that May morning in 1935, it may have been because no newspaper had spotted some important connections—between the tooth research at the Mellon Institute and the corporations funding the Air Hygiene Foundation lobby group, which was also run, of course, out of the Mellon Institute. By the early 1930s a tidal wave of new information about the health risk from low-level fluoride exposure was also filling medical libraries. Several members of the Air Hygiene Foundation were paying particularly close attention. As with silicosis and asbestos claims, big corporations were potentially at risk for massive corporate legal liability—for the harm caused to workers and communities by industrial fluoride exposure.[40]

One Foundation member had particular reason to worry. Tall and athletic, the chief scientist for the aluminum manufacturer Alcoa, Francis Frary, had studied in Berlin, was fluent in several languages, and would personally translate Kaj Roholm's fluoride research.[41] Conditions inside Alcoa's smelting plants were brutal, with "exposure to chemical agents (especially fluorides and carcinogens and, to a lesser degree alumina dusts and asbestos insulating materials)" a frequent hazard for workers, according to the historian George David Smith. "The effects of fluoride emissions was a particular concern of Frary's," Smith noted.[42] During the 1920s and 1930s, African American workers were imported from the Deep South for the "killing potroom labor" inside one plant in the company town of Alcoa, Tennessee. And at the Massena plant in upstate New York, where Alcoa's mostly immigrant workers were shipped in by train, a health study would later confirm that crippled workers were the result of a fluoride dust hazard that had existed at the plant for years.[43]

Francis Frary was a member of an elite fraternity of officials running corporate research labs, a fraternity that would chart the nation's scientific progress during the period between the two World Wars. Other members of this close-knit group included Charles Kettering, director of research for General Motors, and the research directors of U.S. Steel and DuPont.[44] "Those people all knew each other; it was a small, relatively select group who headed research labs," noted the historian Margaret Graham.[45]

Fluoride's threat to corporate America was laid out in an exhaustive review of the new medical information about fluoride's harmful effects, published in 1933 by the U.S. Department of Agriculture. A senior toxicologist, Floyd DeEds, warned of the growing risk from industrial fluoride pollution. "Only recently, that is within the last ten years," he stated, "has the serious nature of fluorine toxicity been realized, particularly with regard to chronic intoxication [a medical term for poisoning]." Like Kaj Roholm, the government scientist singled out the aluminum industry.[46]

DeEds also noted that in 1931 several researchers had, for the first time, linked the ugly blotching or "mottling" seen on teeth in several areas of the United States to naturally occurring fluoride in water supplies.[47] This new dental information appears to have rung an alarm bell for industry. Quietly Alcoa scientists made their own investigations. It was not just nature's fluoride that stained teeth, they discovered; the company found tooth mottling in children living near Alcoa's big aluminum plant in Massena, New York. Crucially, however, Alcoa's chemists reported that there was no naturally occurring fluoride in the local water.[48] A potential source of the fluoride staining children's teeth in Massena was obvious: there was little or no pollution control on many early aluminum plants, and elsewhere around the country the fluoride waste from these industries was routinely dumped in nearby rivers.[49]

Mottled teeth in children had become a potential red flag, warning citizens and workers of industrial fluoride pollution—and pointing directly to a man-made hazard the media had not yet discovered.[50] With public outrage over Gauley Bridge reaching a crescendo in 1935, several powerful industrial corporations now held their breath, hoping to avoid a fresh epidemic of worker lawsuits that this time were for fluoride exposure. The potential for litigation against industry was mapped for all to see by blotchy marks on children's teeth, evidence of "neighborhood fluorisis" in action.[51]

Alcoa's research director, Francis Frary, took action. In September 1935 he approached Gerald Cox, a Mellon Institute researcher, at the American Chemical Society's Pittsburgh meeting. Frary now had a suggestion that would ultimately transform the public perception of fluoride.[52] Though Frary was preoccupied with the "killing" hazards facing his Alcoa employees, and the aluminum industry faced lawsuits from farmers whose cattle had been injured in the vicinity of the smelters, Frary took it upon himself to make a generous suggestion to the Mellon researcher. Had Cox ever considered that good teeth might be caused by fluoride?

Cox understood that Frary was suggesting that he include fluoride in his tooth-decay study. Although this suggestion flew in the face of the results from the dental study at Johns Hopkins a decade earlier—which had showed that fluoride hurt teeth—nevertheless the Alcoa man's proposal was "the first time I ever gave fluorine a thought," Cox later told historian Donald McNeil.[53]

The great makeover of fluoride's image had begun. By August 1936 the Mellon researcher had given laboratory rats some fluoride and announced that the chemical was the mystery "factor" protecting teeth. In 1937 Ray Weidlein and Cox published details of their fluoride "discovery" in the scientific press. And the following year Cox declared in the *Journal of the American Medical Association* that "the case [for fluoride] should be regarded as proved."[54] Virtually overnight, the Mellon Institute rats had put a smiling face on what had been a scientifically recognized environmental and workplace poison.[55]

The Kettering Laboratory

FRANCIS FRARY WAS not the only industry scientist who had grown interested in children's teeth during those Depression years. In April 1936 his colleague Charles Kettering, vice president and director of research at General Motors, quietly held a meeting in GM's Detroit offices with a delegation from the American Dental Association (ADA) and Captain C. T. Messner of the U.S. Public Health Service.[56] Kettering seemed an unlikely candidate for an interest in teeth; he had become famous and wealthy by inventing the electric starter for the automobile. But Kettering's laboratory in Dayton, Ohio, was also the birthplace of two industrial chemicals that would haunt the twentieth century. And like Alcoa's Francis Frary, Kettering was in a unique position to see the health risk that fluorides posed to American workers—and the potential liability facing DuPont and General Motors.[57]

Fluoride and lead were twin pillars on which the great wealth of both DuPont and General Motors was built. In 1921 Kettering's scientists had discovered that lead added to gasoline increased engine efficiency. And in 1928 they patented the fluoride-based Freon gas, which was much less toxic at room temperature than were earlier refrigerants. But those twin pillars had shaky foundations. Tetra ethyl lead (TEL) was so toxic that it killed several of DuPont's New Jersey refinery workers, attracted a rash of ugly newspaper headlines, and almost resulted in the lucrative product's being banned from the market.[58] Similarly, Freon sales quickly stalled following protests from the American Standards Association and the New York City Fire Department, when it was discovered that when Freon was exposed to flame, it decomposed into the nightmarish phosgene and hydrogen fluoride gases.[59] (Phosgene was the same poison gas that had been used to monstrous effect in the trenches of World War I.)

GM and DuPont moved quickly to protect their new products. They hired a young scientist at the University of Cincinnati, Robert Arthur Kehoe, to perform safety studies on lead at GM's in-house laboratory. Kehoe's research—which asserted that lead was found naturally in human blood and that there was a "threshold" level below which no ill effect would be caused—helped to placate the U.S. Surgeon General and "single-handedly spared the leaded gasoline industry from federal regulation in the 1920s," according to the historian Lynne Snyder.[60] "Kehoe's first contract had salvaged a billion dollar industry," wrote another Kettering scientist, Dr. William Ashe.[61] The thirty-two-year-old was rewarded in 1925 with an appointment as the medical director of the Ethyl Corporation, which marketed leaded gasoline.[62]

In 1930 Kehoe rode to the rescue again, performing toxicity studies on Freon. That same year the Ethyl Corporation, DuPont, and the Frigidaire Division of General Motors founded a laboratory at the University of Cincinnati with a $130,000 donation. It was named the Kettering Laboratory of Applied Physiology; a new building was erected, and Kehoe was installed as director.

The dangers of using a potential poison gas in the home—and the risk to firefighters in particular—may have seemed obvious, but Kehoe argued that a blaze would rapidly disperse any poison that might be created, presenting little risk. "Thus even from a fire fighting point of view . . . the decomposition of [Freon] is not to be regarded as of great consequence," he stated.[63] (More than sixty years after his clash with New York firefighters Kehoe's toxic shadow haunted them in the aftermath of the World Trade Center terror attack.[64] Following the building's collapse, rescue workers feared that two enormous tanks of Freon gas that had once fed the towers' air-conditioning system would rupture and burn in the still-smoldering rubble, spewing acid and poison over downtown Manhattan.[65] Although there have been numerous previous reports of phosgene poisoning from Freon, mercifully the refrigerant never burned at Ground Zero.[66])

Kehoe's assurances helped to win the day. A joint venture between GM and DuPont, known as Kinetic Chemicals, quickly erected two massive Freon manufacturing facilities at DuPont's plant in Deepwater, New Jersey. Although Kettering scientists soon measured "high" levels of fluoride in DuPont's New Jersey workers, Freon sales soared from 1.2 to 18.7 million pounds between 1931 and 1943. Freon became the main refrigerant in homes and industry and grossed an estimated $35 million in revenue during this period.[67]

But new experiments soon discovered just how precarious DuPont's exploitation of fluorides might be. The Kettering Laboratory found that hydrofluoric acid—the raw material needed to make Freon and the same gas produced when the refrigerant was burned—was toxic in very low doses.[68] The scientists did not report a level below which toxic effects were *not seen*. The danger to workers who breathed the gas on a daily basis was clear. The gas was stealthy. Even at a level that could not be detected by smell, it caused "exceptional" injury, including lung hemorrhage, liver damage, and "striking evidences of kidney damage." Animals died when exposed to a dose of just 15.2 milligrams per cubic meter (about 19 parts per million).

That toxicity data was published in September 1935. Six months later Charles Kettering met with the American Dental Association. The Freon magnate quickly became a member of the ADA's three-person Advisory Committee on Research in Dental Caries. That Committee, in turn, shepherded publication of *Dental*

Caries—a compendium of dental research from around the world that included several references to Gerald Cox's work at the Mellon Institute as well as that of other fluoride promoters. Neither Charles Kettering's interests in selling industrial fluorides nor the potential health risk from fluorides to U.S. workers were ever disclosed to readers of *Dental Caries.* Nor were dentists told that the General Motors' vice president might have personally funded a portion of the ADA's activities.[69] In a letter dated March 16, 1937, the ADA's chairman, P. C. Lowery, somewhat cryptically promised "Kett" that he will "secure sufficient information" so that the General Motors vice president could, in turn, "furnish the $25,000." In other words, the millionaire industrialist with one of the greatest personal stakes in the commercial exploitation of fluorides was quietly donating to the dental organization that would shortly become one of the most aggressive boosters of fluoride's use in dentistry.[70]

A third connection between industry and some of the earliest attempts to link fluoride with dental health can be found in the actions of Andrew W. Mellon, who was U.S. Treasury Secretary from 1921 to 1932. The silver-haired smelter and Pittsburgh banker was also a founder of Alcoa and one of its biggest stockholders. In 1930 he intervened in efforts to have the Public Health Service support researchers at the University of Arizona who were then surveying naturally occurring tooth mottling.[71] (The U.S. Public Health Service [PHS] was then a division of the Treasury Department.) Mellon's economic interest was clear. Fluoride's legal threat to industry could now be seen, literally, in children's smiles. However, linking dental mottling to naturally occurring fluoride, in areas far from industry, helped to deflect attention from the bad teeth and the myriad other health effects caused by industrial fluoride pollution.[72] A young PHS researcher named H. Trendley Dean was promptly "ordered" to study fluoride. He soon confirmed that natural fluoride in water supplies produced dental mottling.[73] But like the industry scientists before him, Dean also developed "a hunch" that fluoride prevented dental cavities.[74] (Following this hunch, Dean later found that natural fluoride in the local water supplies apparently correlated with fewer cavities; these findings, although much criticized for their scientific method, eventually became a foundation for artificial water fluoridation.)[75]

Dean departed from Washington in the fall of 1931 to study fluoride and tooth decay throughout communities in the South and Midwest. His departure planted a seed for the government's fluoride policies. Several years later, another seed would take root. On September 29, 1939, Gerald Cox, the researcher at the Mellon Institute, made his most radical suggestion yet at a meeting of the American Water Works Association in Johnstown, Pennsylvania. His suggestion took place at a historic moment. The world stood on the precipice of another world war. German tanks had just entered Poland. Aluminum aircraft and steel armor plate would be critical in the coming conflict. Pittsburgh's great blast furnaces and aluminum pot lines, grown cold during the Depression, were being stoked anew, throwing a fresh funereal smoke against the autumn sky. Workers were already flooding war factories, eager for work. Cox proposed that America should now consider adding fluoride to the public water supply.

Until then, health authorities had sought only to remove fluoride from water; now, the Mellon man told the Water Works Association, "The present trend toward complete removal of fluorine from water and food may need some reversal."[76]

It would take a global conflagration, a nuclear bomb, and an Olympian flip-flop by the Public Health Service for water fluoridation to take hold—yet Gerald Cox's 1935 rat study and Dean's population invastigations would be the germ for a vaccine providing a marvelous new immunity in the postwar years. Touted as a childhood protection against dental cavities, water fluoridation would also secretly help to inoculate American industry against a torrent of fresh lawsuits from workers and communities poisoned by wartime industrial fluoride emissions.

4

General Groves's Problem

On the edge of the marsh water, near the monumental K-25 factory at Oak Ridge, Tennessee, stands a solitary blue heron, its head angling for prey. "Danger. No Fishing. Radiation," reads a sign. Across the pond, the gray walls of the plant glitter in the late evening sun. The smokestacks are cold now, the big machines silent and patient as the heron, waiting to be dismantled and hauled away. Close your eyes and the ghosts return. Mausoleum now, this half-mile-long steel colossus was once among the biggest industrial buildings in the world. Here, in the spring and summer of 1945 and throughout the cold war, tens of thousands of women and men worked through the night in a cacophony of heat and smoke, their backs bent to the purpose of a nation. Here, in the shade of Tennessee's Black Oak Ridge, lay America's biggest wartime secret, where nature was rendered in man's image more powerfully than ever before. Here, on the banks of the Clinch River, exotic ore and minerals from the corners of the globe were transfigured with an elemental genius by scientists, farm laborers, and migrants from across the United States, punching time clocks, sculpting the future, and enriching uranium for the Hiroshima atomic bomb.

IT WAS A cold December morning in 1943 in northwest Washington, DC, and Brigadier General Leslie C. Groves had another problem on his desk. The portly, tough-talking engineer was in charge of the United States' biggest and best-kept wartime secret. He was the army's chief of the Manhattan Project, and its staff was building an industrial infrastructure to manufacture the world's first atomic bomb.

It was a gargantuan task. In complete secrecy Groves and the Army Corps of Engineers were overseeing the work of tens of thousands of laborers, scientists, and engineers who in just three years would create factories and laboratories rivaling the size of the entire U.S. automobile industry. The budget of the Manhattan Engineer District, as the project was officially known, eventually would run to over $2 billion and would be concealed almost entirely from the U.S. Congress.[1]

The General's days were a blur of covert action. There were secret flights to mysterious giant new factories being carved from virgin sites in Tennessee, New Mexico, and Washington State; huddled conferences in the Manhattan Project's New York and Washington, DC, offices; and endless telephone calls, trouble-shooting with top military lieutenants. The United States was in a nuclear arms race with Germany, Groves believed. Yet some of the key industrial processes needed to make the U.S. weapon had not even reached pilot-plant stage. Much of the nation's atomic program, he knew, was still mired in laboratory development.

Groves had a new headache that December morning. There were disturbing reports of workers and scientists being gassed and burned in the bomb project's laboratories and factories. Colonel Stafford L. Warren, chief of the Manhattan Project's Medical Section, needed help. He wanted General Groves to use his authority to pry loose some secret information from the army's Chemical Warfare Service. Warren wanted to know what the military's poison-gas experts could tell the Manhattan Project about the toxicity of fluoride.[2]

General Groves immediately agreed to help. Getting more information about fluoride toxicity was vital. Despite the many uncertainties facing the Manhattan Project that bleak winter of 1943, Groves was sure of one thing: fluoride was going to be essential in making the United States' atomic bomb. Manhattan Project scientists were planning to use a "gaseous diffusion" technology to refine uranium. In that process uranium is mixed with elemental fluorine, forming a volatile gas called uranium hexafluoride, which is then "enriched" by diffusing that gas through a fine barrier, or membrane. The lighter molecules containing fissionable uranium needed for a nuclear explosion pass though the membrane more quickly and are captured on the other side. But because only a handful of the lighter molecules make it through the membrane each time, many hundreds of tons of fluorine, and thousands of stages of progressive enrichment, would be needed to produce enough uranium for a single atomic bomb.[3] By January 20, 1945, when the K-25 gaseous diffusion plant on the banks of the Clinch River was loaded with fluoride for the first time, the plant's fantastic appetite would include a work force of 12,000, a hunger for electricity that rivaled the city of New York, and a diet of some 33 tons of uranium hexafluoride each month.[4]

The hunger for fluorine was one of the most closely guarded military secrets of World War II. A special office of the Manhattan Project in New York City, known as the Madison Square Area, coordinated much of the fluoride work. Elemental fluorine was designated simply "the gas" or "fresh air." Scientists at the University of Chicago were advised in a secret 1942 memo that "all fluorides are to be disguised . . . in that they give definite clues to the chemistry involved."[5]

Dragooning fluoride into military service was also one of the central technological challenges of the war, requiring the full resources of academia and industry.[6] While the idea behind gaseous diffusion was simple, elemental fluorine and uranium hexafluoride were extraordinarily corrosive and toxic.[7] Fluorine was easily the Earth's most reactive element, scientists knew, often combining violently with other chemicals even at room temperature, vaporizing steel in a flash of white heat, for example, and presenting bomb-program engineers with extraordinary challenges and nightmarish hazards. So dangerous was the pure element that industry had avoided fluorine before the war, regarding it as "a laboratory curiosity."[8]

Wartime necessity became the mother of invention. Thousands of researchers in crowded laboratories worked to enlist fluoride in the fight against fascism. Scientists from Columbia, Princeton, Johns Hopkins, Purdue, Ohio State, Penn State, Duke, the University of Virginia, MIT, Cornell, and Iowa State studied the chemical, alongside engineers from some of the biggest industrial companies in wartime America. The companies included DuPont, Chrysler, Allis-Chalmers, Westinghouse, Standard Oil, the American Telephone and Telegraph Company (AT&T), Mallinckrodt, Eastman Kodak, the Electro Metallurgical Company, Linde Air Products, Hooker Chemical, Union Carbide, and Harshaw Chemical.[9]

Columbia University scientists made an early technological breakthrough. In December 1940 a tiny two-cubic-centimeter capsule of a liquid, code-named "Joe's Stuff," was delivered to the campus in New York City. Researchers handled it with care. Inside was virtually the entire world's existing supply of a radical new chemical compound known as a "fluorocarbon"—in which carbon atoms were bonded not with hydrogen, as in conventional "hydrocarbon" oil, but entirely with fluorine atoms.[10] The Columbia researchers soon confirmed that the liquid had Herculean strengths. The fluoride atom was bound to the carbon atom so tightly that even the hyperaggressive elemental fluorine gas was held at bay. The discovery was crucial. Inside the Oak Ridge gaseous-diffusion plant, hundreds of huge compressors and blowers would be needed to push the uranium hexafluoride gas through the multiple "enrichment" stages. If regular oils were used to grease these engines, however, the predatory fluorine atom stripped the hydrogen from the hydrocarbon, destroying the lubricant and the machinery.[11]

The bomb-program scientists could now fight fire with fire. Fluoride, bonded to carbon atoms in fluorocarbons, would protect the machinery from the fluoride in the uranium hexafluoride gas. In other words, fluoride would protect the machinery from fluoride's uniquely corrosive powers. A crash research program at Columbia—led by a brilliant Russian immigrant, Aristide V. Grosse—soon found a way of mass-producing the top-secret compounds.[12] By 1945 thousands of pounds of fluorocarbon oils and seals were being delivered to Oak Ridge.[13]

DuPont mass-produced the fluorocarbons. Their prewar expertise in manufacturing Freon was vital to the U.S. nuclear program. Thousands of pounds of similar refrigerants were now needed to cool the K-25 diffusion plant. DuPont's fluoride-based plastic called Teflon also gave the United States a key wartime advantage. Japan's atomic scientists had struggled to manufacture and handle small amounts of the corrosive uranium hexafluoride. But Teflon—which had been first fabricated in a DuPont lab in 1938—allowed U.S. companies to move enormous quantities of fluoride around the country.[14]

"The basic problem" in making the bomb, General Groves wrote, "was to arrive at an industrial process that would produce kilograms of a substance that had never been isolated before in greater than sub-microscopic problems."[15]

Solving that problem required fluorine scientists. Without their inventions, the United States' atomic bomb "would have been impossible," noted the Manchester University scientist and historian Eric Banks. Most historians have focused on the physics of the atomic bomb, chronicling how the atom was split. The vast contribution of chemical engineers to the Manhattan Project—and the radical debut of a powerful chemical element onto the global stage—has largely been ignored. "It is a striking omission," pointed out Banks. "American fluorine chemists had a huge impact on the production of the bomb."

But exploiting fluoride was a double-edged sword, as the bomb program's scientists soon discovered. On January 20, 1943, the senior Manhattan Project doctor, Captain Hymer L. Friedell, paid a visit to the sprawling New York campus of Columbia University, where a small-scale gaseous diffusion plant had already been built. Almost a thousand researchers would eventually work on bomb-related projects at Columbia's War Research Laboratory.[16] After his visit Captain Friedell warned of possible health problems: "The primary potential sources of difficulty may be present in the handling of uranium compounds, as noted above, and the coincident use of fluorides which are an integral part of the process."[17]

His warning was accurate. A fluoride-gas release at Columbia later that year produced "nausea, vomiting and some mental confusion"; in 1944 another researcher, Christian Spelton, developed pulmonary fibrosis after repeatedly fleeing clouds of uranium hexafluoride gas.[18] Other health problems were also reported. Dr. Homer Priest, a leading Columbia University fluoride scientist, complained that his "teeth seemed to be deteriorating rapidly." Dr. Priest told a doctor that he bled more freely and that "there has been a progressive increase in the degree of slowness of healing and of pain in the period he has been doing this work."[19]

The epidemic spread. At Princeton leaking fluoride gas left scientists feeling "more easily fatigued." There were multiple reports of illness at Iowa State and of fluoride acid burns at Purdue, where two researchers were badly gassed with carbonyl fluoride in 1944.[20] Health problems hit industry scientists too. At DuPont "rather severe weakness" was reported in 1943 by three chemists who had received "heavy exposures" to fluorine. "The symptoms were ascribed by them to the oxyfluorides formed," a report said.[21]

Accounts of fluoride injury mushroomed as the laboratory work moved into full-scale industrial production. At Oak Ridge in September 1944, 190 pounds of hexafluoride gas escaped into a room, drifted outdoors, and formed a chemical cloud "20 yards by 20 yards." Nine workers were exposed "for periods of twenty seconds to five minutes," injuring "the mouth, salivary organs, pharynx, skin, eyes and lungs."[22] The news got worse: that same year, 1944, General Groves got shocking new reports of multiple deaths in the nuclear program. Details of those fatalities and fluoride's role have remained hidden, often for a half-century or more.

The stories of the DuPont workers, who may have been fluoride's first wartime fatalities, have not been made public until now. (And they remain anonymous: once-secret military documents describing the deaths do not record their names.) On January 15, 1944, a laboratory assistant, a chemist, and "a girl technician" producing the fluorinated plastic Teflon for the bomb program were exposed to waste gases. Shortness of breath followed twelve hours later and "by the end of 36 hours, all three were in the hospital," Colonel Warren was informed.[23] The chemist recovered but the other two died terrible deaths, turning purple and unable to breathe.[24] When the twenty-three-year-old female "expired at the end of ten days," her autopsied lungs resembled a victim of a World War I poison gas attack. Colonel Warren's deputy, Captain John L. Ferry, suspected that the DuPont fumes contained "certain oxyfluorides" and suggested the military "investigate the possibilities of this material being used as a poisonous gas."[25]

Although the army ordered up fresh toxicity studies, fearing "similar compounds may be formed in some of the other fluoride manufacturing operations," DuPont dragged its feet, investigators suggested, perhaps seeking to protect Teflon's postwar commercial potential. "The manufacturer considers that we were buying a 'packaged product' and is not interested in our investigating the toxicity of the materials involved," reported Captain Ferry. "Several of the components thus far identified give good promise for commercial uses other than that contemplated here," explained a second army official.[26] (Subsequently there were additional reports of sickness associated with Teflon. British scientists visiting a DuPont factory just after the war confirmed that heated Teflon fumes were linked with "excessive weakness, tiredness, nausea and sore throat.")[27]

A Philadelphia Story

THE SECRET DEATHS continued. Arnold Kramish is tormented by injuries sustained in perhaps the worst fluoride accident of World War II. Sitting in a New York hotel eating breakfast one October 2001 morning, pastry crumbs sprinkling his shirt, Kramish described how he still endures "painful" fluoride skin eruptions on his legs—fifty-seven years after surviving an explosion that killed two of his colleagues. In the 1970s he sought medical help for the recurring sores. A Navy doctor explained to him that fluoride "stalks you the rest of your life."

He is stalked, too, by memories of the chemical "hell" that erupted in South Philadelphia in September 1944. After the war Kramish became a top nuclear scientist and government diplomat, well-versed in the ways of government secrecy. But half a century after the fluoride accident, in a bid to gain recognition for the victims, Kramish broke his silence and revealed details of that disaster, including the names of the men who were killed and why General Groves kept the deaths secret.[28]

On the morning of September 2, 1944, twenty-one-year-old Private Kramish and engineers Peter Bragg and Douglas Meigs reported for duty at the sprawling Philadelphia Navy Yard. The Yard housed a super-secret facility using hot liquid fluoride and pressurized steam to enrich uranium for the atomic bomb.[29] Kramish was one of ten volunteers who had arrived to train on the new equipment. Just three days earlier, at the Manhattan Project's vast construction site at Oak Ridge, Tennessee, Harvard University president James Conant had gathered the men and asked for volunteers. Conant warned them that their work in Philadelphia would be "one of the more dangerous parts of the Project," remembers Kramish.

James Conant was acutely aware of the dangers the men faced from fluoride. The chemist was one of President Roosevelt's top atomic advisers. He knew about the DuPont Teflon deaths. And he had seen the secret army reports on fluoride toxicity that General Groves had requested in December 1943.[30] The reports explained that the military was carrying out wartime human experiments with fluoride gases at the army's Edgewood Arsenal in Maryland, searching for chemical warfare agents.[31] The army had received data about fluoride experiments on humans in England that had produced powerful central-nervous-system effects.[32] And there were reports from captured prisoners of war suggesting that the Nazis, too, were investigating fluoride as a war gas.[33] Harvard's president was so disturbed by the "extraordinary" toxicity of certain fluoride compounds, especially those used in the human experiments, that he issued a secret warning to a senior U.S scientist about the atomic industrial fluoride work. "As an organic chemist," Conant wrote, "I think I should point out to you . . . it is conceivable that similar effects would occur with any fluorinated organic acid, although probably the compounds would be less striking in their action. It is further conceivable that these compounds could be formed in small amounts by the action of fluorine gas on the acids or related compounds."[34]

That fall day at Oak Ridge, however, as he asked for volunteers, Conant did not mention fluoride. All ten men raised their hands. "Any mildly inquisitive guy was not going to opt out," said Kramish.

At first the Philadelphia mission was more Keystone Kops than cloak and dagger. When they arrived at the Thirtieth Street train station, a military official in street clothes ordered them into Wanamaker's department store to replace their uniforms with anonymous civilian garb. But the Navy did not give them enough money, and all the men could find were cheap Hawaiian shirts, says Kramish. He remembers ten men furtively changing into their new "outfits" in a nearby subway station, emerging into the sunlight wearing brightly colored shirts and GI boots.

Two days later Kramish, Bragg, and Meigs were at the Navy Yard, working on the secret machinery. At lunch Kramish received a two-dollar bill in his change. "Give it back," his friend told him, warning that it was an omen of bad luck. Kramish pushed the bill into his pocket.

That afternoon, back at the plant, at 1:20 PM a massive explosion suddenly tore at the machinery. Boiling steam and fluoride jetted onto Kramish's legs and back, clawing at his lungs and eyes. He fell backward, temporarily blinded. A trained scuba diver, Private John Hoffman ran into the smoking chaos holding his breath, pulling the injured men from the room and slicing Kramish's clothes from his burned body. This act of bravery would win Hoffman a Soldier's Medal, although the award was kept secret. "I pulled three guys out. Everybody was shell-shocked," Hoffman told me. "Fluorine gas had gotten loose—it was pretty pungent. I had to watch what the hell I was doing."[35]

The afternoon detonation echoed across South Philadelphia. A giant white plume of uranium hexafluoride gas drifted over the dockyard and into the nearby battleship USS *Wisconsin*. Douglas Meigs and Peter Bragg lay in their death throes. A priest attempted last rites on Kramish, whose wife was told that he had been killed. A once secret report of the disaster makes gruesome reading: twenty-six men had been exposed to 460 pounds of fluoride and uranium in a "huge chemical cloud." Douglas Meigs was "sprayed with live steam containing liquid, solid and gaseous material in large quantities"; he died after sixteen minutes. Peter Bragg expired an hour later with third-degree burns over most of his body. He "seemed in a great deal of pain," the report noted, and "became violent shortly before death and resisted all attention."

The remaining men survived, although many had serious and slow-healing wounds. Some experienced "intense pain in the scrotum, penis, or about the anus, probably because of the hydrolysis of the chemicals in these moist areas," the report notes. Survivors also suffered unusual "nervous system" effects. One man was temporarily rendered "almost incoherent." This "altered mental state" was "more than could be explained on a purely fear reaction basis," the report said. "In all probability the injurious effects observed on the skin, eye, mucous membranes of upper respiratory tract, esophagus, larynx and bronchi were all directly caused by the action of the fluoride ion on the exposed tissues," concluded a military doctor.[36]

Kramish reports that at a closed wartime inquiry, he learned that part of his suffering had been unnecessary. The head of the Navy project, Dr. Philip H. Abelson, had known how to treat fluoride burns, according to Kramish. But fluoride and uranium were considered so secret that Abelson refused to give the medical facts to the arriving doctors, telling them, "I'm not sure you guys are cleared," Kramish recalls. As a result, he adds, the doctors walked among the injured and dying men that afternoon "guessing what the burns might be." (Fifty years after the accident, Kramish reports he cornered Abelson one lunchtime in the Cosmos Club in Washington. Abelson refused to talk about the accident, Kramish says. "It was clearly a trauma for him.")

The Philadelphia explosion traumatized the entire Manhattan Project. In addition to the fluoride strewn over south Philadelphia, it was perhaps the largest release of man-made radiation that had ever occurred. General Groves feared that a nuclear fission accident had taken place. The military quickly suppressed media coverage. The Philadelphia coroner was not told the cause of the men's death.[37]

That disaster night, roused by Groves, the Manhattan Project's top doctor, Colonel Stafford Warren, drove through the darkness from Oak Ridge, Tennessee. He arrived at the Philadelphia Navy Hospital in time to seize the organs of the dead men, stuffing the heart and lungs of Meigs and Bragg into his briefcase before returning home, he later told Kramish. (Warren and Kramish became friends after the war.) Warren explained to him that the organs "had become classified material," Kramish recalled, and that they were sent to the University of Rochester for examination. "The deceased were buried without them," Kramish added.

Family members, such as Elizabeth Meigs, who was on her way to meet her husband in Philadelphia for Labor Day, would never learn that fluoride may have killed their relatives. General Groves kept silent about the fatalities. In his book about the Manhattan Project, *Now It Can Be Told*, Groves tells only that several persons "were injured" in Philadelphia and that the investigation "held up the work for a while."[38] Groves's fear of admitting the deaths, Kramish says, was "not only that the atomic bomb project might be compromised, but that if project workers learned of the true hazards of working with uranium, they might balk."[39] Suppressing toxicity information "would extend to fluoride," added Kramish. "Working with it was dangerous."

Arnold Kramish still has the two-dollar bill he received that lunchtime. He keeps it wrapped in lead; it remains contaminated.

Although fluoride played a nearly fatal part in Arnold Kramish's wartime experiences, he believes that few people have any idea of the chemical's wartime importance. "It is not as exotic as the atom," he says. For most historians, radiation is "all they want to talk about."

The Fear Mounts

FEAR NOW GRIPPED wartime fluoride workers across the U.S. atomic complex, and with good reason.[40] Thousands of them were entering an abominable work environment, beyond even Victorian horror, with daily exposure to a witch's brew of fluoride chemicals—including, for the first time in human history, the ferociously reactive elemental fluorine gas.[41]

"When a jet of pure fluorine strikes most non-metallic materials," began one 1946 secret memo detailing occupational hazards, "the surface of the material is instantly raised to an incandescent white heat. Personnel may be severely burned by heat radiated from the surface even when they are not directly exposed to fluorine at all. . . . NO PERSONAL PROTECTIVE EQUIPMENT HAS BEEN DEVISED TO DATE WHICH WILL RELIABLY AFFORD EVEN TEMPORARY PROTECTION AGAINST A HIGH PRESSURE JET OF PURE FLUORINE," emphasized the memorandum.[42]

Incredibly, fluorine was *not* the most toxic gas to which workers risked exposure. When excess fluorine was vented to the atmosphere (a common procedure, as we shall see) a truly venomous family of even deadlier compounds—"oxyfluorides"—were formed. One of these chemicals, oxygen fluoride, "a bi-product of fluorine disposal," was probably "the most toxic substance known," bomb program researchers bluntly reported.[43]

Another common workplace hazard was hydrogen fluoride acid (HF), which had the fiendish property, if splashed on skin, of initially escaping detection but then slowly and painfully eating into a victim's bones.[44] One especially fearsome compound called chlorine trifluoride, which was used to "condition" or clean machinery, was so reactive that Allied intelligence agents suspected Hitler's SS had also experimented with it, as an incendiary agent.[45] U.S. atomic worker Joe Harding, who used chlorine trifluoride at the Paducah gaseous diffusion plant in Kentucky, described the compound as a "violent monster that makes [pure] fluorine look mild by its side."

Working with chlorine trifluoride was "more dangerous than handling TNT while you was climbing a tree," said Harding.[46]

Fluoride posed another hazard. It dramatically boosted the toxicity of other cold war chemicals. The biological havoc wreaked by beryllium, for example—a key metal that makes nuclear weapons more powerful—was at least doubled by the synergistic presence of fluoride, bomb program scientists found.[47] By 1947 there had been nineteen or more deaths reported in the nation's beryllium plants, with the carnage spreading rapidly.[48] (When newspaper reporters got wind of the fact that families living near the beryllium plants were also getting sick, the Atomic Energy Commission tried to suppress the story.)[49]

Beryllium smelters were felled with an especially devastating one-two punch, said the Manhattan Project scientist Robert Turner. Men became ill with a "foundry fever" marked by shivering, high temperatures, and "profuse perspiration." The knockout blow from fluoride fumes followed sometimes days later, the scientist noted, with workers turning purple, gasping for breath, and coughing up blood. [50] Turner was critical of other scientists. Investigators studying fluoride had shown "a disregard of the fundamental principles of modern toxicology." Discovering how workers were being hurt required considering a range of factors, including the size of the particles involved, ways the poison entered the body, and awareness "that the action of a compound is not equivalent to the sum of the action of its component parts," he wrote.[51] Turner described the pathways by which tiny fume-sized particles of beryllium oxyfluoride penetrated deep into lungs "with missile-like force." When the molecules arrived inside the alveoli, the atoms of fluorine and beryllium separated "like a charge bursting." Both beryllium and fluoride were poisonous, the scientist said, but it was the liberation of fluoride deep inside the lung that produced the most catastrophic health problems, destroying tissue, choking breath, and leaving permanent lung scarring.[52]

Similarly, when uranium was converted into hexafluoride gas, that poisonous metal also got a deadly new punch. This enhanced toxicity of uranium presented nuclear planners with perhaps their most diabolical quandary. Enormous quantities of uranium hexafluoride "process gas" were required for even a single atomic bomb. But when the "hex" was exposed to air, it rapidly formed a dense white cloud of HF gas and fume-sized particles of a "highly toxic" compound known as uranyl fluoride or uranium oxyfluoride (chemical symbol UO_2F_2). The compound injured laboratory animals in microscopic quantities, while even "a few milligrams" ingested daily proved fatal, bomb program doctors reported.[53]

Exposure to these two chemicals would be a daily fact of life in the diffusion plants.[54] In the hidden chambers of the massive K-25 plant, where precious uranium for the Hiroshima atomic bomb was first captured, "there will be a continuous escape of UO_2F_2 in the cold trap rooms," officials warned. Those workers would be exposed "8 hours per day regularly," explained Medical Captain John Ferry in a secret June 16, 1944 letter to an Oak Ridge contractor.[55]

"Just Watch Anyone That Has a Tie On"

AS PREDICTED, WHITE fluoride smoke became a familiar sight and smell to generations of workers in America's gaseous diffusion plants. "I have never seen it that there wasn't a thick haze of process gas smoke in the air," said Joe Harding, remembering his almost thirty years inside the gaseous diffusion plant at Paducah, Kentucky.[56]

"It does have a pungent odor," confirmed another worker, Sam Vest, who in 1970 followed his father and two uncles into the Oak Ridge nuclear factories. In a 2001 interview in his home near Oak Ridge the fifty-four-year-old Vest tugged on a never-ending cigarette, recalling his own three decades at America's first gaseous diffusion plant. His soft Tennessee drawl transported a visiting writer back inside the cacophonous K-25 building and to the apprentice electrician's first encounter with uranium hexafluoride gas. Vest watched one morning as clouds of smoke belched from equipment he was replacing. He asked a more experienced worker about the strange white fog.[57] "I said, 'What is that stuff?' And he said, 'That is process gas.' And I said, 'Should we be here? I don't see anybody with respirators on.'" The older worker explained an Oak Ridge safety rule: "Just watch anyone that has a tie on." He added, "And if he leaves hurriedly, you leave behind him." "That was my first indoctrination," Vest said. "I was just a kid."

Medical advice given to men who had been in a chemical release, said Vest, was to "go home and drink a six pack of beer."[58] Vest remembered thinking, "I don't know anything about chemicals or uranium hexafluoride or anything like that. But none of this looks on the level to me. These men are standing in this fog with no respirators. I thought 'My God, what kind of a place is this?'"

On another occasion Vest found himself high above the plant in the "pipe gallery," replacing electrical heaters. "We were wading though this yellow powder," he recalled. "I asked [a colleague] Clyde, I said, 'Clyde, what is all this yellow lying around here?' And he said, 'That is product.' I said, 'What do you mean?' And he said, 'Well, that is UO_2F_2. After it cools down, it solidifies and that is enriched uranium.' And I said, 'Shouldn't we have some kind of breathing apparatus or something?' And he said, 'Hell no, we work in this all the time. It won't hurt you.'"[59]

Similar official safety reassurances, from the highest levels of the United States government, were given to tens of thousands of fluoride workers throughout the cold war. The assurances were false. Fluoride was a state secret. Workers were neither told what chemicals they were handling nor of the warned dangers. "The people hired by the contractors were not, because of security, told of the hazards involved in their work," Colonel Stafford Warren wrote to a deputy, Dr. Fred Bryan, in September 24, 1947.[60]

Despite an early awareness that cancer and occupational injuries were extraordinarily frequent at the gaseous diffusion plants, workers could never prove that such was the case. "All medico-legal and insurance statistics which refer directly to process hazards" were classified "secret," an AEC document noted.[61] In data that were declassified only in 1997, for example, it was revealed that during the earliest months of the K-25 plant's operation, from June 1945 to October 1946, there were 392 "chemical injuries" from uranium hexafluoride, 58 injuries from fluorine, 21 from hydrogen fluoride, and six injuries from fluorocarbons.[62]

Area C

WORKERS QUICKLY GREW suspicious at the endless medical testing. Behind a barbed wire fence at a secret plant in downtown Cleveland, Ohio, known as Area C, segregated young African Americans—who loaded a chalky "green salt" into furnaces—gave regular urine samples to government doctors.

"You had to be tested all the time," said Allen Hurt, an employee of the Harshaw Chemical Company, which ran the secret plant under contract for the Manhattan Project. He was one of five former workers who agreed to talk about his experiences.

The industrial complex on the Cuyahoga River was one of the Manhattan Project's most important sites. Harshaw engineers had invented a way to add extra fluoride molecules to uranium tetra fluoride—the "green salt" the workers were handling—manufacturing the vital hexafluoride "process" gas needed for uranium enrichment. ("Hex" means six and "tetra" means four.) By June 1944, the plant was capable of producing a ton of "hex" each day for shipment by truck to Oak Ridge for the K-25 gaseous diffusion plant.[63]

The government reassured the workers about the tests. In a 1948 visit to Cleveland, for example, a Manhattan Project senior doctor, Bernard Wolf, gathered the workers together to tell them that "all our records indicate that no unusual hazard existed." The truth was very different. Secretly, on August 5, 1947, the AEC's W. E. Kelly had informed Harshaw's senior manager, K. E. Long, that "the status of health protection at Area C is unsatisfactory is several respects." He cited in particular:

1. Contamination of the Area C plant, Harshaw plant area and an unknown amount of contamination of the surrounding neighborhood with uranium and fluoride compounds.
2. Exposure of operating personnel to uranium and fluorine compounds by direct contact and inhalation.[64]

Harshaw workers knew *something* was in the air. "The moment you stepped out of the time clock office, there would be an odor, a burning sensation," recalled Henry Pointer. "It would sting your face, you would inhale it too." Union organizer John L. Smith was sick one day after repairing a pipe. "It was the fumes—next thing I felt breathing difficulty and started vomiting and went to the first aid and started shitting in front of them at the same time," he said. (Although he never knew what had poisoned him, Smith's symptoms were of acute fluoride poisoning.)[65]

There were fluoride fatalities at Harshaw as well. Young black women made up about half of the Area C workforce. Twenty-two-year-old Gloria Porter started at the Cleveland works in 1943, filling hydrogen fluoride tanks. On October 9, 1945, she saw a man eaten alive by the fluoride acid when a storage tank at Area C exploded.[66] "I heard this rumble," remembers Porter, who had just finished her shift. "All of a sudden this cast iron [storage tank] just burst open and the smoke, the fumes from the acid, you just couldn't see nothing, and that stuff was rolling and the more it rolled the further we would run."

A male worker helped Porter to scramble over the barbed wire fence that surrounded Area C. As she stared back, a horrific image was seared in her mind. She watched men struggling through a giant cloud of hydrofluoric acid. "I saw all of them coming out with hunks of flesh just falling off of them, and the stomach, and their arms, and I said 'My God, I can't look at that. That man can't live.' He looked just liked bone, but he fell right then." Two men were killed in the accident, and a good friend was badly burned, recalls Porter, who left Area C the following year.[67] "After the explosion, I just wanted to get out," she added

African Americans may have been hired for fluoride work in order to conceal the chemical's toxic effects. "Most fair complexioned men could not be employed in the production plant," reported a once classified wartime study of Harshaw fluoride workers.[68] Acid fumes produced skin that was "dehydrated, roughened and irritated," the report noted. Some workers had "hyperemia" or acute reddening of the face. When that report was published, however, the black- and-white language of segregation had grown less stark. The chemical sensitivity to the fluoride was now more subtly described as "more severe in fair complexioned men."[69]

Harshaw veterans confirmed that only African Americans were employed inside the heavily guarded Area C plant. Outside, white male supervisors oversaw the big cylinders being hoisted onto trucks for the journey to Oak Ridge, remembered a former worker, James Southern. "Yeah, but they weren't pulling," interjected worker Henry Pointer, "the labor people were all black."

One young white laborer, John Fedor, who joined the company in 1939 with a tenth-grade education, was never permitted to enter the

Area C complex. He had no idea that the plant was performing secret war work for the government. "To work there you had to be cleared and I was not cleared to go in," he explained. Nevertheless Fedor grew worried about fluoride exposure at Harshaw's big hydrogen fluoride (HF) plant, which supplied Area C, and about the "terrible" conditions those workers endured. (He became a union organizer after the war.) His Safety Committee invited state inspectors inside the HF plant. Inside, fluoride levels as high as 18 parts per million were measured, six times the permitted safety standard.[70] "There were men walking around with rags over their noses, there were no respirators, there was no safety program," Fedor remembered. Burns and acid splashes were common. "The good Lord knows what it did to the inside of a person's body. How many people may have suffered fatalities over the years I have no idea," he added.[71]

Allen Hurt carries visible reminders of his years at Harshaw Chemical. He pulled a trouser leg up to reveal fifty-year-old scars he blamed on fluoride. "They didn't give you protection," he said. "It would eat the clothes and it would do the same thing to your skin." Sickness has stalked former employees, survivors claim. By the time the plant closed in 1952, an estimated 400 to 600 workers had been employed at the Area C plant. Cancer and heart ailments have been especially frequent among former workers, John L. Smith claims. "The people who worked there are dead. Those that ain't dead, there's five of them in the nursing home." The remaining veterans smolder with anger. Mostly, they wish they had been given the dignity of choosing their wartime fate. "At least we should have been properly informed," said Smith. "What few is left is as pissed off as they can be."[72]

"Hazards to the local population could occur"

WHEN HE WAS shown several declassified documents describing how fluoride and uranium were regularly vented from the Harshaw smokestacks, union organizer John Fedor was suddenly concerned. "I wonder about the immediate area," he remarked, "whether there were illnesses caused by that, or whether it just dissipated when it got in the air?"

Fedor is right to be concerned about the effects of fluoride on the area around Harshaw. It was not, of course, just the atomic workers who were secretly at risk from fluoride. From the beginning of the nation's nuclear program, officials worried about families living near bomb factories. "Hazards to the local population could occur if large amounts of fluorine or if fluorides were to be discharged in effluents," wrote the medical director Colonel Stafford Warren.[73]

Again, the fears proved accurate. Fluoride was secretly vented, and it spilled across communities in New Jersey, Pennsylvania, Kentucky, Tennessee, and Ohio.[74] Those releases increased as the United States expanded its cold war atomic arsenal and built two mammoth new gaseous diffusion plants, at Paducah, Kentucky, and Portsmouth, Ohio.[75]

Environmentalists often cite Cleveland's Cuyahoga River—which burst into flames in June 1969—as the lurid spectacle that helped bring about the Clean Water Act. The shocking sight of a waterway ablaze precipitated a moment of national clarity, focusing attention on the dumping of chemical wastes into the environment. Less well remembered, however, is a $9 million lawsuit brought in 1971 by the local Sierra Club against the Harshaw Chemical Company for fluoride pollution, which, the organization charged, had eaten and corroded the main Harvard Dennison Bridge over the same Cuyahoga river.[76] That bridge had to be rebuilt.

The government had watched the situation in Cleveland nervously. Following "complaints" in 1947, a team from the University of Rochester's Atomic Energy Project was quietly dispatched to measure fluoride pollution. The scientist Frank Smith secretly reported levels of 143 parts per million of HF venting from the Harshaw smoke stacks. (By contrast, 3 parts per million is the standard considered safe today for workplace exposure.) "The results are on the low side," Smith wrote, "since the efficiency of the sampling procedure we used is not too good for [elemental] fluorine and oxygen fluoride; if considerable quantities of these two gases were present in the air, we probably missed a part of them."[77] The AEC was worried about lawsuits. Dr. Smith pointed to several lower fluoride readings in his data. Those measurements, he said, might prove "the most valuable . . . [as they] in no case exceed the level declared legally permissible in Massachusetts, California and Connecticut."

Storm clouds continued to gather over Cleveland. A July 1949 AEC report warned that "although the complaints from civic organizations have been concerned with general atmospheric pollution, and neither fluoride nor uranium have been mentioned specifically, it is likely that as time progresses, the extent of air pollution by fluorides will receive attention."[78] The AEC ran more secret tests after a consultant, Philip Sadtler, was hired in 1949 by the local community to investigate Cleveland air pollution. While uranium releases were within permissible levels, they concluded that "the fluoride data, however, satisfied none of the criteria."[79]

Several of the former Area C workers confirmed that pollution was rampant. Allen Hurt parked his car downwind from the plant whenever he worked the night shift. "Overnight, fallout would come, and my black car was full of gray dust, and I washed if off and I could see little fine pits where it had ate into the paint. If it does that in metal, what would it do to us?" he wondered. Hurt recalled that local residents complained: "They had a problem with the people up on the hill, because it was coming up there and bothering their homes."

Environmental damage around atomic bomb plants was often widespread. At Oak Ridge, officials planned, in 1945, to dump 500 pounds of fluorides each day into the nearby Poplar Creek; a decade later, airborne fluoride emissions had scarred a fifty-square-mile area of wounded and dying trees, officials stated, and posed a clear threat to grazing animals. And in 1955, some 615,000 pounds of fluorine was "lost in the vent gases" from a single in-house plant making uranium hexafluoride at Oak Ridge.[80]

Lawsuits alleging fluoride human injury and destruction of crops and farm animals were sparked against DuPont's Chamber Works in New Jersey and the Pennsylvania Salt Company's plants in the Pennsylvania towns of Easton and Natrona.[81] At a second gaseous diffusion plant in Portsmouth, Ohio, which began operations in 1954, fluoride exposure was immediately declared a "significant liability" for "both employees and the general public," a document noted.[82] At the AEC's giant Feed Materials Production Center in Fernald, Ohio, waste fluorides were "the biggest single problem," where some 15,000 pounds of fluorides were being disposed of each month in the nearby Miami River, according to a pollution expert, Arthur Stern.[83]

And as late as the mid-1980s, thirty years after it began operation, the gaseous diffusion plant at Portsmouth, Ohio, was still dumping 15.6 tons of fluorides each year into the atmosphere.[84]

Darkness hid fluoride releases at the K-25 plant in Tennessee, according to former supervisor Sam Vest. "I could pull into the parking lot at night and smell it. I could tell they were releasing fluorine from the fluorine plant. They waited until after dark to release it, because it was just a horrendous cloud." Some workers found a strange beauty in the nighttime releases at Oak Ridge, Vest added. "Operators described it as being just beautiful, to just stand there and watch crystals on a clear cold night go up [into the air]."

5

General Groves's Solution

Dr. Harold Hodge and the University of Rochester

The Manhattan Project had seen the danger from fluoride early. Before the war private industry had contained the legal dangers from factory pollution by forming the Air Hygiene Foundation at the Mellon Institute. Also fearing lawsuits, in 1943 General Groves established the Manhattan Project's Medical Section at the University of Rochester to "strengthen the government's interests," placing Dr. Harold C. Hodge in charge of a secret unit studying fluoride and the other chemicals being used to make the atomic bomb.

FROM HIS CORNER office window in the medical school at Strong Memorial Hospital that summer of 1943, Dr. Harold Hodge could see construction workers placing the finishing touches on a half million-dollar building at the University of Rochester known as the Manhattan Annex.[1] The heavily guarded structure, funded by the U.S. Army, would be home to the Manhattan Project's Medical Section. Orders had been placed for hundreds of experimental animals: Puerto Rican monkeys, dogs, mice, rabbits, and guinea pigs.[2] And an umbilical cord–like tunnel linking the military annex with the university hospital was urgently being readied.

As the new Annex foundations were put down, so too was the keystone laid for the postwar practice of toxicology in the United States—and for the future career of the thirty-nine-year-old biochemist, Dr. Harold Hodge. The Annex would soon house the largest medical laboratory in the nation, with a staff of several hundred scientists testing the toxicity of the chemicals being used to build the atomic bomb.

Military pilots flew the exotic new compounds directly from the bomb factories to Hodge's team at Rochester. "Harold would actually meet the pilots under [cover of] dark to get the material to test," said toxicologist Judith MacGregor, who befriended Hodge at Rochester, where she was a graduate student in the 1960s, and who was mesmerized by her mentor's tales. "It was unbelievable."

That spring of 1943, Hodge had been placed in charge of the bomb program's Division of Pharmacology and Toxicology and given control of a secret biomedical research unit known as "Program F" to study fluoride toxicity.[3] The Manhattan Project had "a whole section working on uranium and a whole section working on fluoride," explained Jack Hein, who worked with Hodge at Rochester during the early cold war as a young graduate student and remembers the scale of the fluoride studies. "The toxicology studies were very comprehensive. They were looking for toxic effects on the bone, the blood, and the nervous system. . . . Without the Manhattan Project and the atomic bomb, we wouldn't know anywhere near as much as we do about the physiological effects of fluoride," Hein added.[4] "His research suddenly blossomed into an immense program," noted Paul Morrow, a uranium expert who also joined Hodge at Rochester in 1947 and who worked on some of the earliest experiments.

Hodge's war work germinated into a career as the nation's leading expert on fluoride. Over more than half a century the tall, black-haired researcher published several books and some three hundred scientific papers. He was chairman of the National Research Council's Committee on Toxicology and first president of the Society of Toxicology. And a generation of Hodge's Rochester colleagues and students—men such as Herbert Stokinger, Paul Morrow, and Helmuth Schrenk—went on to occupy leading positions in government agencies and universities after the war.[5] "He was unarguably the dean of American toxicology," stated a former colleague and Rochester alumni, Ernest Newbrun, now a professor emeritus at the University of California at San Francisco.[6]

To several generations of colleagues, the soft-spoken scientist with the slicked-back hair was a gentleman scholar and tutor, advising them to "play it straight," and regularly, in his early seventies, trouncing graduate students at squash.[7] But Harold Hodge—grandfather, soft-spoken friend, and "dean of American toxicology"—shouldered dark secrets for much of his professional life.

That summer of 1943, as Dr. Hodge stood at his office window, he confronted a terrible dilemma. Speed was essential in beating the Germans to full-scale production of the atomic bomb.[8] The fate of tens of thousands of American workers lay in his hands. His laboratory's evaluation of the toxicity of chemicals needed for the bomb, such as fluorine, beryllium, and trichloroethylene, would fix work conditions for the women and men inside the Manhattan Project's bomb factories, help determine how quickly the plants could achieve full production—and whether employers would be successfully sued for damages if those workers claimed injury from chemical exposure.[9] "The questions were many and the answers few," wrote Hodge. "There was no time to wait for months, or even weeks, while the accepted laboratory tests established the toxicological facts. Production had to proceed with no delays."[10]

"People working in the atomic energy production plants were going to be chronically exposed," said Jack Hein. "We didn't know too much about the toxicity of fluoride, other than the early studies saying a little too much in the water causes damage to teeth," he added.[11]

General Leslie Groves understood the dangers of such pell-mell production. He feared that personal injury lawsuits would be an Achilles heel for the entire nuclear program. Leading insurers, such as Aetna and Travelers, were providing health coverage for workers in the new bomb factories.[12] Successful claims for fluoride injury or for neighborhood pollution might hemorrhage compensation payments, create a public-relations disaster, risk jeopardizing the embryonic nuclear industry—and threaten the United States' unprecedented new military power.[13]

The army moved quickly to protect itself. Its first weapon was secrecy. The second weapon was seizing control of basic science. In particular the crucial toxicity studies on bomb program chemicals performed at the University of Rochester were sculpted and shaped to defend the Manhattan Project from lawsuits.[14] Those marching orders—conscripting science and law for military service—were drummed home in a July 30, 1945, memorandum titled "Purpose and Limitations of the Biological and Health Physics Research Program," written by the head of the Medical Section, Colonel Stafford Warren. According to Warren, "The Manhattan District, as a unit of the U.S. Army . . . has been given a directive to conduct certain operations which will be useful in winning the war." As such, "medico-legal aspects" were accorded a clear priority for scientists, he added, "including the necessary biological research to strengthen the Government's interests."[15]

Scientists soon delivered courtroom ammunition. "Much of the data already collected is proving valuable from a medical legal point of view," noted a February 1946 memo to General Groves's deputy, Brigadier General K. C. Nichols. "It is anticipated that further research will also serve in this manner," the memo added.[16]

Colonel Warren had chosen his top fluoride expert carefully. The son of an Illinois schoolteacher, Harold Hodge was a biochemist whose specialty was the study of bones and teeth. He had arrived at the University of Rochester in 1931, where he was one of an elite cadre of men selected by the Rockefeller Foundation as dental research fellows. The Rockefeller Foundation was then funding basic research at selected dental schools in a bid to lift the standards of dental care in the United States. Hodge was also a pharmacologist and toxicologist who by 1937 had forged close links with corporate America.[17] By the summer of 1943 some of those corporations and institutions were taking a lead role in developing America's first nuclear weapon. Eastman Kodak, a Rochester company where Hodge had investigated chemical poisoning before the war, was now a leading industrial contractor at Oak Ridge, Tennessee.[18] Rockefeller interests were also using fluoride to refine uranium at an undisclosed site in New Jersey and funding their own biomedical research at the University of Rochester.[19]

Harold Hodge's role as gatekeeper at the wartime crossroads of law and medical science was spelled out in a 1944 letter introducing the Rochester scientist to the DuPont company. The letter, stamped "confidential," again lays out a fundamental scientific bias in the Manhattan District's medical program—a bias against workers and communities, and in favor of corporate legal interests.

"The Medical Section has been charged with the responsibility of obtaining toxicological data which will insure the District's being in a favorable position in case litigation develops from exposure to the materials," Colonel Stafford Warren told Dr. John Foulger of DuPont's Haskell Laboratory in a letter dated August 12, 1944. Harold Hodge was to insure that information about the toxicity of certain fluoride compounds was coordinated between the government and its contractors, Warren explained. "It would be desirable," he told Foulger, "to have the work on the toxicity of fluorocarbons being done in your laboratory parallel the investigations being made on similar compounds elsewhere. For that reason it would be appreciated if Dr. Harold Hodge of the University of Rochester could visit your laboratory in the near future and an exchange of ideas be effected."[20]

Harold Hodge, Devil's Island, and the Peach Crop Cases[21]

Harold Hodge's diligence in defending the war industry can be seen in a 1946 court challenge from farmers living near a DuPont fluoride plant in New Jersey. Although not mentioned in any history of the Manhattan Project, the lawsuits were regarded by the military as the most serious legal threat to the U.S. nuclear program, requiring the direct intervention of General Leslie Groves. A closing chapter in the Manhattan Project, the aggressive use of secrecy, science, and public relations by Groves and Hodge, and at least a half dozen federal agencies battling the farmers, is an opening scene in the story of how fluoride was handled by our government following World War II.

The gently rolling alluvial soil along the shore of the Delaware estuary in Southern New Jersey is some of the most bountiful farmland in the United States. Its historic harvest of fruit and vegetables won New Jersey the accolade of The Garden State. The orchards downwind of the DuPont plant in Gloucester and Salem counties were especially famous for their high-quality produce; their peaches went directly to the Waldorf Astoria Hotel in New York. Campbell's Soup bought up their tomatoes. But in the summer of 1943 the farmers began to report that their orchards were blighted and that "something is burning up the peach crops around here."

Poultry died after an all-night thunderstorm, they reported. Fields were sometimes strewn with dead cattle, residents recalled, while workers who ate the produce they had picked vomited all night and into the next day. "I remember our horses looked sick and were too stiff to work," Mildred Giordano, who was a teenager at the time, told reporter Joel Griffiths. Some cows were so crippled that they could not stand up, and grazed by crawling on their bellies. The injuries were confirmed in taped interviews, shortly before he died, with the chemical consultant Philip Sadtler of Sadtler Laboratories in Philadelphia. On behalf of the farmers' crusading attorney, Counselor William C. Gotshalk of Camden, New Jersey, Sadtler had measured blood fluoride levels in laborers as high as 31.0 parts per million. (Blood fluoride is normally well below 1 part per million. These levels are potentially lethal doses.)[22]

"Some of the farm workers were pretty weak," Sadtler noted. The New Jersey farmers organized a Fluorine Committee. They patriotically waited until the war was over, then sued DuPont and the Manhattan Project for fluoride damage. Thirteen claimants asked for a total of $430,000 in compensation.

Little wonder the farmers reported health problems. Conditions on the other side of the DuPont fence were extraordinarily dangerous. More than a thousand women and men were employed on Manhattan Project contracts at the Chamber Works during the war, secretly manufacturing elemental fluorine, uranium hexafluoride, and several exotic new fluorocarbons.[23] Chemical exposures were frequent, making the DuPont employees perhaps the most endangered and fearful of the wartime fluoride workers. By the end of January 1944 at least two DuPont laboratory workers had been killed and several scientists injured. Work conditions at the secret fluoride-producing East and Blue Areas of the Chamber Works were especially dreadful, with "gross violations of safety," inspectors noted.[24]

One unit was especially notorious, the government reported. "The plant frequently caught on fire, and the activators often burned out so the employees were frequently exposed to rather large amounts of fluorine compounds," Captain Mears of the Manhattan Project noted in October 1945. "Medical hazards were attributed to fluorine in a gaseous state, silver fluorides in a powdered state and liquid 2144 [code for fluorocarbon]."[25]

Injured workers paraded into the DuPont hospital. Doctors often reported "a fibrotic condition of both lungs" on X-rays; serious chemical burns were seen "very frequently." The mounting injury toll was blamed on fluoride.[26] In February 1945 doctors at the East and Blue Areas reported seventy-nine "sub-par or so-called chronic cases." Sixteen of those workers had their condition detected in the last two months.[27]

A Manhattan Project medical investigator, Captain Richard C. Bernstein, warned his boss, Colonel Warren, that workers now feared assignment to the DuPont fluoride processing areas as "an exile to Devil's Island."[28] Another report warned of brewing labor unrest. "Fear of the physical consequences was becoming prevalent in the Areas," wrote Manhattan Project investigator First Lieutenant Birchard M. Brundage in February 1945. "This fear was being used by certain agitators to cause trouble in the personnel," he added.[29]

The farmers' lawsuits electrified the Manhattan Project. There had been no disclosure of the diabolical work conditions at DuPont. Now, a public lawsuit pointed a finger directly at the Chamber Works and fluoride. A once secret November 1945 memo measures the government's concern: "The most serious claim to neighboring properties of any operations of the [Manhattan Engineering] District is the litigation known as the 'peach crop cases.' These are cases claiming damages to the fruit crop and to the peach trees themselves in and around the operation of the Chambers Works of the DuPont Company at Kearney, New Jersey. This damage is allegedly caused by the release into the atmosphere, both unintentional and necessary as a result of the process [*sic*] of hydrogen fluoride. The claims against the District approximate $430,000. Part of the loss would be due to the private contractor and part to the operation of the contractor on behalf of the District."[30]

The military sprang into action. Dr. Hodge was dispatched to New Jersey to marshal the medical response to the farmers' rebellion. Although DuPont's smokestack fluoride had long been spilled into the environment and a great volume of new fluoride compounds were being made inside the wartime plant, he quickly reported back to Colonel Stafford Warren at Oak Ridge that the mottled teeth seen in the school near the DuPont plant could be attributed to natural fluoride in the ground water.[31] Such natural fluoride in the water supply meant that the dental markings could not be used as unequivocal proof of industrial poisoning. "The situation was complicated by the existence of mottled enamel as a result of fluoride in the drinking water," Hodge told Warren.

Dr. Hodge had an idea for calming the citizen panic. His prescription gives an early meaning to the term *spin doctor*—and provides a clue that the promotion by the U.S. government of a role for fluoride in tooth health has a powerful national-security appeal. "Would there be any use in making attempts to counteract the local fear of fluoride on the part of residents of Salem and Gloucester counties through lectures on F toxicology and perhaps the usefulness of F in tooth health?" Hodge inquired of Colonel Warren.[32] Such lectures, of course, were indeed given, not only to New Jersey citizens, but to the rest of the nation throughout the cold war.

A good cop–bad cop assault was launched against the farmers. Almost immediately their spokesperson, Willard B. Kille, a market gardener, received an extraordinary invitation: to dine with none other than General Leslie R. Groves, then known as "the man who built the atomic bomb," at his office at the War Department on March 26, 1946.[33] Although Kille had been diagnosed with fluoride poisoning by his doctor, he departed the luncheon convinced of the government's good faith. The next day he wrote to thank the general, wishing the other farmers could have been present, he said, so "they too could come away with the feeling that their interests in this particular matter were being safeguarded by men of the very highest type whose integrity they could not question."

Behind closed doors however, General Groves had mobilized the full resources of the federal government and the Manhattan Project to defeat Kille's farmers and their Fluorine Committee. The documentary trail detailing the government's battle against the farmers begins with a March 1, 1946, memo to top Manhattan Project doctor Colonel Stafford Warren, outlining the medical problem in New Jersey. "There seem to be four distinct (though related) problems," Colonel Warren was told.

1. A question of injury of the peach crop in 1944.
2. A report of extraordinary fluoride content of vegetables grown in this area.
3. A report of abnormally high fluoride content in the blood of human individuals residing in this area.

4. A report raising the question of serious poisoning of horses and cattle in this area.

Under the personal direction of General Groves, secret meetings were convened in Washington, with compulsory attendance by scores of scientists and officials from the U.S. War Department, the Manhattan Project, the Food and Drug Administration, the Agriculture and Justice departments, the U.S. Army's Chemical Warfare Service and Edgewood Arsenal, the Bureau of Standards, and DuPont lawyers.[34] These agencies "are making scientific investigations to obtain evidence which may be used to protect the interest of the Government at the trial of the suits brought by owners of peach orchards in . . . New Jersey," stated Lieutenant Colonel Cooper B. Rhodes of the Manhattan Project in a memo dated August 27, 1945, and cc'd to General Groves.[35] The memo stated:

> SUBJECT: Investigation of Crop Damage at Lower Penns Neck, New Jersey
> TO: The Commanding General, Army Service Forces, Pentagon Building, Washington D.C.
> At the request of the Secretary of War the Department of Agriculture has agreed to cooperate in investigating complaints of crop damage attributed . . . to fumes from a plant operated in connection with the Manhattan Project.
>
> Signed L. R. Groves, Major General U.S.A.[36]

"The Department of Justice is cooperating in the defense of these suits," General Groves subsequently wrote in a February 28, 1946, memo to the Chairman of the U.S. Senate Special Committee on Atomic Energy.[37]

General Groves, of course, was one of the most powerful men in postwar Washington, and the full resources of the military-industrial state were now turned upon the New Jersey farmers. The farmers' expert witness, scientist Philip Sadtler, was singled out by the military. A handwritten note in General Groves's files in the National Archives demands to know: "Col. Rhodes, Who is Sadtler"?[38]

Groves learned that the Sadtler family name was one of the most distinguished and respected in American chemistry. The firm of Samuel P. Sadtler and Son was established in 1891 and routinely consulted for top industrial corporations, including Coca-Cola and John D. Rockefeller.[39] Philip Sadtler's grandfather, Samuel P. Sadtler, had been a founding member of the American Institute of Chemical Engineers, while his father, Samuel S. Sadtler, was one of the first editors of the venerable science publication *Chemical Abstracts*. (Today Philip Sadtler's "Standard Spectra" are a diagnostic tool used in laboratories around the world.)

But back then, in New Jersey, counterespionage agents followed him and accused him of "dealing with the enemy," stated Sadtler.[40] He recalled one confrontation with two U.S. Army captains that ended in a South Jersey orchard when Gotshalk, the farmers' lawyer, asked the military officials, "Since when are the farmers of the United States the enemy?"

Why was there such a national-security emergency over a few lawsuits by New Jersey farmers? In 1946 the United States had begun full-scale production of atomic bombs. No other nation had yet tested a nuclear weapon, and the A-bomb was seen as crucial for U.S. leadership of the postwar world. The New Jersey fluoride lawsuits were a serious roadblock to that strategy. In the case of fluoride, "If the farmers won, it would open the door to further suits, which might impede the bomb program's ability to use fluoride," remarked Jacqueline Kittrell, a Tennessee public-interest lawyer specializing in nuclear cases, who examined the declassified fluoride documents. (Kittrell has represented plaintiffs in several human radiation experiment cases.) She added, "The reports of human injury were especially threatening, because of the potential for enormous settlements—not to mention the PR problem."[41]

Indeed, DuPont was particularly concerned about the "possible psychologic reaction" to the New Jersey pollution incident, according to a secret 1946 Manhattan Project memo. Facing a threat from the Food and Drug Administration (FDA) to embargo the region's produce because of "high fluoride content," DuPont dispatched its lawyers to the FDA offices in Washington, where an agitated meeting ensued. According to a memo sent the following day to General Groves, DuPont's lawyer argued "that in view of the pending suits

. . . any action by the Food and Drug Administration . . . would have a serious effect on the DuPont Company and would create a bad public relations situation."

After the meeting adjourned, Manhattan Project Captain John Davies approached the FDA's Food Division chief and "impressed upon Dr. White the substantial interest which the Government had in claims which might arise as a result of action which might be taken by the Food and Drug Administration."[42] There was no embargo. Instead, new tests for fluoride in the New Jersey area would be conducted—not by the Department of Agriculture but by the Chemical Warfare Service—because "work done by the Chemical Warfare Service would carry the greatest weight as evidence if . . . lawsuits are started by the complainants." The memo was signed by General Groves.[43]

The farmers kept fighting. On February 2, 1946, Willard Kille wrote to the influential Senator Brian McMahon, Chairman of the Special Committee on Atomic Energy, on behalf of the Fluorine Committee, telling him about the peach trees and poisoning. General Groves quickly interceded, informing the Senator, "I do not believe it would be of any value to your committee to have Mr. Kille appear before it." Groves assured Senator McMahon that "I am keeping in close personal touch with the matter from day to day in order that I may be personally certain that while the government's interests are protected no advantage is taken of any injured farmer."[44]

The New Jersey farmers were ultimately pacified with token financial settlements, according to interviews with descendants still living the area.[45] Joseph Clemente says that his father told him the family had been "paid off" by DuPont after the cattle died suddenly during the war. The Clemente farm lay just across the road from the Chamber Works. His grandfather had been a wartime manager inside the Chamber Works and his family owned a construction firm that had helped to build the plant; accordingly, his father accepted DuPont's cash settlement. "It wouldn't have been very good if my family had caused a lot of stink about the episode," Clemente said.

"All we knew is that DuPont released some chemical that burned up all the peach trees around here," a second resident, Angelo

Giordano, whose father James was one of the original plaintiffs, told the medical writer Joel Griffiths, who visited the orchard country in 1997. "The trees were no good after that, so we had to give up on the peaches."

Their horses and cows also acted sick and walked stiffly, recalled his sister Mildred. "Could any of that have been the fluoride?" she asked. According to veterinary toxicologists, various symptoms she went on to detail are cardinal signs of fluoride toxicity. The Giordano family has been plagued by bone and joint problems, too, Mildred added. Recalling the settlement received by the Giordano family, Angelo told Griffiths that "my father said he got about $200."

The New Jersey farmers were blocked in their legal challenge by the government's refusal to reveal the key piece of information that would have settled the case—the amount of fluoride DuPont had vented into the atmosphere during the war. "Disclosure . . . would be injurious to the military security of the United States," wrote Manhattan Project Major C. A. Taney Jr.[46]

Gotshalk, the farmers' attorney, was outraged at the stonewalling. He called it "a callous disregard for the rights of people" and accused the Manhattan Project of using the "sovereign power of the government to escape the consequences of what undoubtedly was done."[47]

Gotshalk was right. A once-secret memorandum sent to General Groves in Washington—which Gotshalk and the farmers never saw—reveals that the wartime DuPont plant was belching out mass quantities of hydrogen fluoride: at least 30,000 pounds, and perhaps as much as 165,000 pounds, was expelled over the adjacent farmland each month.[48]

The scale of the pollution was explained to General Groves. DuPont was then producing 1,500,000 pounds of HF each month for its commercial Freon-producing [Kinetics] plant, according to his deputy Major C. A. Taney. "Assuming that the losses were only 1 percent at Kinetics, the amount vented to the atmosphere would be about equal to the average loss from the Government facilities at the Chamber Works during the worst months of 1944," Major Taney wrote. But the pollution might be much worse, he added, in which case the lion's share of the blame would be attributable to DuPont's commercial operations. "If the losses at Kinetics ran as high as 10 percent, which is possible, the fumes produced at the Chamber Works would obviously be caused to the greatest extent by DuPont's own operations and not by the Government facilities," the memo stated.

The memo to Groves is probably the smoking gun tying DuPont to the reported injuries. The emissions data would certainly have been crucial courtroom ammunition for the plaintiffs, according to the scientist Kathleen M. Thiessen, an expert on risk analysis and on the health effects of hydrogen fluoride.[49] She notes that the amount of fluoride spilled over the orchards and farms in 1944 from the Chamber Works—at least 30,000 pounds monthly—is "consistent" with the injuries reported within a ten-kilometer radius around the DuPont plant. "The air concentrations could easily have been high enough to cause vegetation damage, and if they are high enough to cause vegetation damage they are high enough to cause damage to livestock eating that pasture," the scientist estimated.

Could the fluoride have hurt the local citizens too?

"It is going to depend on where they lived and how much of that local produce [they ate]," Thiessen explained. The reports of high blood fluoride levels in local citizens, and of badly contaminated local produce, were again "consistent" with human fluoride injury, she added.

Denied the government data, the farmers settled their lawsuit, and their case has long since been forgotten. But the Garden State peach growers unknowingly left their imprint on history. Their complaints of sickness reverberated through the corridors of power in Washington and triggered Harold Hodge's intensive secret bomb-program research on the health effects of fluoride.

"Because of complaints that animals and humans have been injured by hydrogen fluoride fumes in [the New Jersey] area," reads a 1945 memo to General Groves from a deputy, Lieutenant Colonel Cooper B. Rhodes, "although there are no pending suits involving such claims, the University of Rochester is conducting experiments to determine the toxic effect of fluoride."[50]

6

How the Manhattan Project Sold Us Fluoride

Newburgh, Harshaw, and Jim Conant's Ruse

For half a century assurances from the Public Health Service that water fluoridation is safe have rested on the results of the 1945 Newburgh-Kingston Fluorine-Caries Trial, in which the health of children from the fluoridated town of Newburgh, New York, were compared for ten years with children from neighboring nonfluoridated Kingston. But recently declassified documents link the wartime Public Health Service's interest in fluoride to the Manhattan Project. And a trail of papers showing how bomb-program scientists from the University of Rochester secretly monitored the Newburgh experiment, studying biological samples from local citizens—and crudely manipulating at least one other wartime study of fluoride's dental and toxic effects—suggests that Newburgh was simply another cold war human experiment, serving the interests of the nuclear industrial state.

THE VIEW FROM the Old Firehouse on Broadway in the city of Newburgh, New York, is one of the more majestic in the Empire State. The boulevard climbs purpose-straight through the center of town from the valley below, and whipped by a January wind, a lone pedestrian can see east across the mighty Hudson River to a spine of rolling hills in the Connecticut distance. In the spring of 1945 the wind carried the laughter of hundreds of excited school children as they chattered their way to a free public-health clinic inside the Old Firehouse. Doctors wanted to examine the children.

Newburgh had become only the second place in the United States to artificially add fluoride to public water supplies.

"Last week came news that fluorine is to be tried out with whole towns as guinea pigs," *Time* announced approvingly in April 1944. The magazine suggested that, where fluoride was found naturally in the groundwater, "dentists' chief occupation is holding citizens' mouths open to display their perfect teeth."[1]

It wasn't just teeth the doctors were interested in. The Newburgh-Kingston Fluorine-Caries Trial, as it was formally known, was considered the most extensive of the several fluoride experiments then being planned around the United States. Over a period of ten years a team from the New York State Department of Health would conduct a battery of psychological exams and X-rays on the Newburgh children, plus measuring their blood, urine, height, and weight. The information would be compared with data from children in the neighboring fluoride-free town of Kingston, New York. The news that Newburgh would host the experiment created a buzz among local citizens. The gritty, blue-collar industrial town was home to a large population of immigrant Italian Americans as well as African Americans who had come from the South. Most considered themselves fortunate to be early recipients of a new public-health measure.

"I can remember a lot of excitement as a young child," remembered a lifelong Newburgh resident and former Mayor, Audrey Carey, who regularly attended the Broadway clinic in 1945 as a child. Carey's parents were poor, she explained. Her father became only the second African American on the Newburgh police force, and the family was grateful for the daughter's free health check-ups.

"In the front room there was a dental chair and someone would check your teeth and you would see the nurse," Carey recalled. "You would have your height, your weight [measured, and] they would do some urine. I can remember that occurring every month of the year for a very long time."

The tests were designed to answer a simple safety question—whether the chemical produced *nondental* health problems (a medical agenda that, of course, was not publicized to local citizens). "Are there any cumulative effects—beneficial or otherwise, on tissues and organs other than the teeth—of long-continued ingestion of such small concentrations . . . [of fluoride]?" the doctors explained to their colleagues in various academic publications and conferences on the topic.[2]

Some of the most powerful voices in the nation were asking similar questions about fluoride's toxicity—with wartime urgency. Earlier in the fall of 1943 President Roosevelt's science adviser, James Conant, had organized a major Conference on Fluoride Metabolism, secretly convened on behalf of the Manhattan Project.

The conference was held on January 6, 1944, in New York City, and conference transcripts and letters from Conant are among the first documents that connect the atomic-bomb program to water fluoridation and to the Public Health Service (PHS).[3] Weapon makers wanted to use the health service as a wartime camouflage, a fig leaf for the atomic bomb. In a letter dated September 25, 1943, Conant explained to the chief of the Division of Industrial Hygiene, J. J. Townsend, that a "consultant" Dr. Stafford Warren would secretly provide the conference financing. This consultant, of course, was none other than Colonel Stafford Warren, the Manhattan Project's Medical Director.

"It is sincerely hoped that the Public Health Service will be willing to sponsor the conference and to send out the invitations to the contributors under its own letterhead," Conant wrote to Townsend. "All the arrangements such as the selection of the speakers will be taken care of by Dr. Warren. The purpose of this letter," Conant added, "is to assure you of the importance of this symposium and of the real need for the information in connection with the war effort. However, this picture of the purpose of the meeting is for your information only, and it is desirable that the impression be given that the interest is in industrial hazards only."

Dr. Townsend replied that if the Public Health service could review the agenda and "the qualifications of the individuals who might be invited to attend . . . the Surgeon General would be very glad to call such a conference."[4]

On January 6, 1944, a Who's Who of the wartime fluoride industry passed through the doors of New York's Hotel Pennsylvania. Mingling were the top medical men from the army and from the companies and universities building the atomic bomb, including DuPont, Union Carbide, Columbia, and Johns Hopkins. Also attending were Alcoa's top fluoride expert, Francis Frary; Helmuth Schrenk from the Bureau of Mines; the biochemist Wallace Armstrong from the University of Minnesota; and Edward J. Largent from the Kettering Laboratory.

Dr. Paul A. Neal of the National Institutes of Health outlined the critical importance of fluoride to the war economy—and emphasized how little doctors knew about health effects on workers. Aluminum, magnesium, refrigerants, aerosol propellants, insecticides, phosphates for animal feeds, hydrofluoric acid ("especially its use as a catalyst in oil refining"), and the employment of fluoride fluxes among an estimated 150,000 welders were just some of the burgeoning uses for fluoride in the war effort, Neal reported. There was a "definite need," he added, "for careful, thorough investigation on workmen who have been exposed for many years to fluorides. However, it has been postponed until after the war since such an investigation could hardly be made at this time without undue interruption of the output of these industries."[5]

The conference organizers had made what *seemed* to be a surprising addition to the guest list: Dr. David B. Ast, chief dental officer of the New York State Health Department. Dr. Ast was then preparing to add sodium fluoride to the drinking water of Newburgh, New York, in a stated bid to improve dental health in children. Although the conference had been secretly arranged by the Manhattan Project—whose industrial contractors were concerned that workers in bomb factories would be poisoned by fluoride—the dental researcher quickly justified his attendance at the conference. Military officials and industrial contractors heard a conference report that "animal tests were of doubtful value" in studying fluoride toxicity in humans, and that there was confusion over amounts that "may cause deleterious effects in adults." Dr. Ast then boldly volunteered a solution.[6] He suggested that researchers could examine whether fluoride in drinking water was harmful to people, and thereby help to determine whether the chemical posed a risk to workers in factories. The "accumulated effects of small doses of fluoride in drinking water [could] be studied in the U.S. . . . [and that] evidence of the effects of consumption of fluoride over that period of time might [become apparent]," Ast told the conference.[7]

Until such human fluoride studies could be done, however, a

temporary workplace standard had to be fixed. Following the morning conference session, the Manhattan Project had arranged a "luncheon for ten persons who will meet to set standards." It is not clear if the ten men who met for lunch that day—including the Public Health Services' H. Trendley Dean, the researcher who had reported that fluoride found naturally in water in some areas of the country was associated with fewer cavities—knew that their meal was paid for by the Manhattan Project. But Harold Hodge knew: he paid the tab with bomb-program funds. "It would be convenient if cash can be provided and delivered here by Dr. Harold Hodge," the Manhattan Project's Captain Ferry had ordered.[8]

A sacrifice was needed from war workers, the lunch team decided. Although earlier that morning DuPont's Dr. A. N. Benning had described how 1 part per million of hydrogen fluoride in air etched glass in two hours, the diners determined that 6 parts per million of fluoride breathed in factory air would be the wartime fluoride standard for an 8-hour workday, six days a week. The existing 3-ppm threshold in several states was "an arbitrary figure not based on any specific evidence," stated Dr. Carl Voegtlin of the University of Rochester, who chaired the lunch session. "We do not want to set up standards that are so extreme on the lower side that it makes it hard to operate the plants," Voegtlin added, "We can say that *in the absence of definite evidence, we feel . . .*" [emphasis in original].

Francis Frary of the Aluminum Company of America doubted whether standards were even necessary. "The best guide is the individual response," suggested Frary, explaining that "I doubt in the case of man whether there is enough hydrofluoric acid in the air that is comfortable to breathe that would cause any damage."

Hodge finessed the problem, suggesting that "We can also say that men working in plants where we know the atmosphere is varied at all times, should by certain screening methods, be protected."

A lone dissent drifted across the lunch table. "I should think that someone is going to be hurt by the long exposure to the irritant," interposed Dr. Wallace Armstrong from the University of Minnesota.[9]

Following the New York conference, as the giant gaseous diffusion plant secretly rose amid the virgin woodland at Oak Ridge, Tennessee, planning for the public-water-fluoridation experiment in Newburgh also proceeded apace. A Technical Advisory Committee was selected to guide the New York Health Department. The chairman of that expert committee, it was announced, would be a "pharmacologist" from the University of Rochester, Dr. Harold Hodge. "Possible toxic effects of fluoride were in the forefront of consideration," the Advisory Committee stated.[10]

On May 2, 1945, the Hudson River city became the second community in the world to be artificially fluoridated. Over the next ten years its residents were studied by the New York State Health Department. Secretly, in tandem with the state's public investigation, Hodge's classified "Program F" at the University of Rochester conducted its own studies, measuring how much fluoride Newburgh citizens retained in their blood and tissues—key information sought by the atomic bomb program.[11] Health Department personnel cooperated, shipping blood and placenta samples to the Rochester scientists. The samples were collected by Dr. David B. Overton, the Department's chief of pediatric studies at Newburgh.[12]

Hodge was not the only scientist associated with the Newburgh experiment who had ties to the bomb program. Dr. Henry L. Barnett, who joined the Technical Advisory Committee after the war, was described as a pediatrician. But Barnett had also been a Manhattan Project medical captain, sent to Japan following the nuclear bombings as a leading member of the Atomic Bomb Casualty Commission.[13] And Dr. Joe Howland, who drew control samples of blood from residents of Rochester, New York, where no fluoride had been added to water supplies—for comparison with fluoride levels in the blood of Newburgh citizens—was an especially practiced human experimenter.[14] On April 10, 1945, for example, as chief of Manhattan Project medical investigations searching for information on the health effects of bomb program materials, Captain Howland had driven a plutonium-laden needle into the arm of Ebb Cade, an unsuspecting victim of a Tennessee car accident, who had the simple misfortune of landing in the Oak Ridge hospital.[15]

Although Dr. David Ast of the New York State Health Department clearly realized that water fluoridation could give industry useful information about fluoride's health effects on humans—as evinced by his testimony at the Manhattan Project's 1943 Conference on Fluoride Metabolism (above)—today he maintains that he did not know about the Manhattan Project's involvement at Newburgh. "If I had known, I would have been certainly investigating why, and what the connection was," Dr. Ast told me.[16]

The final report of the Newburgh Demonstration Project, published in 1956 in the *Journal of the American Dental Association*, concluded that "small concentrations" of fluoride were safe for U.S. citizens. The biological proof—"based on work performed . . . at the University of Rochester Atomic Energy Project"—was delivered by Dr. Hodge.[17]

Publicly the safety verdict boosted federal efforts to promote water fluoridation. Privately the data was also helpful to the nuclear weapons industry, explained Hymer L. Friedell, the Manhattan Project's first medical director. Workers alleging harmful exposure to fluoride would now find it more difficult to sue the government or its industrial contractors, Friedell stated.[18] "Any claim about fluorides—here was the evidence that it was of no consequence," said Friedell.[19]

"Anything that was evidence of a 'no-effect' level was important information," agreed the former Rochester scientist and historian, J. Newell Stannard.[20]

Although he claimed no knowledge of the Medical Section's role in the Newburgh experiment, Hymer Friedell was not surprised that bomb-program scientists had been involved. "There may have been some things done that were not ever in the record," he admitted.

But there were records. In the once-secret archives of the Manhattan Project's Medical Section, there exists an entire file on Newburgh. Inside the file—coded "G-10" by the U.S. Army—is a startling revelation: The top fluoride scientist for the U.S. Public Health Service, Dr. H. Trendley Dean, the man who later became famous as "the father of fluoridation," had secretly opposed the Newburgh fluoridation experiment, fearing fluoride's toxicity.[21]

Dean's opposition was a potential disaster. News that the leading fluoride scientist from the PHS was against adding fluoride to Newburgh's water—on the grounds of toxicity—would certainly have frightened the Newburgh citizens, perhaps aborted the nation's water-fluoridation program entirely, and eventually have alerted nuclear workers to the danger of handling fluorides.

But Dean's dissent was never made public. Instead, Harold Hodge passed the troubling news on to Colonel Stafford Warren at Oak Ridge:

"Dear Staff:" Hodge wrote on September 15, 1944. "Here is a copy of the current file relating to the Kingston-Newburgh study. If desired, I would be glad to come down to your place and talk this problem over. Sincerely, Harold." (Scrawled on the letter in what may be Warren's handwriting is a note: "Return to Medical Section files.")[22]

Enclosed with Hodge's letter are key documents detailing the planning and protocol for the Newburgh experiment. The Manhattan Project was, indeed, deeply interested in public water fluoridation. The papers include letters from Hodge to Newburgh planners requesting additional "bone" studies—key information sought by the bomb program—and an agenda for a meeting of the Newburgh Technical Advisory Committee, with the word "Warren" scrawled across the top.[23]

The G-10 file also records Dean's opposition to water fluoridation. His showdown with the Newburgh planners occurred at 2:00 PM on April 24, 1944, at the Department of Health's offices at 80 Centre Street in New York City, according to the Advisory Committee meeting minutes sent to Colonel Warren.[24] Dr. Harold Hodge chaired the meeting. Almost immediately, "a question of cumulative poisoning was raised. This is the crux of the whole problem of toxicity as it relates to this study," meeting minutes record.

Dr. Dean took the floor. The PHS expert explained that in parts of the country with high levels of groundwater fluoride (8 ppm) he had seen evidence of "toxic effects" in local residents, including "bone changes" and "cataracts." He wanted more time "to study lower concentrations to see at what level the effects disappear," he told the committee. Dean worried that fluoride posed a special risk to the elderly; he told the committee that he feared Newburgh's citizens might experience "cumulative effects past middle age." The government expert explained that if, for example, a person's kidneys did not work well, that person would be at greater risk for poisoning as more fluoride accumulated in their body. According to the Technical Advisory Committee meeting minutes, an unanswered question about the pending experiment "was what to look for in the way of evidences of early intoxication. Dr. Dean recommended that both the child and the past middle age groups be considered. With the renal impairment common to older age groups, fluorine intake and output even in small concentrations may not be balanced."

But Hodge and his Newburgh team were anxious to proceed. "Much publicity" had already been given to the proposed experiment, recalled Dr. Edward S. Rogers of the New York State Department of Health. Similarly, another Advisory Committee member, Dr. Philip Jay from the University of Michigan, "felt this was the propitious time for such a study from a psychological standpoint." Another Committee member alluded to pressure from Washington policy makers. "While her own feeling was conservative," noted Dr. Katherine Bain of the U.S. Department of Labor's Children's Bureau, "the project had the approval of the Children's Bureau." (The Children's Bureau was financing the Newburgh experiment.)

Chairman Hodge called a final Advisory Committee vote at 4:15 PM, on whether to proceed with the experiment. Dean was the lone voice in opposition. "Dr. Dean did not agree that the proposed program could be considered a perfectly safe procedure from a public health point of view," the meeting minutes record.[25] Nevertheless, the committee voted in favor of the experiment to fluoridate Newburgh's water.

Shortly afterwards, as wartime pressures mounted in that summer of 1944, Dean performed an unreported but spectacular flip-flop, transforming himself from foe to friend of water fluoridation. Just three months after giving Newburgh the thumbs-down, Dean announced that he now favored adding fluoride to public drinking water in the city of Grand Rapids, Michigan. He would be one of the lead investigators, comparing children's teeth for ten years with another neighboring nonfluoridated city, Muskegon. Six months later, on January 25, 1945, America's great fluoride experiment began. One hundred and seven barrels of sodium fluoride were delivered to Grand Rapids, where, at 4:00 PM city technicians gingerly began tipping it into the city's drinking water supply.[26]

Dean's wartime gyration was well rewarded. In 1948 he was appointed the first director of the National Institute of Dental Research, and in 1953 he took a senior position with the American Dental Association. Until now Dean's dissent on Newburgh has never been made public. The government has long dismissed claims that any of its scientists ever endorsed water fluoridation despite reservations regarding its safety.[27]

When the scientist and historian Newell Stannard was told of the once-classified correspondence between Hodge and his Manhattan Project bosses on Newburgh—as well as the military's involvement in the public water fluoridation experiment—he was surprised but saw the logic. "I don't think [the military] was really interested in water fluoridation. I think they were looking for information on toxicity on fluorine, and fluorides," he said.

But former Newburgh Mayor Audrey Carey is appalled at the news that medical officials from the atomic weapons establishment secretly monitored and studied her fellow citizens during the cold war. "It is reprehensible; it is shocking; it reminds me of the experiments that were done regarding syphilis down in Alabama [in which African Americans were not told that they had the venereal disease, so government doctors could study them]," she said in an interview.[28] Now Carey wants answers from the government about the secret history of fluoride and about the Newburgh fluoridation experiment. "I absolutely want to pursue it," she said. "It is appalling to do any kind of experimentation and study without people's knowledge and permission."

Did Harold Hodge and the Rochester bomb scientists suppress or censor adverse health findings from the Newburgh study? There is some indication that they did; however, as we shall see, prying information from the University of Rochester's cold war archive is no easy task, confounding the best efforts of a Presidential Commission in 1994. (For a further discussion of censorship and of Newburgh health effects today, see chapters 7 and 17.)

Evidence that military censors *did* remove information about fluoride's harmful effects can be seen in another study performed by Rochester bomb-program scientists, published in the August 1948 issue of the *Journal of the American Dental Association.* A comparison with the original, unpublished secret version found by the medical writer Joel Griffiths in the files of the Manhattan Project's Medical Section illustrates the ways cold war authorities censored damaging information on fluoride, to the point of tragicomedy.

In these files Manhattan Project Captain Peter Dale at the University of Rochester reported in the second half of 1943 on the preliminary results of two "dental investigations," a study of oral conditions among laboratory fluoride workers at Columbia University, and a study of dental conditions among workers exposed to dilute and anhydrous hydrofluoric acid in production.

The results from Columbia, where scientists at the War Research Laboratories were using fluoride to enrich uranium, were disappointing, even worrying. Fluoride did not prevent cavities, Captain Dale suggested. Of the ninety-five laboratory workers examined, "the total number of tooth surfaces filled and attacked by caries was not significantly altered by exposure to hydrofluoric acid vapor," Dale reported.[29] The fluoride might have been producing a harmful effect. Dr. Homer Priest, a leading fluorine scientist, reported that his "teeth seemed to be deteriorating rapidly." Dr. Priest also told the Medical Section that his gums bled more freely and that "there has been a progressive increase in the degree of slowness of healing and of pain in the period he has been doing this work."[30]

The Columbia data were never published in the scientific literature. But the results of the second dental study, on the laborers at the Harshaw Chemical Company in Cleveland, became an important piece of "evidence" for the idea that fluoride reduced cavities.[31] The study is particularly illustrative. As we saw earlier, work conditions at Harshaw Chemical Company were appalling. Two workers had been killed by fluoride acid in 1945. So much fluoride and uranium was escaping from the plant that the FBI had been called in. And the Atomic Energy Commission proposed secretly tracking former workers, to discover the incidence of lung cancer.[32] None of that was made public, however. All that the medical community learned about Harshaw and fluoride was from a study published in the 1948 issue of the *Journal of the American Medical Association*—a study "based on work performed . . . for the Manhattan Project at the University of Rochester at the suggestion of Harold C. Hodge"—that reported that the men had better teeth. When compared with the original secret study, the published version reveals crude censorship and data distortion, according to the toxicologist Phyllis Mullenix, who read both versions.[33]

- The secret version states that most of the men had few or no teeth; they were "in large proportion edentulous [toothless] or nearly edentulous." This information, however, was left out of the published version. The published study merely notes that the fluoride workers had fewer cavities than did unexposed workers.

- The published version omits the suggestion that fluoride was actually harming the men's teeth. While the secret version states, "There was some indication [teeth] may have been etched and polished by [the acid]," and that "exposure of the teeth to the acid may have contributed to the attrition observed," the public version, instead, concocts an observation seen nowhere in the original. It states that "strangely enough, dental erosion or decalcification of enamel and dentin commonly seen in workers exposed to inorganic acids [fluoride] was not seen." The published version omits information about the harmful effect that fluoride may have had on teeth, ignoring physical evidence that indicated otherwise.

"A lie," commented Mullenix. The published version had simply reversed the original medical observation that fluoride may have corroded and consumed the men's teeth, she said.

- The published version implies that the men were at fault for refusing to wear protective masks, instead "preferring to chew tobacco or gum for protection." The secret study makes no mention of masks (and a later Ohio State study criticized Harshaw for not giving its workers protective masks).[34]

- The published study states that men "with clean mouths" had good teeth. Men "with neglected mouths" had "a peculiar brownish deposit which seemed to cover the enamel of the anterior teeth in large quantities." The secret version, however, *makes no distinction* in the men's oral hygiene, noting that "all men, as a group, neglected their mouths." The published report therefore makes the bad, or discolored, teeth appear to be the worker's fault. "The dirty brown teeth were now a function of the men's hygiene," Mullenix remarked. "In other words, [the censored study is] blaming the victim for not having a clean mouth."

The published Harshaw study helped to shift the national medical debate over exposure to industrial fluoride. Several studies during the 1940s had already shown that acid in an industrial environment hurt workers' teeth, and Dr. Priest's experience at Columbia University suggested that the same was happening with wartime fluoride workers. Now, said Phyllis Mullenix, instead of blaming fluoride for eroding teeth, with the help of "a clever editing job" the published study became a piece of dental propaganda that "buries the American fluoride worker."

"It totally changes the viewpoint," Mullenix told me. "This makes me ashamed to be a scientist." Of other cold war–era fluoride safety studies, she asks, "Were they all done like this?"

Recently, in Cleveland, a roomful of surviving Harshaw fluoride workers erupted in grim laughter when told about Harold Hodge's censored dental study. I showed Allen Hurt the once-secret results of the long-ago measurements of fluoride in his urine, analyzed by AEC doctors at the University of Rochester; the fluoride was recorded at the extraordinarily high levels of 17.8 mg/liter.[35] Today he is plagued with arthritis, he says, while many of his Harshaw friends died young of cancer. Nevertheless, smiling a largely toothless grin, Hurt commented on the published dental study: "They had to come up with something."

7

A Subterranean Channel of Secret-Keeping

AFTER THE WAR Harold Hodge became the leading figure promoting water fluoridation in the United States and around the world, while the University of Rochester served as a kind of queen bee for cold war–era dentistry, hatching a generation of dental-school researchers who were unanimous in support of a central role for fluoride in their profession.

"If you look at the credentials of the people who have been important in academic dentistry, you will find that Hodge's interests here at Rochester were responsible for many of those people getting their expertise," noted the toxicologist Paul Morrow, who worked alongside Hodge for almost twenty years. The fluoridation of public water supplies was the crowning glory of Harold Hodge's career. "He pioneered [fluoridation] very adamantly," Morrow pointed out. "That was one of the most difficult things he did. There was an extraordinary resistance to the use of 'rat poison' in public water supplies."

Today, however, revelations that Hodge concealed wartime information about fluoride's central nervous system effects in atomic workers, secretly studied the health of the subjects of the water fluoridation experiment at Newburgh, New York, on behalf of the Manhattan Project, and gave information on fluoride safety to the U.S. Congress that later proved inaccurate (see chapter 11), all call into question Hodge's agenda as the grand architect of America's great postwar fluoride experiment.

Even during his lifetime, researchers had begun to examine his career more closely. In 1979 a journalist, John Marks, reported that

Hodge had helped the U.S. Central Intelligence Agency (CIA) in its search for a mind-control drug. In his book, *The Search for the Manchurian Candidate*, Marks described how the CIA had given the hallucinogenic drug LSD to unsuspecting Americans. He wrote that Hodge and his Rochester research team had been "pathfinders" in that research program, figuring out a way to radioactively "tag" LSD.[1]

"I knew he had something to do with the CIA, but that is all," recalls the scientist and historian J. Newell Stannard, who worked alongside Hodge at Rochester in 1947.

Marks may have only scratched the surface of Dr. Hodge's work for the CIA. The journalist filed Freedom of Information Act requests and received scores of heavily redacted files. Although the names of people and institutions have mostly been blacked out, Marks identified several of the files as referring to CIA contract work at the University of Rochester. The letters, reports, and accounting statements make chilling reading. They are the bureaucratic account of a laboratory and its scientists eagerly hunting for chemicals to "selectively affect the central nervous system" and to produce symptoms "even more bizarre" than LSD.

The CIA studied fluoride as a potential mind-controlling substance. A March 16, 1966, memo from the "TSD" (most likely Technical Services Division) titled "Behavioral Control Materials and Advanced Research" reports on the "disabling" effects of "dinitrofluoride derivatives of acetic acid" that are "currently undergoing clinical tests."[2]

For many, Harold Hodge's image of respectability collapsed completely in the late 1990s. The reporter Eileen Welsome found a once-classified memo that implicated Hodge in perhaps the most diabolical human experiments ever conducted in the United States. On September 5, 1945, he attended a University of Rochester planning meeting with several other scientists. Their purpose: to discuss the research "protocol" for injecting plutonium into unsuspecting and uninformed patients at the University of Rochester's Strong Memorial Hospital.[3] A second AEC document, reporting on the experiments, thanks "Harold Hodge . . . [who] participated in the early planning of the work and frequently made general and specific suggestions which contributed much to the success of the program."[4] In the 1990s the federal government settled a lawsuit with family members of those plutonium experiment victims, paying approximately $400,000 to each family.[5]

Hodge oversaw additional injections in Rochester hospital patients during the late 1940s, to find out how much uranium would produce "injury."[6] In the fall and winter of that year seven people would be injected with uranium in the "Metabolic Unit" at Rochester's Strong Memorial Hospital. A tunnel connecting the Army Annex to the Hospital permitted the uranium and plutonium to be transported to the ward in secrecy.

On October 1, 1946, "a young white, unmarried female, aged 24" was "injected with 584 micrograms of uranium." She was "essentially normal except for chronic undernutrition which probably resulted from emotional maladjustments," the report stated. In early 1947 a sixty-one-year-old white male alcoholic was admitted to the hospital with a suspected gastric lesion. Although the patient did "not appear ill," the scientists noted, "as he had no home, he willingly agreed to enter the Metabolic Unit." Like the other patients, the man did not know he was the subject of an experiment. Nor was there any attempt to argue that the uranium would have any therapeutic effect on his condition. Injections were explicitly given "to find the dose of . . . uranium which will produce minimal injury to the human kidney," a summary noted. The Rochester scientists believed that a human subject "should tolerate" 70 micrograms of uranium per kilogram of body weight. Accordingly, on January 10, the same "cooperative . . . short, gray-haired man" was injected with 71 micrograms of uranium per kilogram.[7]

In the 1950s Dr. Hodge was a key figure in the Boston Project. In this series of experiments, Hodge arranged for Dr. William Sweet of the Massachusetts General Hospital to inject "the highest possible dose" of various uranium compounds into patients hospitalized with brain cancer. The researchers wanted to learn the quantity of uranium to which atomic workers could safely be exposed.[8]

In 1995 a former senior government physicist, Karl Z. Morgan, described Hodge during these cold war years as a particular enthusiast of human experiments. Morgan had visited Hodge's laboratory and years later told government investigators that Dr. Hodge had been one of the Rochester scientists "itching, you might say, to get closer to Homo Sapiens."[9]

The Trapezius Squeeze

TWO FORMER ROCHESTER students, Judith and James MacGregor, were able to get a close look at the unique influence Hodge exerted over the U.S. medical establishment. The pair had followed Hodge to San Francisco in 1969, when the sixty-five-year-old became professor emeritus at the University of San Francisco Medical School. His office door was frequently open, and they listened in awe as the old man clutched the telephone, reaching across the country, making decisions on faculty appointments at medical schools, on the composition of scientific boards and panels, and on the various national committees that set standards for chemical exposure in the workplace.[10]

"He would be talking to leaders all over the country. Herb Stokinger [the former head of occupational medicine at PHS], people that chaired public health committees for the government would be asking for comments or recommendations on appointments on senior committees, and things like that," stated Judith MacGregor. "He was just incredible at getting things done," she added.

"A great persuader," noted J. Newell Stannard, who worked with Hodge in the 1940s at the University of Rochester. "He had people that would be grateful to do most anything if Harold asked them to do it."

While Hodge wielded the cold steel of political power in the medical world, he generally did so by staying behind the scenes. According to colleagues, his influence was subtle and covert. "He was supremely apt at getting difficult decisions made in the way that he thought they should be without ever raising his voice or appearing to be confrontational," remarked James MacGregor, now a senior official at the U.S. Food and Drug Administration. "He was perhaps the world's master at that," he added.

"He could leave the fewest ripples on the water," said Judith MacGregor. More than a decade after his death, she can still feel the old man's fingers slipping around her shoulder and neck, her resolve buckling. She called this Hodge's "trapezius squeeze"—his signature greeting, which involved taking hold of the shoulder muscle called the trapezius and slowly tightening his fingers, all the while looking into your eyes. MacGregor called Hodge "Grandpapa" behind his back—but she was powerless at the old man's touch. "He would kind of squeeze your muscle a little," she remembered. "It was like a handshake. You knew that when he gave you the trapezius squeeze he was going to ask for something. And you knew that you were going to do it. You couldn't refuse the guy."

Dr. Harold Hodge, it now seems, performed a trapezius squeeze on us all.

"A Whole Song and Dance"

PROBING HODGE'S SECRET fluoride work at the University of Rochester is difficult. Hodge died in 1990. His archive remains closed. And even the multimillion dollar resources of a U.S. Presidential Committee in the 1990s could not breach Rochester's cold war defenses, according to the attorney Dan Guttman, a top investigator in that effort.

Guttman has a quick sense of humor and a sharp mind. He needed both in 1994 for his new job as executive director of President Bill Clinton's Advisory Committee on Human Radiation Experiments (ACHRE, also known as the Clinton Radiation Commission). The attorney had gone to law school with Hillary Clinton. He was tapped by the president to investigate the hundreds of radiation experiments that scientists had performed on unsuspecting U.S. citizens during the cold war—including some on pregnant women, retarded children, and prisoners.[11]

Perhaps the most notorious were the experiments described above with plutonium and uranium that Hodge had helped to plan at the University of Rochester. Guttman therefore wanted access to the University's cold war–era files. He had attended the school as an undergraduate in the 1960s but was "stunned" to learn that his alma mater had been "the Grand Central Station of biomedical research" for the Manhattan Project.[12] The former student approached Rochester's President Thomas A. Jackson at an alumni gathering. On President Clinton's behalf he asked for Jackson's cooperation in obtaining documents from the university archives. Jackson seemed "completely uninterested," Guttman recalled. "I was very disturbed by the University's reaction which was, for practical purposes, obstructing fact finding."

It was not just the University of Rochester who stiffed the U.S. President's Human Radiation Commission. Guttman found himself sitting at a table with Pentagon bureaucrats and lawyers, demanding secret military documents about medical experiments performed on U.S. citizens. At first the Defense Department seemed helpful, Guttman explained; but when the Commission stumbled upon the existence of an inner-sanctum military organization—which appeared to have been in charge of cold war–era human experiments by both military and civilian agencies—the Pentagon suddenly froze. Guttman remembers a specific meeting with top military officials. He asked for all existing records of the Joint Panel on the Medical Aspects of Atomic Warfare, as the secret group had been known. The Joint Panel had included representatives of the CIA, the military, the PHS, the NIH, and the AEC.

"The reaction of the Defense people was, 'We are not supposed to give you that,'" Guttman recalled. "We said 'Excuse us? This was the whole point [of the Clinton Radiation Commission]!'" Guttman asked for the documents nicely. He asked in writing. He asked for six months. He was stiffed. "It was stunning," he said. "All the documents were allegedly destroyed, shredded," he says he was finally told. "We went through a whole song and dance."

Guttman hoped that the Joint Panel documents would shed light on so-called cut-out or "work for others" arrangements, in which the true sponsor of a medical research project is concealed. For example, Guttman explained, "is the CIA having its work done by some innocuous entity that is then funded by some other agency? We were hoping that some of the 'work for others' might have become more apparent through the documents of this interagency group." (Dr. Harold Hodge's work for the CIA at Rochester had been done using precisely such a cut-out arrangement, according to the journalist John Marks. The Geschickter Fund for Medical Research—a Washington, DC, foundation sympathetic to the CIA—had nominally provided Hodge funds, although money secretly came from the government intelligence agency.)

The shredding of public documents about human experiments and military involvement with civilian health agencies during the cold war left Guttman scratching his head. "You ask as a citizen, what was that about?" he said. But the Clinton Radiation Commission *was* able to make a historic discovery. Guttman's team learned that documents had been classified during the cold war, not just to protect secrets from the Russians, but also to hide medical information from U.S. families. "When the Radiation Commission got started," Guttman explained, "people thought that [the government] kept too many secrets but that was for national security reasons. What we discovered was that there was a subterranean channel of secret-keeping, where those on the inside knew that this was not national security, and could not be kept secret for national security reasons, and they had a whole other category, 'embarrassment to the government, resulting damage to the programs, or liability to the government and its contractors.'"

Censorship of the health claims of injured atomic workers, and of medical reports produced by bomb program scientists, was performed by the Insurance Branch and by the Public Relations section of the AEC and the Manhattan Project.[13] Guttman's team found explicit instructions to medical censors, written by the AEC's medical advisor at Oak Ridge. They are worth citing at length:

> There are a large number of papers which do not violate security, but do cause considerable concern to the Atomic Energy Commission Insurance Branch and may well compromise the public prestige and best interests of the Commission. Papers referring to levels of soil and water contamination surrounding Atomic Energy Commission installations, idle speculation on the future genetic effects of radiation and papers dealing with potential process hazards to employees are definitely prejudicial to the best interests of the government. Every such release is reflected in an increase in insurance claims, increased difficulty in labor relations and adverse public sentiment. Following consultation with the Atomic Energy Commission Insurance Branch, the following declassification criteria appears desirable. If specific locations or activities of the Atomic Energy Commission and/or its contractors are closely associated with statements and information which would invite or tend to encourage claims against the Atomic Energy Commission or its contractor such portions of articles to be published should be reworded or deleted.

> The effective establishment of this policy necessitates review by the Atomic Energy Commission Insurance Branch, as well as by the Medical Division, prior to declassification.[14]

Guttman was baffled by what he discovered. Harold Hodge and his Rochester team had been given the job of monitoring workers' health across the entire bomb-program complex—collecting and measuring fluoride, uranium, and other toxic chemicals in the workers' urine—and acting as a repository for their complete medical records.[15] It had been a massive undertaking. Tens of thousands of men and women were employed in the factories making the atomic bomb. Rochester and DuPont each acquired a new IBM punch-card tabulating machine, a forerunner of the computer, to tabulate and analyze the data.[16] Dan Guttman discovered "boxes" of this raw information. But something was missing. The "big unanswered question" about the Rochester data, Guttman explained, was the absence of any epidemiological analysis of worker health.

"What was happening with all that worker safety data that was going to Rochester, and what were they doing with it?" wondered Guttman. "I was really hoping we would find more than just lots of charts, [that] we would find somebody analyzing this stuff. Rochester was an arm of the government, so there should have been some summary, something [like a letter to the AEC stating]: 'Dear Head of the Division of Biology and Medicine, this is what we are finding.' Where is all that stuff?" Guttman asked. "Rochester was extremely uncooperative."

Guttman's committee was asked to uncover information about human-radiation experiments. It had not asked questions about fluoride, however. Was it possible the team had missed other human experiments performed by the Manhattan Project and the AEC?

"Sure," Guttman told me. "On fluorine I would not be surprised if there were missing experiments. I would be surprised if there were missing radiation experiments, but fluorine, I wouldn't be surprised."

The University of Rochester did perform human experiments using fluoride. We may never know exactly how many experiments, nor the souls experimented upon. Nevertheless, a paper trail of now-yellowing documents once again leads back to the "Manhattan Annex" and the passageway to the Strong Memorial Hospital. Rochester scientists gave fluoride to "patients having kidney diseases" to determine how much fluoride their damaged kidneys could excrete.[17] And in a single, cryptic fragment of a declassified Rochester document, a chemical compound, "boron trifluoride," is listed as being "inhaled" for thirty days. Scientists took measurements, including dental studies and weight response. One measurement—item "H"—reads simply: "Human excretion of F."[18]

Postscript: The New World

A MONTH AFTER the Hiroshima bombing, in September 1945, the Danish health expert Kaj Roholm made his first trip to the United States. He wanted to meet America's fluoride researchers and to study wartime advances in American medicine.[19] Top doctors regarded him highly. The Rockefeller Foundation offered financial support and arranged introductions. Roholm traveled widely along the East Coast, visiting hospitals and the medical schools at Yale, Harvard, and John's Hopkins. After the horror and deprivation of wartime Europe, the Dane found the country "inspiring and hospitable," though he did note that the absence of public-health care made him think that "it would be a catastrophe to get sick in the United States."[20]

At the National Institutes of Health in Bethesda, Maryland, Roholm met with the senior dental officials Frank J. McClure and H. Trendley Dean. There they discussed the "fluoride problem." Before the war the American Medical Association and the U.S. Department of Agriculture had warned of the health risk from small amounts of fluorides, and the American Dental Association had editorialized against the idea of water fluoridation.[21] But in his meetings Roholm discovered that the years of conflict had wrought a profound change in Washington's views. "In the United States it is common to associate fluoride as a less toxic element than previously known," he reported.[22]

In 1944, for example, the Department of Agriculture had increased its maximum accepted contaminant level for fluoride pesticides from 1.43 milligrams of fluoride per kilogram, to 7 mgs F per kgm.

And in the water-fluoridation experiments involving thousands of U.S. citizens, fluoride was being added to public-water supplies in Newburgh, New York, and Grand Rapids, Michigan.[23]

Roholm saw the danger. He examined X-rays the PHS had taken from a region of the United States where there were high levels of natural fluoride in the water. The black-and-white images looked familiar. As he had observed in the men and women poisoned by fluoride in the Copenhagen cryolite factory, Roholm detected "numerous cases of typical osteosclerosis" in the X-rays. The promise of better teeth appeared to be worth a great deal to U.S. officials, the Dane mused with dry understatement.

While the "therapeutic concentration for this outcome [better teeth] is close to the toxic limit," Roholm stated, "this, however, has not prevented the Americans from performing several studies."

"The mood was that of great optimism in Bethesda," he wrote. "It will be very interesting to see the results within the next five to ten years."[24]

Roholm returned to Denmark. Although he did not know it, his days were numbered. He was appointed professor of public hygiene at the University of Copenhagen on January 1, 1948. In February he gave his inauguration lecture to students on the history of Danish public-health measures. Although his pithy style made the material "come alive," observers noted that the professor looked pale.[25] Roholm's first lecture as a professor would be his last; stomach cancer had begun its deadly march. One month later Roholm entered the hospital.

The disease tore through his strong body like a wildfire. Each day his best friend, Georg Brun, visited him in the Copenhagen hospital. Throughout that grim March of 1948, as the scientist lay close to death at the age of forty-six, he seemed unable to accept that his life was almost over. Both men avoided the truth. "I tried to say to him that he would be all right," Brun said. "He wouldn't accept anything else." Roholm died of cancer of the large intestine on March 29, 1948. He left a wife and two young children.

Kaj Eli Roholm's death was a tragedy for his family and friends and for the twentieth century—for all who rely on scientists to tell them the truth about the chemicals they handle in the workplace and the risk from industrial pollution.

8

Robert Kehoe and the Kettering Laboratory

FROM THE DARKNESS it can be difficult to determine the source of a shadow. Dr. Robert Arthur Kehoe of the Kettering Laboratory cast such a shadow over us all, one of the darkest of the modern era.

For more than sixty years Americans breathed hundreds of thousands of tons of raw poison wafted into the atmosphere from leaded gasoline. This toxic air contributed to a medical toll of some 5,000 annual deaths from lead-related heart disease and an almost incalculable toll of tragedy in the neurological injuries and learning difficulties imposed on children. One estimate, based on government data, suggests that from 1927 to 1987, 68 million young children in the United States were exposed to toxic amounts of lead from gasoline, until the additive was finally phased out in the United States.[1]

For this in good measure we can thank Dr. Kehoe. Dark-haired and dark-eyed, Kehoe described himself as a "black Irishman" and claimed to be descended from Spaniards who had been shipwrecked on the Irish coast during Elizabethan times. The scientist possessed boundless energy, and a keen mind, and he could also tell "one hell of a dirty joke," colleagues remembered. Others who confronted him professionally, however, remembered Kehoe as arrogant and aloof.[2]

For almost fifty years Kehoe occupied some of the commanding heights of the nation's medical establishment. He was at various points president of the American Academy of Occupational Medicine and president of the American Industrial Hygiene Association; he served as a consultant to the Public Health Service, the International Labor Organization, and the Atomic Energy Commission.[3] Kehoe also exercised a powerful influence on the publication of medical reports, since he sat on the editorial boards of leading scientific publications.[4] He preached the gospel of leaded gasoline's safety from his pulpit at the Kettering Laboratory for the duration of his entire scientific career.[5]

Kehoe did much the same for fluoride, with health consequences of a potentially similar magnitude.

The Fluorine Lawyers and the "Infectious Idea of Easy Pickings"

SPOOKED CORPORATIONS STAMPEDED Kehoe's laboratory following World War II.[6] The great factories that had throbbed and roared for the long years of national emergency had spewed unprecedented volumes of poisonous gas and smoke into the skies over numerous American cities and manufacturing areas. There were aluminum plants on the Columbia River and at Niagara Falls; uranium plants in New Jersey, Cleveland, and Tennessee; steel mills in Pittsburgh; gasoline refineries in Los Angeles; and phosphate plants in Florida. These were just some of the industrial operations that had won the war for the United States, but from which a steady rain of fluoride and other pollutants now fell, endangering the health of workers in factories and people living nearby.

Patriotic U.S. citizens tolerated the smoke of war. When peace arrived, they turned to the courts. Perhaps the first to file suit were the injured peach farmers from the Garden State, downwind from DuPont's Chamber Works. They were quickly followed by numerous additional lawsuits alleging fluoride damage to crops, farm animals, and citizens.[7]

"Soon we had claims and lawsuits around aluminum smelters from coast to coast," recalled Alcoa's leading fluoride litigator, Frank Seamans. "Once this sleeping giant was awakened, claims and lawsuits were brought against all types of plants involving fluoride emissions—steel plants, fertilizer plants, oil refineries, and the like," he added.[8]

To battle this "awakened giant," Seamans and attorneys for other beleaguered corporations organized themselves into a self-described Fluorine Lawyers Committee, which met regularly through the cold war years.[9] The Committee would eventually include attorneys representing several of America's top corporations, including Aluminum Company of Canada, U.S. Steel, Kaiser Aluminum and Steel, Reynolds Metals Company, Monsanto Chemical, the Tennessee River Valley Authority (TVA), Tennessee Corporation and subsidiaries, Victor Chemical, and Food Machinery and Chemical Corporation. Those corporations, guided by the needs of the Fluorine Lawyers, and directed by a Medical Advisory Committee of doctors from the corporations, funded the fluoride research at the Kettering Laboratory.[10]

The gathering storm clouds were surveyed after the war at a confidential conference at the Mellon Institute on April 30,1946. Among the guests filing through the ornately decorated aluminum doorways of the bunkerlike structure on Pittsburgh's Fifth Avenue were representatives from several of the companies facing fluoride lawsuits and complaints, including Alcoa, Pennsylvania Salt, and Harshaw Chemical.[11]

Robert Kehoe dispatched a loyal young Kettering lieutenant to the conference. Although Edward Largent's only degree was a BA obtained in 1935 from Westminster College in Fulton, Missouri, his willingness to sacrifice his own body and the bodies of others on behalf of the Kettering Laboratory's corporate clients, had already propelled him to the front line of industry's defense against fluoride litigation.[12] Starting in 1939, the giant Pennsylvania Salt Company and the Mead Johnson food company paid for a special experimental diet for the Kettering researcher. Pennsylvania Salt manufactured numerous fluoride products, including a cryolite pesticide spray, while Mead Johnson made a children's food, called Pablum, containing animal bone meal. (Bone meal can contain high amounts of fluoride.) Largent "converted to a human guinea pig" for the Kettering sponsors, eating, drinking, and breathing large quantities of fluoride for several years.[13] Under the direction of a Kettering toxicologist, Francis Heyroth, the eager young researcher consumed fluoride in various forms: as cryolite, calcium fluoride, hydrogen fluoride, sodium fluoride, and sodium fluoroborate. As with similar experiments, in which human volunteers breathed lead fumes in a Kettering Laboratory gas chamber, the data were subsequently used to promote industry's position that "moderate" levels of fluoride—or lead—in the body were in "equilibrium" with the environment and, if kept below certain thresholds, were both natural and safe. Such a hypothesis was immensely practical, of course. Following Largent's wartime experiments eating cryolite, for example, the Department of Agriculture raised the amount of cryolite pesticide residue permitted on agricultural produce, an obvious windfall for the Pennsylvania Salt Company.[14]

Now, in April 1946, Largent was one of those sitting in the audience at the Mellon Institute as the grand old man of prewar fluoride science, Alcoa's director of research, Francis Frary, took the stage. Frary explained to the Mellon audience some of industry's worries: how fluoride accumulated in the human skeleton and how coal had recently been identified as an "important" new source of airborne fluoride.[15] Largent was well aware of the legal risks that fluoride posed to corporations. He had been battling farmers who had launched court cases against several big chemical companies in New Jersey and Pennsylvania, alleging damage to crops and herds in a postwar barrage of litigation in the Philadelphia and Delaware Valley area. Largent described these as "almost epidemic."[16]

Industry confronted a potentially devastating cold war domino effect—that America's industrial workers would follow the farmers into court. Largent had been monitoring the fluoride exposure inside the Pennsylvania Salt Company's two big plants in Natrona and Easton, Pennsylvania. The X-rays showed "bone changes" in workers' skeletons and pointed to a clear and present danger, he stated. "These X-ray data could easily be misused by dishonest people to conduct a probably successful attempt to obtain compensation," Largent told a colleague from the Harshaw Chemical Company in an April 1946 letter that discussed the importance of the pending Mellon conference. "The infectious idea of 'easy pickings' may spread to include damage claims regarding occupational injuries," he added.[17]

The Mellon Institute audience was captivated by the bold new medical theory of a second speaker. According to the roentgenologist (X-ray expert) Paul G. Bovard, much of the bone damage seen on workers' X-rays was probably *not* caused by fluoride, and the Danish scientist Kaj Roholm had been a needless worrywart.[18] Dr. Bovard's fresh perspective was terrific news, Largent reminded the Pennsylvania Salt Company. "Several of [your] employees show bone changes which might be successfully, even if it were dishonestly, made to appear like fluorine intoxication. The possibility of a roentgenologist being led by a dishonest lawyer to make such an error is not too far-fetched; it shows with great emphasis how fortunate we are to have the help and interest of a man with Dr. Bovard's capabilities."[19] Bovard's fresh thinking would prove "invaluable assets to the defense against dishonest claims for compensation," Largent concluded.[20]

Largent passed on more good news. Following the Mellon conference, other U.S. companies had also expressed "intense interest" in the fluoride problem. Alcoa's Francis Frary had told Largent that the aluminum company might support an expanded research program at Kettering. Other companies soon contacted Robert Kehoe directly. The DuPont medical director, Dr. G. H. Gehrmann, told Kehoe that DuPont, too, might be interested in joining the fluoride research at Kettering.[21] Such collaboration became a reality that summer and fall. On July 26, 1946, industry representatives met again, this time in the Philadelphia headquarters of the Pennsylvania Salt Company. And by the end of the year DuPont, Universal Oil Products, Reynolds Metals, and Alcoa had all agreed to pay for expanded fluoride studies at Kettering. Of special interest to sponsors: the willingness of the Kettering team to procure additional humans for experimentation. "This program should allow for new human subjects and should materially contribute to this subject," noted Pennsylvania Salt's S. C. Ogburn Jr., in a November 1946 letter to Edward Largent.

More Human Experiments, and a Suspicious Scientific Study

THE EXPANDED RESEARCH program quickly bore fruit, both in fresh human experiments and in an influential scientific paper attacking Kaj Roholm. In January 1947, as industry checks for the fluoride research started to arrive in the Kettering Laboratory mailroom, Edward Largent looked around for more human subjects. He did not have to look far. Largent sometimes ate in the Kettering lunchroom with members of a local African American family, the Blackstones, several of whom worked for the University of Cincinnati as laboratory assistants and animal handlers. "A group of black boys—a wonderful family, Elmo and Peanut and Gentry," remembered Edward Largent years later.[22]

The Blackstone brothers had helped Dr. Robert Kehoe in his lead experiments.[23] In 1947 a new item appeared on the Blackstone's menu—extra-dietary fluoride. In May of that year, forty-one-year-old Elmo Blackstone began eating fluoride and carefully collecting his urine and excreta. The industrial experiments would continue for three and a half years, during which time he would consume a startling 12,047 mg of fluoride in the form of sodium fluoride and sodium fluoroborate, considerably more fluoride than even Largent had ingested. In one experiment, begun in June 1948, Elmo was given 84 mg of sodium fluoride each week in his food for 130 weeks.[24] There is no surviving record of whether Elmo Blackstone experienced injury as a result of these experiments, but the historians Gerald Markowitz and David Rosner describe similar Kettering human experiments with lead as "particularly pernicious because their objective was not the discovery of a therapy for those with lead poisoning but was to gather evidence that could be used by industry to prove that lead in the blood was normal and not indicative of poisoning by industry."[25]

In 1951 Edward Largent mounted a major assault on the research of Kaj Roholm, describing health effects of fluoride exposure in American workers that were much less severe than those reported by the Danish scientist.[26] His paper laid a medical keystone for America's cold war industrial enterprise.[27] The war had hugely increased U.S. industrial dependence on fluoride, a hunger that grew voraciously as the American economy began its spectacular cold war expansion, with entire new enterprises, such as fluorocarbon plastics, aerosols, refrigerants, uranium enrichment, rocket fuels, and agricultural chemicals, all requiring that employees breathe and absorb fluoride.[28] By 1975 the government estimated that 350,000 men and women in 92 different occupations were exposed to fluoride in the workplace.[29] Yet the consequences of that chemical exposure would be largely overlooked, in part because of Largent's 1951 paper, published in the influential *American Journal of Roentgenology.*

Roholm had reported that fluoride produced a host of medical symptoms in factory workers. Most distinctly, fluoride could visibly disfigure a worker's bones, disabling them with a painful thickening and fusing of spinal vertebrae, a condition Roholm called "crippling skeletal fluorosis." Largent now contradicted the Dane, reporting that *no* disabilities had been caused by fluoride in the U.S. workers he had studied. Instead, he argued that fluoride "deposition" only highlighted a preexisting condition, making it more "apparent" to X-rays. "One wonders if Roholm may not have overemphasized the part that fluorides may play in causing limitation of mobility of the spine," Largent wrote. Perhaps the crippled spinal columns of the Danish workers were mostly the result of "hard labor," he suggested.[30]

Largent's 1951 paper was influential among those for whom it was meant to be influential, so that in 1965, for example, the nation's leading fluoride expert, Harold Hodge, could state that "crippling fluorosis has never been seen in the United States."[31] But Largent's paper also appears to have been a grim scientific hoax. At the end of his paper the Kettering researcher had ostentatiously posed a question: why did fluoride appear to affect American and European workers differently? "Just why disability has not been recorded in American workers remains unanswered," Largent wrote.

The answer is simple. The facts were hidden by a Kettering cover-up that misled a generation of medical researchers about the consequences of industrial fluoride exposure and sentenced many thousands of U.S. workers to undiagnosed fluoride injury. Just three years earlier Kettering's Robert Kehoe had privately told Alcoa that 120 workers at its Massena aluminum smelting plant had "bone fluorosis" and that 33 were "severe" cases that showed "evidences of disability ranging in estimated degree up to 100 percent."[32] Similarly, while Largent publicly reported no fluoride disability, privately three doctors had told him that workers' X-rays showed evidence of fluoride-linked medical injury, according to his personal correspondence and long-concealed records.

Largent's 1951 paper was based on X-rays of workers at the Pennsylvania Salt Company. Fluoride was burrowing inside the employees' bodies, deforming and crippling their bones, according to a radiologist, Dr. Thomas Smyth. Ira Templeton, one worker from the company's plant in Easton, Pennsylvania, "showed marked increase in the density of the pelvis, upper portion of the femur, vertebrae, ribs, clavicle, scapula and forearm. Dr. Smyth considered these [effects] to be indicative of marked fluorine intoxication," Largent told management. At another Pennsylvania Salt plant at Natrona, Pennsylvania, X-ray images of a worker, Elmer Lammay, revealed that "bone growths on some of the vertebrae were extensive enough to indicate that some of the bones of the spine were becoming solidly fused together," Largent reported to management.[33] A second Natrona worker, Ross Mills, also revealed a "clear-cut increase in the density of the lower ribs and the lower thoracic and lumbar spine, typical of fluorine absorption," according to radiologist Paul Bovard, who classified Mills a "probable case of fluorosis."[34]

Although the Kettering researchers hid the incriminating X-ray pictures from the workers, on January 31, 1947, a mix-up occurred and Ira Templeton's results were sent directly to the Easton plant. "All of the films show osteosclerosis previously described and considered to be as a result of fluoride poisoning. . . . Very truly yours, Russell Davey, M.D.," read the mailed analysis.[35] Pennsylvania Salt's management was furious at the misdirected letter. Its workforce might learn of the danger from fluoride exposure, the company worried. "You can appreciate the seriousness of this situation to us," wrote a senior official, S. C. Ogburn Jr., to Dr. Robert Kehoe, Largent's boss at the Kettering Laboratory. "Doubtless, this letter has been widely discussed at our Plant and is evidence of extremely poor tact, to say the least, on the part of Drs. Pillmore and Davey," Ogburn added.[36]

Kehoe asked the offending radiologist, Dr. Davey, to send future X-rays directly to the Kettering Laboratory and thereby "absolve the management of the Easton plant of any responsibility." He added, "We wish to avoid any situations that would result in undue suspicions or anxiety on the part of any of these men." And Kehoe swiftly reassured Pennsylvania Salt's management that any apprehension or concern by workers about their health was the result of a semantic misunderstanding. In Europe the terms "fluorine poisoning" and "fluorine intoxication" might suggest disability and even worker compensation. In the United States, however, Edward Largent and the radiologist Dr. Paul Bovard were using these terms differently, infusing medical language with new meaning, Kehoe insisted. Poisoning was "merely an unfortunate choice of verbal expression," he added.[37]

Dr. Kehoe and Edward Largent now delivered their sponsors some good news. Dr. Bovard had reversed the earlier diagnoses of fluoride poisoning by Drs. Smyth and Davey. He now claimed that, "with the exception of spinous ligament changes seen in films of Ira Templeton," the bone changes were "so commonly seen in laborers as to have no necessary or likely relation to fluorine deposition." Pennsylvania Salt should therefore "differentiate between the terms, 'fluorine intoxication,' which carries with it the implication of illness and disability, or impending disability, and 'fluorine deposition,' which signifies demonstrable change but without implying, necessarily, that illness or disease has occurred or is imminent," suggested Largent.[38]

The Kettering researcher's published verdict of "no disability" was manifestly suspicious. All three radiologists had diagnosed some degree of fluoride-induced spinal thickening, "ligament changes," or "fluorosis" in the Pennsylvania Salt workers. A careful reader of Largent's published paper might also note an important distinction between the way Largent had arrived at his medical conclusions and how Kaj Roholm had investigated the same problem. The Dane had listened closely to the health complaints of the Copenhagen employees. He had concluded that fluoride poisoning was insidious and hydra-headed and that several groups of symptoms—including stomach, bone, lung, skin, and nervous problems—often presented themselves at different times in different people, making fluoride injury both serious and sometimes difficult to diagnose.[39] Largent's 1951 published finding of "no disability" in the Pennsylvania Salt workers, however, was made without ever talking to the employees themselves. Nor had the Kettering team performed any medical examinations beyond studying bone X-rays in a distant office. "Detailed clinical examination of the workmen in these plants could not be carried out and therefore no other data are available for consideration," Largent wrote.[40]

Sins of the Father

EDWARD LARGENT'S WILLINGNESS to perform human experiments was remarkable. In the haste of World War II, he had helped the Manhattan Project fix fluoride inhalation "safety" standards at 6 parts per million for U.S. war workers who breathed in fluoride in factories.[41] Following the war Largent even turned to his own family to obtain additional scientific data.[42]

"He couldn't get experimental subjects," explained his son Edward Largent Jr., who today is a classical composer and professor emeritus at the Dana School of Music at Youngstown State University in Ohio. "A lot of people were just antifluoride for whatever reasons," he added.

His son, then a high school student, was selected by his father because he "was available and he was willing," his father told the medical writer Joel Griffiths. "Willing human subjects are not that easy to find," he explained. Largent told his son that he needed more data for whatever research he was doing, Largent, Jr. remembered. "It was really sort of a cursory knowledge. I wouldn't have understood a lot of what he was talking about because I was only a sophomore in high school."

The Manhattan Project's Rochester division had already reported earlier experiments with hydrogen fluoride gas on dogs. At concentrations of approximately 8.8 parts per million of hydrogen fluoride, the lungs of one out of five dogs hemorrhaged.[43] Largent, Sr., had read the study but appeared skeptical about the results. "When I read it I wasn't impressed with what it meant in terms of potential human exposure," he told Griffiths. There was no review committee for the Kettering inhalation experiment and no formal consent forms. "I was the review committee," he said. He did not anticipate health problems in the experimental subjects. "As far as we were concerned, there were no such risks," he added.

In order to perform these new experiments, Largent had to have a gas chamber built. The process was a challenge. HF gas is corrosive, and the acid attacked the metal cylinders and valves. "It was found to be very difficult to maintain a specific concentration of HF in air inside the inhalation chamber," he reported.

Once the gas chamber was built, Largent reserved the greatest amount of fluoride for one of the Kettering laboratory's African

American laboratory assistants, forty-six-year-old male Gentry Blackstone. For fifty days in the early spring of 1953 Blackstone sat in the Kettering gas chamber six hours a day, breathing an average dose of 4.2 parts per million of hydrogen fluoride acid. But Largent did not experiment on Gentry Blackstone alone. Largent also exposed his own wife, Kathleen, to a lower dose of 2.7 parts per million. And although Gentry Blackstone received the largest amount of fluoride over the longest period of time, the single highest exposure values were given to Largent's son. On June 22, 1953, Edward Largent Jr., aged seventeen, entered a Kettering gas chamber for the first time. Cold cosmetic cream was applied to his face. The experiment would continue for twenty-eight days, six hours at a time, with weekends off.

"I had to sit in this cage," the son remembered. A small fan was placed in front of the boy to improve the gas circulation. Outside, his father operated the controls and watched. The walls of the chamber were made from transparent plastic sheeting. The gas whispered in. At first, it caught the teenager's lungs and burned his nostrils, he said. His skin reddened and flaked. He read fiction to relieve the tedium, eyes stinging and smarting. The average dose for the six weeks that Edward Largent Jr. sat in the chamber was 6.7 parts per million—almost two and a half times what his mother received. For one remarkable week in early July 1953, however, with a break for Independence Day, the scientist gassed his son with doses of hydrogen fluoride that averaged 9.1 parts per million and climbed as high as 11.9, almost four times the maximum allowable concentration then set by federal authorities and twice what the father had tolerated himself. The son's urine levels spiked at 40 parts of fluoride per million. The highest doses given to his son were accidental, the father said in retrospect; "It was our inability to keep it from going higher than we wanted it to."

Largent's experiments rang alarm bells for industry. At a 1953 Symposium on Fluorides at the Kettering Laboratory, he described his inhalation studies and spelled out the potential dangers they had revealed.[44] The gathered officials—including the head of the Fluorine Lawyers' Committee, Alcoa's Frank Seamans—knew that American workers were regularly exposed to 3 parts per million of fluoride in their factories and workplaces. They also knew that when fluoride urine levels rose above 8 milligrams per liter, there was real danger that fluoride was building up in the skeleton and might soon become visible to X-rays. Largent delivered the bad news. Fluoride levels in his experimental subjects had spiked sharply immediately after their gas chamber exposures, even at lower "acceptable" exposure levels. "Urinary concentrations averaged about 10 mg. per liter," he told the industry men, "although the atmospheric concentrations of HF were near to 3 ppm, which is generally accepted as satisfactory for prolonged occupational exposure."[45] In public Largent continued to maintain that fluoride was safe in low doses.[46] Privately he told the industry representatives at the 1953 Symposium, "One wonders (whether) . . . prolonged exposure to HF at such a level may not give rise to medico-legal controversies."[47]

Despite his private warnings to industry, Largent's experiments on his family and on the Blackstones are now considered a scientific foundation for today's official safety standard for the tens of thousands of workers who each day breathe the gas in their factories. The other source for safety assurances? Experiments done in 1909 on rats.[48]

Even though the family experiments seem shocking, Edward Largent Jr. refuses to judge his father for placing him in a hydrogen fluoride gas chamber. Although the music professor has experienced knee problems in recent years, he blames a youthful passion for soccer; he doubts that it had anything to do with his summer spent breathing fluoride in the basement of the Kettering Laboratory, where he remembers only moderate discomfort. Mostly, he told me, "It stank and it was very boring. Be careful about criticizing," he warned, referring to the 1950s experiments. "Those were different times. The criteria and the sensitivities to such things were very different." He added, "It is like trying to judge a Beethoven symphony today. You have to look at the circumstances, the instruments he was writing for, the audience situations."

After the experiments Edward Largent Jr., abruptly changed his career plans. He had passed his entrance exams for medical school at Ohio State, but suddenly plumped for music. Science no longer seemed so appealing. "I just decided I didn't want to do that," he said.

His father would be haunted in later life by his own service as a human laboratory animal. Painful "osteofluorosis" led to a knee replacement and a reliance on medication for relief, the former Kettering researcher told medical writer Joel Griffiths in a taped interview in the mid-1990s. "Both knees were hurting," Largent explained, because of "the deposition of fluoride." Ironically, he seemed to have wound up suffering from the very type of skeletal disability his industry-funded scientific studies said did not exist. (In a second interview, however, Largent reversed himself and denied to Griffiths that he had ever suffered osteofluorosis.)[49]

Edward Largent Sr. died in December 1998, five days after an operation for a broken hip, suffered after a nighttime fall: gripped by Alzheimer's dementia, Largent had forgotten to use his walker to get to the bathroom. At the end of his life, his son recalled, Edward Largent "was angry and frustrated and very frightened because he knew there was something that wasn't right and that he couldn't figure out how to deal with it." The son wondered whether his father's bone pain in later life was because of his fluoride experiments. Edward Largent Jr.'s mother also suffered from ill health in her final years. Kathleen Largent had a leaking heart valve and a nerve disorder known as myasthenia gravis. (Arthritis, increased risk of hip fracture, Alzheimer's, and other central-nervous-system disorders have all been linked by scientists to fluoride exposure.)[50]

In recent years Edward Largent Jr. has spent hours reading about the Manhattan Project, wondering if his father was involved. An elder brother said their father had worked at Oak Ridge. And as a boy, Edward Largent Jr. remembers his father arriving from Tennessee at their Cincinnati home on a Friday night during the 1940s, driving a black car with government plates. "The car would go in the garage and I would say 'Let's go for a ride,' and Dad would say 'No, no we can't use that car.' And then he would leave Sunday afternoon in the government car."

9

Donora: A Rich Man's Hocus Pocus

I have felt the fog in my throat—
The misty hand of Death caress my face;
I have wrestled with a frightful foe
Who strangled me with wisps of gray fog-lace.
Now in the eyes since I have died.
The bleak, bare hills rise in stupid might
With scars of its slavery imbedded deep;
And the people still live—still live—in the poisonous night.

Attributed to area resident John P. Clark, whose mother-in-law, Mrs. Jeanne Kirkwood, aged seventy, died at Clark's home at 2 AM on Saturday, October 30, 1948.[1]

THE MOST VISIBLE U.S. air pollution disaster after the war was in Donora, Pennsylvania, where twenty people were killed and many hundreds were injured following a smog that blanketed the mill town over the Halloween weekend of October 1948. Philip Sadtler, the chemical consultant and antipollution crusader, had gone to Donora immediately afterward and written a report blaming fluoride. However, his conclusions were soon drowned out by the subsequent official Public Health Service investigation that blamed a temperature inversion and "a mixture" of industrial pollutants.[2] Robert Kehoe and Edward Largent also investigated the disaster and prepared medical evidence against the Donora survivors who sued the U.S. Steel Company for damages. Kehoe's files shine a stark new light upon these historic events.

Halloween 1948: Donora

WHEN PHILIP SADTLER stepped from the train platform onto Donora's cobbled streets that November morning in 1948, he carefully made his way up McKean Avenue and past the many churches and Slavic working clubs of the industrial Pennsylvania town.

Grief and fear still clung to the air. It was only five days after what had been the worst recorded air pollution disaster in U.S. history.[3] Bodies stiffened in Rudolph Schwerha's funeral home. Scores of citizens had been hospitalized and many hundreds lay seriously ill.[4]

Sadtler nodded a greeting at a knot of Donora's grim-faced citizens. He studied them closely, already gathering clues. Over that Halloween weekend twenty people had been killed in Donora and the nearby town of Webster. Two more would die that same week, and many more would succumb to their injuries in the weeks and months ahead.[5] An estimated 6,000 men, women, and children had been sickened, out of a population of 13,500. They were choked and poisoned in their homes and beds by a toxic gas from the metal-smelting plants along the banks of Monongahela River, which cut between the two towns. The deadly effluent was trapped in the river valley by a seasonal temperature inversion. A layer of warm atmosphere had pressed down on the cold dense air below and a blanket of industrial filth had smothered Donora and Webster for almost five days.

The townspeople were unaware at first that a disaster was unfolding. Their Halloween parade on the Friday night down McKean Avenue was a ghoulish farce. "They were just like shadows marching by," the mayor's wife said. "It was kind of uncanny, especially since most of the people in the crowd had handkerchiefs tied over their nose and mouth to keep out the smoke. But, even so, everybody was coughing. The minute it was over, everybody scattered. They just vanished. In two minutes there wasn't a soul left on the street. It was as quiet as midnight."[6]

As midnight struck, death began to stalk the brightly painted wood-framed homes that climbed the hills surrounding Donora.

Perhaps the first to die was Ivan Ceh, a seventy-year-old retired steelworker. When he was twenty-two, Ceh had set sail from Yugoslavia to work in the Donora mills. At around 8:30 PM that Friday evening, as the toxic fumes crept though the town, the unmarried Ceh began hacking with a dry cough, struggling to breathe. His torment worsened through the night. With his lungs fighting for oxygen, the steelworker's heart suddenly failed at around 1:30 AM. "It was observed that a white frothy fluid was coming out of the patient's mouth during the last moments of life," noted one medical report.[7]

Ceh's violent demise would be typical that night. A Scottish widow who had lived in Donora for twenty-four years since arriving in the United States had also fallen ill on Friday. The town's smogs had frequently left her breathless but this was much, much worse. She coughed through a sleepless night, her lungs scrambling for air. Two hypodermic injections brought no relief and, at 2:00 AM on Saturday, she also died of heart failure.[8]

The undertaker Rudolph Schwerha may have been the first to realize that a tragedy was unfolding. A telephone call announced the arrival of a new death, just as his assistant returned to the morgue with Ivan Ceh's body. "Now I was surprised," Schwerha told *The New Yorker* magazine. "Two different cases so soon together in this size town doesn't happen every day."

Donora's longest night would be etched in the memory of its residents. Almost fifty years later Gladys Shempp gestured to the curtains in her Donora home and described that long-ago Friday of October 29, 1948, as she struggled through air "as yellow as the color of those drapes. You couldn't see. Your eyes were burning, and the tears were running down your face."

The following morning, Saturday, October 30, her husband, Bill Shempp, was called out to the Donora fire station to give oxygen to residents. The smog had thickened. The volunteer firefighter crept through empty streets he no longer recognized. "It was like a claustrophobia," he said. "You didn't know where you were. It would take us at least two or three hours to get to one home."

A vision of hell greeted the firemen. Frightened citizens clamored for oxygen. Shempp released the elixir into a homemade oxygen tent made out of a sheet or blanket. It helped, he said, but when the firemen tried to leave, panic ensued. "They were in great fear of not being able to breathe," Bill Shempp remembered. "They were getting some relief temporarily, and then to shut it off on them, we had quite a problem."[9]

Fire chief John Volk discovered men and women whose lungs clawed for air but whose grip on life was slipping. "I found people laying in bed and laying on the floor," he remembered. "Some of them didn't give a damn whether they died or not. I found some down in the basement with the furnace draft open and their head stuck inside, trying to get air."[10]

A doctor's receptionist, Helen Stack, continued to answer a telephone that had rung endlessly throughout Friday night with cries for help. "Everyone who called up said the same thing," Stack told *The New Yorker.* "Pain in the abdomen. Splitting headache. Nausea and vomiting. Choking and couldn't get their breath. Coughing up blood."

On Saturday morning Stack called her good friend Dorothy Hollowitti to check on Dorothy's father, who'd also fallen sick from the smog. She wanted to reassure her friend that the doctor was on his way. "Dorothy was crying when she answered the phone," said Stack. "I'll never forget what she said. She said, 'Oh, Helen—my dad just died! He's dead!'"

Dorothy's father, the retired steelworker Ignatz Hollowitti, was the sixth victim of the smog.[11] Incredibly, even by that Saturday afternoon many Donora residents still had no idea that a disaster was upon them. Allen Kline was a twenty-two-year-old sportswriter for the *Daily Republic*, covering the Donora high school football games. Donora had a passion for sports. Hometown hero Stan Musial had just completed another fabulous season with the St. Louis Cardinals, batting a league high .376 average. But that Saturday at the football game, it was impossible to see the players from the press box and there was a great deal of "coughing and hacking" from spectators, Kline remembered. "It was almost unbelievable," he added. "It seemed to be nighttime in the middle of the day."[12]

During the football game an announcement was made: the children of Bernardo Di Sanza should return home. The announcer did not mention the reason, but the sixty-seven-year-old Di Sanza was dead. The Donora death fog had now claimed eleven victims.[13]

On the sideline reporter Allen Kline heard firemen "telling stories about how many people they had administered oxygen to, and how people were dropping over here and there." A temporary morgue had been set up in the Community Center. Kline quickly called the Pittsburgh offices of the Associated Press and UPI wire services. He discovered that, ironically, while Donorans were just learning of the disaster, the Pittsburgh wire services were already reporting the deaths to the nation, sealing Donora's place in history.

Donora residents now heard the news over the radio. Walter Winchell broadcast a report on his nationwide show on Saturday evening. Panic quickly gripped the town, phone lines jammed with incoming calls from worried relatives and friends, and hundreds of residents attempted to flee the valley for higher ground. Poor visibility and choked roads, however, meant that for many evacuation was nearly impossible, reported the *New York Times*.[14]

Reports of the unfolding horror quickly reached U.S. Steel's corporate headquarters in Delaware. Its subsidiary company, American Steel and Wire, ran Donora's zinc and steel works. On Sunday morning at 3:00 A M, with the death toll at nineteen, U.S. Steel general counsel Roger Blough made a frantic phone call. He reached the zinc works superintendent M. M. Neale in Donora and ordered him to shut the smelter down.[15] The call may have prevented a much greater disaster. A local doctor, William Rongaus, later testified that if the smog had lasted just one more evening, "the casualty list would have been 1,000 instead of 20."[16]

U.S. Steel had reason to be concerned. Donora was a company town, entirely dominated by the mighty steel and zinc plants that stretched for three fuming and clamorous miles along the town's riverfront. By 1948 five thousand of Donora's men sweated in those mills, turning out record profits that year for the company.[17] Even the town's name betrayed its corporate roots. "Donora" was an amalgam of the first name of Nora Mellon, the wife of Pittsburgh industrialist Andrew Mellon, and the surname of a former company president William Donner.[18] U.S. Steel had long ago purchased the Donora Works from Mellon, but the town's corporate character remained; the steel company's accounting department even drafted Donora's town budget.[19]

Donora was famous for its culture. Many workers were immigrants from Eastern Europe, Slovenia, northern Spain, and Italy.

They had seen newspaper advertisements placed by steel barons Andrew Carnegie and Andrew Mellon in the European papers and had arrived in Donora in the early part of the twentieth century, an excited chorus of foreign tongues bubbling up the valley, mingling with earlier Scottish and Irish immigrants and African Americans from the southern states. The zinc workers—whose toil at the white-hot furnace face was some of the dirtiest in Donora—were mostly from northern Spain.[20]

"Donora was a great Spanish town," remembered Bill Shempp. "They used to have a festival out at Palmer Park every year and people came from as far away as California and it would last for a week or so, and they would practically camp out."

Today a stroll through a wooded Donora cemetery whispers a memory of the new industrial world those immigrants found. Birdsong spills upon the gravestones, some marked with distinctive twin-horizontal Coptic crosses, etched with Slavic, Spanish, and Italian names. Coal barges still push up the Monongahela River. A train whistles in the valley below. On one gravestone an engraved photograph of a young man in an uncomfortable-looking suit stares out from behind a glass panel like an icon, this grave a final resting place for a long-ago dream of that Promised Land in western Pennsylvania.

In Philadelphia that disaster weekend Philip Sadtler's father, Samuel Sadtler, flipped through the pages of his Sunday newspaper. It was full of speculation that Harry Truman would lose the coming November election to Republican presidential challenger Thomas Dewey. But as Sadtler read, his eyes lit on a short description of the terrible events in Donora. *Time*, *Newsweek*, and the *New York Times* all carried similar accounts of the tragedy. Scores of Donora's sick and injured were being evacuated by air to Myrtle Beach in South Carolina.

As he read about the Donora events, Samuel Sadtler became suspicious. He recalled a similar disaster in Belgium some eighteen years earlier, when fumes from metal-smelting and fertilizer factories had been trapped by a temperature inversion and had killed sixty-three people in the Meuse Valley. Thousands more had been left ill with respiratory and heart problems. Kaj Roholm and other scientists had reported that fluoride emissions from industrial plants in the Meuse Valley had caused the disaster.[21] There had been three zinc plants in the valley. Roholm's book sat in Sadtler's library. He wanted his son to go to Donora and investigate the situation.

"Father said, 'That's fluorine,'" remembered Philip Sadtler. "I said, 'Well, so what Dad? I can't afford to go out there.'"

But five days later Philip Sadtler stepped off the Donora train. The six-foot-tall Sadtler already had his own reputation as a talented scientist and air-pollution investigator. He had examined several big fluoride pollution cases just after the war in Ohio, Florida, New Jersey, and Pennsylvania, including the so-called Peach Crop cases, linked to the Manhattan Project (see chapter 5). Sadtler had also measured fluoride content in vegetation along the industrialized Delaware Valley and found damage endemic and widespread.[22] "There were at least ten thousand square miles of damage from fluorine. Most people did not know that was going on," he said.

Sadtler's train ticket to Donora was paid for by a group of crusading Florida farmers. They were suing phosphate fertilizer plants near the town of Bradenton, on Florida's southwest coast, claiming that fluoride air pollution was destroying their crops and their health. Thirty-eight-year-old Sadtler was their courtroom scientific expert. The Florida farmers hoped that a verdict of fluoride poisoning in Donora might help their own court case and worried that the Donora deaths would be blamed instead on sulfur dioxide, a much less toxic pollutant that at the time was being generated in large volumes by the coal used to heat homes.

"The Bradenton farmers called and said, 'Don't let them call it sulfur dioxide,'" Sadtler told me. They feared that if Pennsylvania's industrialists could point the finger at sulfur dioxide produced by Donora's coal-burning citizens, instead of industry's fluoride emissions, then there would be no one to blame for the disaster. "All the culprits in the country at that time wanted to call it sulfur dioxide," Sadtler recalled. By blaming air pollution on sulfur dioxide, the industrial polluters were safe; fluoride, on the other hand, was much more likely to be blamed on metal smelters and manufacturing plants, and could lead to convictions in court.[23] (Today the fluoride researcher and activist Mike Connett describes sulfur dioxide as the Lee Harvey Oswald of air pollution. Like Oswald, sulfur dioxide is a convenient scapegoat and, like Oswald, it is highly unlikely that sulfur dioxide could accomplish all that it is blamed for.) Sadtler thought that the farmers were probably right. He had earlier investigated some big sulfur dioxide pollution incidents, and he felt that the damage in Donora "sounded a lot worse than sulfur dioxide ever caused," he said.

Now, treading Donora's cobbled streets, Sadtler continued gathering clues. When the Donora townspeople talked, he watched their mouths. Many had teeth that were badly mottled, he said. Sadtler knew that the mottling—the white blotches and chalky marks that appeared on teeth—was known as dental fluorosis. He knew that such dental fluorosis was an indication that a community had been exposed to fluoride over a long period of time and was a cardinal sign of fluoride poisoning. Scientists call such long-term and moderate exposure chronic. Larger acute exposures, on the other hand, such as burns or serious lung damage, are the sort of fluoride poisoning that might occur during an industrial accident. Sadtler even joked about the dismal dental situation he found in Donora, where many workers were entirely toothless. "They did not have any tooth problem with the employees in the smelter," Sadtler said, "because when they went to work they put their teeth in the locker. No tooth problem. But people outside [the smelter] did have the mottling."

As Sadtler approached the Donora town hall, more people passed. He heard several ugly hacking coughs. Respiratory disease such as pulmonary fibrosis, emphysema, and dyspnea (shortness of breath) is another obvious sign of chronic fluoride poisoning.[24] He soon learned that the mill town and the surrounding county had a notorious reputation among local people and doctors, even within smoky, industrial Pennsylvania, for lung problems and respiratory disease.[25]

"There were lots of respiratory problems in the area," said the Donora resident Gladys Shempp. "Everybody was always sneezing and carrying on. But they took it for granted, that was just part of life."

Sadtler soon had a third clue to the health of Donora citizens. He learned that arthritis was unusually common in the town. The scientist knew that fluoride was stored in bones as well as teeth; the Danish scientist Roholm had linked fluoride to arthritis-like symptoms. Steel mills added a fluoride mineral called fluorspar to help flux and draw the steel from the molten ore. Fluoride was among the worst pollutants of the U.S. steel industry and the subject of millions of dollars in legal claims against steel mills around the country.[26] The Donora zinc plants also gave off copious fluoride fumes. Working in the steel and zinc mills, or simply living in Donora where the poison was breathed each day, had produced "very obvious" physical effects, both in the teeth and in the bones, of the local people he met, Sadtler said.[27]

Philip Sadtler was not the only new scientist in Donora that day. News of the disaster had electrified the captains of U.S. industry. They quickly dispatched their top lieutenants to western Pennsylvania. That Sunday night, while Donora's firefighters gave oxygen to suffocating residents, twenty-eight miles to the north telephones started to ring in Pittsburgh—home to the U.S. Steel Corporation and the giant Aluminum Company of America. Industrialists knew that the Donora disaster might get much worse. In the wee hours on Sunday morning, U.S. Steel executives had placed an emergency call to the Mellon Institute, whose director, Ray Weidlein, had answered the telephone that weekend. There was already a growing national agitation against pollution, Weidlein knew. The steel industry had reaped record profits in 1947 and 1948. Yet almost no effort was being made to staunch the torrent of raw chemical pollution spilling into waterways and filling the nation's skies. Just three days before the Donora disaster *Collier's* magazine had reported, with stunning prescience: "It is an American habit to poison our air as flagrantly as we have poisoned our water. . . . Given the right weather conditions enough poisonous fumes are poured into the air every day to produce a great disaster. It happened once in Belgium. Now European nations have air pollution control. Should we wait until some appalling catastrophe happens here?"[28]

An aggressive investigation of pollution from the Donora factories might place legal responsibility for the deaths squarely on the smelters, costing millions in victim compensation and requiring expensive new pollution-control equipment in fluoride-emitting industries—not just in Donora, but across the country. "It would have been very hard on chemical plants. It would have been hard on the steel industry, it would have been hard on the aluminum industry," said Philip Sadtler.

There was another worry. Both the U.S. Army and the Atomic Energy Commission (AEC) had a secret and vital interest in the outcome of the Donora disaster, Sadtler knew. Vast amounts of fluoride gas were now needed by the AEC for the uranium-enrichment factories that were being planned and constructed across the United States in Ohio, Kentucky, and Tennessee. Sadtler had already measured high human blood fluoride levels among poisoned peach farmers living near the DuPont Chamber Works plant in New Jersey, where DuPont made top-secret fluoride compounds for the Manhattan Project. If fluoride were fingered for the Donora deaths, it might bring new scrutiny of worker health safety in those AEC bomb factories, resulting in damage suits and expensive requirements for air-pollution controls.

"It would have been very hard on the Atomic Energy Commission," said Sadtler. "They would have had to pay millions of dollars in damages if [citizens] knew the real story."

Newspaper reporters were already sniffing a possible military connection to Donora. "Death Smog Eyed Closely in Washington," headlined one story in the *Pittsburgh Press.* "Military intelligence officials are watching closely Pennsylvania's investigation into causes of the mystery fog at Donora, Pa.," wrote the newspaper's Washington correspondent, Tony Smith. "The government," he wrote, "has given much attention to possible air contamination around atomic energy projects, and has taken precautions to guard against it. Other types of industry, particularly war industries, may also cause air pollution. . . . A source intimate with the operations of central intelligence said that agency will order one of its own if the results of Pennsylvania's aren't considered satisfactory," Smith continued. "Should central intelligence investigate the Donora smog, it would undoubtedly be an unannounced and secret operation."

The Mellon Institute's Ray Weidlein, who had been a consultant to the U.S. military on chemical war gases during World War I, took swift action. On October 31, as an autumn rain fell that Sunday morning in Donora and washed the worst of the smog away, suited strangers began flocking to the traumatized mill town. One of the first to arrive, at 6:00 A M that Sunday, was Wesley C. L. Hemeon of the Mellon Institute. For the next month Hemeon would walk

Donora's streets, acting as the eyes and ears of Ray Weidlein and the many friends of the Mellon Institute.

Hemeon's first stop was an emergency meeting that Sunday afternoon held by Donora's Board of Health. Although the meeting was closed to the general public, the Mellon man managed to slip in. Passions ran high. Donora doctor and health-board member William Rongaus rose and told mill officials that the smog was "just plain murder." Air pollution that night had affected many other towns, he said, but the deaths had occurred only in Donora and across the river in Webster. Many of the deaths were within blocks of the U.S. Steel zinc works.

Poison gas from the zinc mill had been injuring Donora's residents "silently and insidiously" since the mill opened in 1915, Rongaus told the board members. It was not only asthmatics who had been made sick during the disaster; there were numerous reports of normally healthy people experiencing central-nervous-system effects, such as shaking, chronic fatigue, dizziness, and acting "crazy." Many of those symptoms would last for months. At least one Donora woman suffered a miscarriage that evening as well.[29] "I treated many patients who were young and strong and never had any symptoms of asthma," Dr. Rongaus stated. "All complained of severe pains in the lower chest. It seemed to me like a sort of partial paralysis of the diaphragm."

As he sat through the meeting, Wesley Hemeon of the Mellon Institute grew increasingly nervous. The United Steelworkers' safety director, Frank Burke, blamed the zinc mill for fluoride and sulfur-gas pollution. Then it got worse. The steel workers' representative pointed an accusing finger at the medical experts from the Mellon Institute. Workers trusted neither the Mellon Institute nor health officials from the Commonwealth of Pennsylvania to investigate the disaster, Burke announced. State health authorities had done nothing to protect Donora citizens, despite thirty years of lawsuits and complaints. "This is worse than a catastrophe," Burke told the Donora Council. "Twenty of your citizens are dead. Why weren't washers used in the mill to strain poisons out of the air? We want the facts and we are going to get them."

The president of Donora's Board of Health, Charles Stacy, agreed with Burke—any state investigation of the smog would be "a whitewash." Stacy called for an immediate federal investigation by the U.S. Public Health Service. Like many Americans, Donora residents had emerged from the Depression and World War II with renewed faith in the power of the federal government and its ability to improve living conditions. Initially, however, Washington public-health officials had seemed reluctant to get involved in Donora. Twice during the disaster weekend federal authorities had dismissed frantic calls from Pennsylvania asking for government intervention. On Saturday evening, for example, the mayor of Donora, the badly shaken August Chambon, had declared a state of emergency and called Washington for help. His own mother had been stricken. After returning from shopping, she was discovered "lying on the floor, with her coat on, and a bag of cookies spilled all over beside her, gasping for breath and in terrible pain," newspapers reported. A quick federal response might have enabled authorities to measure the exact chemical content of the air pollution or to draw timely blood samples. On Sunday, however, a second plea to Washington from the state authorities was rebuffed.[30]

But subdued Mellon officials soon saw a silver lining in the proposed federal inquiry. They faced a public-relations disaster. Anger in Donora and Webster glowed hot as molten steel. Daily press accounts of smog victims' funerals fanned public emotion. Each shovel of earth t[illegible] on the lowered coffins was a drumbeat of accusation against [illegible] Steel. The first lawsuits against its subsidiary, American Steel and Wire, were already being composed.

The stakes had suddenly become very high, industry saw. Successful lawsuits could prove crippling to many U.S. corporations, warned Alcoa's medical director, Dudley Irwin. He compared the disaster's potential aftermath to the effects of the Gauley Bridge silicosis deaths in West Virginia during the early 1930s. "The repercussions of the Gauley Tunnel [*sic*] episode on silicosis probably will be dwarfed by the effects of Donora on air pollution," Irwin told the powerful trade group known as the Manufacturing Chemists Association, whose Air Pollution Abatement Committee gathered at the Chemists Club in New York City on January 11, 1950, in the aftermath of the Donora disaster. "The Donora incident has not only made the public air pollution-conscious and unduly apprehensive, but also it has advanced opinion with regard to the imposition of restrictive measures by many years," said Irwin. "The outcome of the legal action arising from the Donora experience may set a pattern that could be followed in other areas."[31]

Although the cards now seemed stacked against it, industry had an ace in the hole: a friend in Washington. Only 170 miles from the grieving mill town, across the Allegheny Mountains in Washington, DC, the Truman Administration was basking in the sunny afterglow of the November election triumph. Plum jobs were going to those who had engineered the upset victory over the Republican Thomas Dewey. One of President Truman's most trusted deputies and a key figure in the election victory was fellow midwesterner Oscar R. Ewing. As acting chair of the Democratic National Committee, the Harvard-trained lawyer had raised millions of dollars for the election campaign and had helped to craft the president's folksy media image of "just plain Harry."[32] After the 1948 election Oscar Ewing was reinstalled as head of the giant Federal Security Agency (FSA), in charge of the U.S. Public Health Service.

Ewing had a very private past. For two decades he had been a top Wall Street lawyer for Alcoa. He strolled to work at his offices on lower Broadway in Manhattan swinging a leather briefcase embossed with the gold letters "One Wall Street." Inside were legal papers from the powerhouse law firm of Hughes, Hubbard, and Ewing. The senior firm member Charles Evans Hughes had been an Alcoa attorney since 1910. Hughes would subsequently be a Republican presidential candidate and a U.S. Supreme Court chief justice, while Oscar Ewing became one of the most powerful attorneys in America, earning a reported Depression-era salary of $100,000.[33]

During the war Ewing had moved to Washington as Alcoa's top legal liaison with the federal government.[34] A key wartime concern of the aluminum manufacturers was, of course, lawsuits from workers and communities for fluoride air-pollution damage to health and property. One of Ewing's legal friends was lawyer Frank Ingersoll, from the same Pittsburgh firm as Frank Seamans, head of the Fluorine Lawyers Committee (see chapter 8).

The old friends kept in touch with Ewing, even after he became a Washington public servant. A "Dear Jack" letter from Frank Ingersoll in June 1947, for example, sought Ewing's help in getting a friend appointed to the Federal Trade Commission (FTC).[35] "Dear Frank," Ewing responded, "I would be only too happy to help anyone in whom you, [Alcoa president] Roy Hunt and George Gibbons are interested."[36]

In the grim days of early November 1948, Ewing's Public Health Service now echoed industry's response to the disaster. The same week of the Donora funerals, the U.S. Steel Corporation had taken out a newspaper advertisement denying responsibility for the deaths. "We are certain that the principal offender in the tragedy was the unprecedentedly heavy fog which blanketed the Borough for five days," the company wrote. That same week federal PHS official John Bloomfield also pinned responsibility on the weather, telling newspapers the smog had been an "atmospheric freak."[37]

The Mellon Institute was backing away from direct involvement in the disaster investigation because it wanted "no legal entanglement."[38] Wesley Hemeon told industry leaders in Donora on November 8 that he now favored an investigation by the Public Health Service. A week later, at the annual meeting of the Mellon Institute's Industrial Hygiene Foundation, the PHS announced that it, too, had reversed course. James Townsend of the PHS announced that Donora would be the first investigation of an air-pollution disaster by the agency and its biggest project since their aftermath studies of the Hiroshima atomic bombing.[39]

The PHS chose Helmuth Schrenk to head its investigation. Schrenk was a senior scientist from the Pittsburgh office of the federal Bureau of Mines, located only blocks from Ray Weidlein's Mellon Institute. And although it was not made public then, nor would the Donora citizens learn of his dual identify for more than half a century, Helmuth Schrenk was a poison-gas expert who had worked as a secret consultant during the war for the Manhattan Project atomic bomb program. His special expertise was fluoride gas.[40]

On November 30 Helmuth Schrenk and his PHS team moved into the municipal Borough Building in downtown Donora.[41] It was not a moment too soon. A day earlier Philip Sadtler had seized newspaper headlines. He had completed his investigation, reporting that "fluorine gas" from industrial plants had killed and injured the Donora residents. Other toxic gases—including sulfur dioxide and carbon monoxide—had been in the air that night and contributed to health problems, he stated, but none of them had been present in quantities to kill.[42]

Numerous mills in the area used large quantities of fluoride-containing raw materials, Sadtler wrote. Blood levels of the dead and injured "showed 12 to 25 times the normal quantity of fluorine," he reported. Another symptom of "acute" fluoride poisoning that night, Sadtler noted, included the widely reported appearance of dyspnea, a shortness of breathing similar to asthma. Fluoride had been polluting Donora for years, Sadtler concluded. He reported mottled teeth in Donora residents, the destruction of farm crops, high fluoride content in vegetation, crippled farm animals, and the etching of windows by fluoride gas.[43]

Sadtler publicly sided with those Donora residents who blamed the zinc works for their long-standing health problems and the environmental damage. The Danish scientist Kaj Roholm had identified zinc ore as being high in fluoride content. Ironically, the same zinc ore used in the Meuse Valley in Belgium, where 63 people had been killed in that industrial disaster in 1930, may also have poisoned Donora's citizens. Sadtler spoke with an official from the New York chemical testing firm of Ledoux & Company, which analyzed metal ores imported into the United States. That official told him that the Donora mill had been "smelting high-fluorine content zinc ore from the Meuse Valley," Sadtler reported.[44] After the Donora mill began using the Belgian ores, U.S. Steel had asked Ledoux & Company to "stop analyzing the ore for fluorides," noted Sadtler. "That was told to me by one of the heads of the company," he added.

But Sadtler still had some lingering questions about the sequence of events in Donora that weekend. Temperature inversions and bad fogs were common during the fall in Donora and along the Monongahela Valley. Why had so many people been killed and injured that weekend? Why had the deaths occurred in such a short period of time? At one point nine people died in six hours. Most deaths happened on Friday night and before noon on Saturday. Yet the weather was just as bad on Saturday evening, and the zinc mill did not cease operations until Sunday morning.[45]

"It was really very queer," said Donora's Red Cross director, Cora Vernon, who was prepared for more deaths on Saturday evening. "The fog was as black and as nasty as ever that night, or worse, but all of a sudden the calls for a doctor just seemed to trickle out and stop. I don't believe we had a call after midnight," she told *The New Yorker.*

Sadtler suspected that something had suddenly produced an extraordinary amount of fluoride that Friday night. He wondered whether top-secret military work had been going on in the Donora mills. "It might have been that they were smelting something for the Atomic Energy Commission," he speculated. Perhaps, he said, the Donora mills were being used that night to roast not zinc ore, but uranium tetrafluoride, to "drive off the fluorine, so that they could get the uranium."

Investigative reports fifty years later by Pete Eisler in *USA Today* and subsequent disclosures by the Department of Energy, all since Sadtler's death, have revealed that private industrial plants were routinely used for secret nuclear work in the 1940s and 1950s. Although none of these disclosures has mentioned Donora, many have revealed that workers were frequently injured by that work and rarely informed about health risks.[46]

Dr. Weidlein Goes to Washington

SADTLER'S VERDICT OF fluoride poisoning in Donora maddened industry. An account of his findings was published on December 18, 1948, in the leading trade magazine, *Chemical and Engineering News*. Retaliation was swift. Sadtler heard immediately from the magazine's Washington editor, who told him that he could not accept any more reports about Donora. Although Sadtler had been a frequent contributor—and his grandfather had been a founding member of the American Chemical Society, which publishes *Chemical and Engineering News*—the editor explained that the director of the Society was now none other than the Mellon Institute's Ray Weidlein. "He told me Dr. Weidlein had been to visit," Sadtler said. "Why would the Mellon Institute, supposedly a nonbiased, nonpolitical organization do such a thing? Well, U.S. Steel, the owners of the zinc works, had an influence with the Mellon Institute, so it only took a telephone call to have Dr. Weidlein go to Washington."

Robert Kehoe also attacked Sadtler. His Kettering Laboratory had been hired by U.S. Steel to conduct a private investigation of the disaster, and it would gather medical evidence to fight lawsuits by victims' family members and smog survivors. Dr. Kehoe fired off a blistering volley to the editor of *Chemical and Engineering News*, Walter J. Murphy, on December 22, 1948. In a letter underlined

"Personal and Confidential," Kehoe called Sadtler's conclusion of fluoride poisoning, which had appeared in the magazine two weeks earlier, "wholly unwarranted," "almost certainly untrue," and a disservice "not least to the families and friends of the unfortunate victims." (Kehoe did not mention in his letter, however, that he was working on behalf of U.S. Steel, which was being sued by those same "unfortunate victims.")

"The analysis of the blood for fluoride is a very difficult procedure," Kehoe wrote, "and even under conditions of severe exposure the concentrations of fluorine in the blood [are] quite low. My associates and I believe that no such results as have been reported here [by Sadtler] are possible of achievement, and therefore we regard the entire story as a deliberate lie or as an irresponsible expression of technical ignorance or incompetence." Kehoe was careful to keep his attack anonymous. "Since I and my associates are engaged in investigations at Donora I do not wish to be quoted in any way in this connection, lest I be suspected of having drawn conclusions before facts are available," he added.

Murphy passed the smoldering letter to his boss, executive editor James M. Crowe, who responded to Kehoe on January 7, 1949. "I have heard from Sadtler recently," Crowe wrote Kehoe, "and he insists that he has made tests on the blood of victims of the disaster and on vegetation, etc., in the area and that he has chemical evidence of unsafe concentrations of fluoride. He claims that he volunteered to check his analytical methods and results with the representatives of the public health agencies, but that they were uncooperative. . . . I note from your letter that the analysis of fluorine in blood is quite difficult and that you feel Sadtler could not have obtained the results indicated. It seems to me that this is the one point, at least, where scientific methods could be checked and agreement reached on whether the results are or are not accurate. It is not our intention to become embroiled in this matter and permit our pages to become a battleground for this case, but for our own information we would be interested to know the results of any analytical findings of your investigation."[47]

Kehoe would send no analytical results to the magazine. Secretly his Kettering Laboratory had now obtained a *similar* blood fluoride result to Sadtler's. Kehoe's first letter attacking Sadtler had been cc'd to Dr. Dudley Irwin, Alcoa's medical director. Alcoa was then sponsoring Kehoe's fluoride research at Kettering and may have been the master puppeteers in the Donora investigation.

Kehoe's Donora deputy, Dr. William Ashe, had reported earlier that summer on the crippling disability fluoride air pollution had caused among aluminum workers inside Alcoa's smelting plant in Massena, New York. Ashe thought that poison gas had caused the Donora deaths. "My assumption that it was a gas which was hydrolyzed in the lung and produced its pathology some little time after it was inspired is based on a very superficial check of the clinical picture as seen by two doctors and two patients," Ashe told Kehoe. (When two PHS officials visited Cincinnati to discuss the disaster investigation, Ashe advised Kehoe to keep this speculation private. "I think that it would be wise to refuse to let them know what our guesses are," he said.)[48]

Following the disaster, Alcoa had quietly obtained a blood sample from one of the first Donora victims, Mike Dorance. On December 30, 1948, in a letter marked "CONFIDENTIAL," Alcoa reported the results of that blood analysis to Dr. Ashe. The letter, which was also cc'd to Dr. Dudley Irwin, was written by the head of Alcoa's analytical division, H. V. Churchill. Alcoa's fears about Donora, and the awful parallel with what Philip Sadtler had found, are wholly evident in this confidential note, written on company stationery:

"Dr. Irwin suggested that we analyze the sample of blood for fluorine content, and we have just completed that analysis. This sample was received by us and contains *20.3 p.p.m.* fluorine," Churchill wrote. "I trust that you will find this information of some use to you" (emphasis in original).[49]

This blood fluoride level is, of course, almost exactly what Sadtler had reported finding in Donora victims—the data that Robert Kehoe had objected so strenuously to seeing published. Dr. Ashe responded to Alcoa on January 3, 1949. He pointed out that no fluoride had been found in Mike Dorance's lung tissue, the only organ tested, and that a volume of "fluid" squeezed from the lung had been too small to test. "Please be assured that we are grateful to you for this data and know that it is completely reliable information. The only problem is: Where did the fluorine come from?" Ashe wrote to Churchill.[50]

"The fluorine finding clearly had some people worried," noted scientist Kathleen M. Thiessen, an expert on risk analysis who reviewed many of the Kettering papers on the Donora investigation for this book. Mike Dorance's fluoride-saturated blood, however, could not be regarded as proof that fluoride was the killer that weekend, Thiessen said. If Dorance had inhaled lethal doses of fluoride that night, she would have expected to see some measurement of fluoride in his lung tissue, she cautioned.[51] Nevertheless, she described the blood fluoride level measured by Alcoa as "excessive" and enough to kill. "That's high," she said. "If that was all you had, you could say it was highly likely that person died of fluoride poisoning."

One more dagger was secretly pointed at Philip Sadtler. When he had first arrived in the mill town, Sadtler met with a deputy from Pennsylvania's Health Department to offer his services as an investigator.[52] But the official quickly attempted to head Sadtler off, he said. "I went to the Borough Hall, it was about 7:30 on a Friday night, met the deputy and he said 'I will see you in my office in Harrisburg [the distant state capital] on Monday,'" recalled Sadtler. "That killed everything. I had nothing to go on. I was quite upset and there was a schoolteacher who heard that, and after a few minutes' conversation he went into the borough council and told [them] they should hear me. So I told the borough council what I knew and they appointed me an official investigator. So when I came back a week later, the union had already appropriated $20,000 [*sic*] to investigate or pay for an investigation, but somebody inserted in pen in the minutes 'at his own expense.' Therefore I was not going to get anything from that $20,000."[53]

Unknown to Sadtler, federal authorities had privately warned the Borough Council not to work with the independent investigator. PHS investigator Duncan A. Holaday reported back to officials in Washington that Sadtler "has broken into print previously in somewhat the same role, as one who could solve complicated problems quickly for a sufficient monetary consideration." Local officials had been given a choice, Holaday added. He explained to them, "The Public Health Service . . . could not work in cooperation with a private individual who had been hired on a fee basis. It was suggested that if they so desired I would submit to them a list of competent industrial hygiene consultants, any of whom would give them an honest appraisal of the situation."[54]

10

The Public Health Service Investigation

The big federal investigation now shifted noisily into gear. From November 1948 and through the following spring Donora residents were bombarded with door-to-door surveys and endless questionnaires from the Washington investigators. Public Health Service air sampling vans criss-crossed the steel bridge between Webster and Donora. The town hall sprouted an air monitor.

Donora residents were elated. They were confident that Harry Truman's Public Health Service would deliver "fair deal" answers about the Donora smog. They also hoped that the federal investigation would help resolve thirty years of community conflict with U.S. Steel. Many residents saw the disaster of 1948 as simply the most recent and violent insult the community had suffered from industry.

When the Donora zinc works opened in 1916 it was the biggest of its kind in the world, and one of the dirtiest. The plant used coal and gas to roast the zinc ore and drive impurities into the air. Ironically, and too late for Donora, that technology was almost immediately superceded in newer plants by much cleaner technology, which used electricity to melt the ore.[1] But U.S. Steel was not prepared to abandon its expensive Donora investment. Zinc was fetching high prices as a vital ingredient in munitions for World War I, which was then raging in Europe.

Each day the Donora works billowed out giant clouds of oily and foul-smelling smoke that drifted on the winds west across Donora or east into the town of Webster. Local families were outraged by their foul-breathed neighbor. Webster's farmers and small holders had chosen the pristine river valley for its natural beauty and the rich soils long before the zinc works had arrived. Some farmers had been on the same land since the Whiskey Rebellion of 1794. Now toxic smoke filled their homes and they watched in horror as the farmland above their town grew barren, rutted gullies slicing at the balding hillsides.

The children of Donora and Webster were born into a near-eternal darkness of smoke and fumes, frolicking on land defoliated by chemical poisons.[2] Even the dead could not rest. Industry's fumes laid waste to Donora's lovely Civil War-era Gilmore cemetery. As the rootless earth eroded down the side of the valley, gravestones toppled and observers reported seeing dogs make off with human bones.[3] A 1941 novel by a former Donora steelworker, Thomas Bell, recalls a view of the zinc works from the Webster side of the river:

> Freshly charged, the zinc smelting furnaces, crawling with thousands of small flames, yellow, blue, green, filled the valley with smoke. Acrid and poisonous, worse than anything a steel mill belched forth, it penetrated everywhere, making automobile headlights necessary in Webster's streets, setting the river boat pilots to cursing God, and destroying every living thing on the hills.[4]

Webster families and some Donora supporters began to organize. The first health-damage suits against the zinc plant were filed in 1918. Marie Burkhardt, a Donora resident since 1904, told a jury that since the plant's opening she had suffered chest pains, a hacking cough, the loss of her voice, and headaches. The jury found her complaints plausible, and so did an appeals court judge. Burkhardt won a judgment of $500 against the zinc plant. Suits like Burkhardt's would continue, angry and unabated, until the plant closed some forty years later. Although claims in the name of 659 plaintiffs had totaled $4.5 million in 1935, court victories were rare and settlements were usually tiny; residents faced an uphill battle against the richest steel company in the world, armed with legions of lawyers to defeat and delay the protests.[5]

"Suits did not get very far," noted a lifelong Webster resident, Allen Kline. He remembered "two or three" small victories like Burkhardt's. "In one case they got an award of $500. Another won $2500. Mostly people got tired of fighting."

The children of Webster were some of America's earliest environmental protesters. Allen Kline's name was listed on a lawsuit against U.S. Steel by his grandfather when Kline was eight years old. His grandfather, an immigrant from Italy, had built their family home in Webster in 1914. He owned farmland in the hills above the town. Two years after he constructed the family's home, the zinc plant was built. For almost fifty years the Kline's home sat directly downwind from the zinc works. Kline remembers a 1938 visit from distant cousins who lived in Allentown, Pennsylvania, on the other side of the state. They were supposed to stay for a week, but instead, "They were here for two days," he recalled. "They didn't know how we lived under these conditions. . . . We didn't know what it was to breathe clean air."

After the 1948 disaster in Donora a protest group called the Society for Better Living took root in Webster's treeless soil. The twenty-two-year-old Kline became the secretary of the Society, which eventually had about 200 members. Its slogan: "Clean Air and Green Grass."

For the next decade the Society waged a David-and-Goliath struggle with U.S. Steel. Tensions ran high in the community. Many Donora workers saw the Society as a threat to their jobs. Several Society officers received death threats, reported Kline. "A lot of people made a good living at the mill," he added. But the tiny group persisted. Its members held rallies, issued Kline's press releases, and even traveled to Washington, DC. Years later Kline remembered this Quixotic lobbying trip to the nation's capital. The self-described "idealistic" young newspaperman and his band of Webster residents had a fantastic notion: why didn't Congress enact nationwide laws against air pollution to protect communities such as their own? Their Washington pleas fell on deaf ears: "I don't think anybody ever knew we were there," said Kline.

The president of the Society for Better Living, Abe Salapino, and deputy Kline grew anxious that spring of 1949. They watched as U.S. Steel public-relations men squired federal health officials around town, wining and dining them at local restaurants. "We were concerned that they were winning the battle on this gastronomical front," said Kline. But Salapino owned a local restaurant. Guests came from Pittsburgh for his delicious meats and pastries, calling first to make sure that the wind was not blowing zinc fumes into the restaurant windows. Salapino and Kline now organized a "sumptuous" meal for the Public Health Service men on their final night in Donora, courtesy of the Society for Better Living. "You couldn't believe this party," said Kline. "We had most of them drunk. We decided there is no way we are not going to get a favorable report out of this group."

That summer, shortly before the much-anticipated PHS report was released, Allen Kline and other members of the Society for Better Living got their own surprise invitation. The president of the American Steel and Wire Division of U.S. Steel, Clifford Hood, wanted them to come to Pittsburgh for a friendly meeting. Kline was stunned. He had spent the last year issuing press releases blaming the company for the Donora deaths and complaining about pollution. At the meeting Hood denied that the zinc works had caused the disaster, but he conceded that U.S. Steel fumes may have damaged some vegetation in the valley. The admission was an about-face from the aggressive position the company had long taken in court. The meeting then became "almost a love session" between the two adversaries, Kline recalled. President Hood gave the twenty-two-year-old a couple of his Havana cigars. "I was terribly impressed by him," said Kline.

The following day the Donora papers reported the goodwill meeting and the steel company's promises to reduce smoke from the mills. The Society for Better Living was "perfectly convinced" of U.S. Steel's sincerity, the newspaper wrote. Kline realized that the meeting had been a public-relations stunt, a "carrot" for his group to improve U.S. Steel's image in Donora. For the remaining decade of the zinc plant's operation, no air scrubbers were installed, according to the Society for Better Living.[6]

While Clifford Hood was passing out cigars to the Webster environmentalists, behind the scenes his company had hired the powerful Pittsburgh law firm of Reed, Smith, Shaw, and McClay, which was headquartered in Andrew Mellon's Union Trust bank building. For much of the century the firm had been fighting citizens in court who claimed that their health and property had been hurt by industrial pollution. The well-heeled Pittsburgh lawyers were given new marching orders after the disaster: defeat the families of the Donora victims in court and escape any legal requirement to clean up the smelter operation.

Robert Kehoe's scientists became the secret weapon of the Pittsburgh lawyers, serving as U.S. Steel's Trojan horse in Donora, nuzzling close to the official PHS investigation, and prying access to the government investigation and its confidential data. As a result PHS investigators gave Kettering officials samples of autopsy material they had gathered immediately after the disaster—information they should not have given out. And when two of the Donora dead were exhumed for additional studies in March 1949, once again Kettering officials joined the PHS doctors around the autopsy table.[7] A former PHS historian, Lynn Page Snyder, calls this manipulation of the public trust by Kehoe the "underbelly" of the Donora investigation. While gaining broad access to the government investigation, Kehoe was privately working with U.S. Steel to shoot down citizen lawsuits.

"Ethically, what was problematic to me was that Kettering officials were given slides with lung tissue, and permission was not requested from the next of kin of the people who passed away," Snyder remarked. "Some of the autopsies were done on people who were dug up after they had been interred. And the PHS and the Borough council and the Board of Health locally worked carefully with the families of the deceased to convince them to dig the bodies back up." Kehoe's access to all this medical data was granted, "without informing area residents of the purpose of Kettering efforts," Snyder added.

Snyder wrote a detailed study of the Donora disaster as a graduate student, and she grew concerned that the federal government's investigation had focused on the weather in Donora that weekend, rather than on the "incredibly filthy" metal-smelting industries. "I am disturbed by the way it is remembered," she said. "I would like to see more discussion of the industrial nature of this disaster."

According to Snyder, PHS officials were willing collaborators in efforts to suppress information about industry's role in the deaths.

When Kehoe prepared U.S. Steel's medicolegal defense against the Donora survivors, for example, he asked his government connections for information on the exact sequence of deaths and the time and location in which they occurred. The chief of the PHS's Division of Industrial Hygiene, J. G. Townsend, wrote back two weeks later giving Kehoe the government data that plotted the onset of the sickness in Donora during the disaster. And a second "special table" of data, correlating smog affliction with preexisting illness, was sent to Kettering and marked by the PHS "This information is *CONFIDENTIAL* and is submitted to Doctor Ashe for his personal use only."[8]

Snyder says that those statistics, which were reworked by Kehoe's team to narrowly define a so-called smog syndrome, helped to discount the role of the disaster in the many hundreds of chronic illnesses or deaths in the smog's long medical aftermath. Many of the lawsuits filed against U.S. Steel involved such cases. "That particular information was helpful to William Ashe," Snyder pointed out, "so that the Kettering people could construct a legal argument that ruled out a number of claims as being unrelated to the smog."

The evidence that the federal government had secretly cooperated with Kehoe disturbed Snyder. "It is collusion," she remarked. "I read that memo [the one marked "confidential"] as evidence of a public health service person collaborating in the case being prepared by Kettering against the plaintiffs—the citizens in Donora and in Webster—without their knowledge." Snyder added, "The information about the illnesses and the times of onset belonged to the citizens, just like the autopsy material. It was not information that ought to have been given to a private interest preparing [to defend a lawsuit] against them."

In October 1949 the PHS report on Donora was finally released. It was an enormous disappointment to the victims' families. They had hoped it would explain what poison killed their relatives that night and where it had come from. The 173-page government document, Public Health Service Bulletin 306, did neither. "They produced a report which looks the size of the Holy Bible," said Allen Kline, "and came to virtually no conclusions."

The government verdict that "no single substance" was responsible for the Donora deaths, however, was a triumph for the U.S.

Steel Company. The report's emphasis on the bad weather effectively endorsed the same argument made by the U.S. Steel lawyers, that the disaster was "not foreseeable" and therefore an "act of god." Blaming the weather had opened the door for a legal escape act. The report's failure to identify which factory or chemical had caused the deaths completed the corporate getaway. "The report did not improve the prospects of the town one whit," noted Lynne Snyder.

Oscar Ewing—Alcoa's former chief counsel, friend of President Truman, and head of the Federal Security Agency—wrote the introduction to the official final report of the Donora investigation. He was silent about his past corporate loyalty to Alcoa. He was silent about the fact that the international aluminum industry had been fighting lawsuits alleging fluoride damage from air pollution for forty years. And he was silent about the sixty-three people who had been killed in 1930 in the Meuse Valley air pollution disaster in Belgium. Instead, Ewing fatuously declared that air pollution was "a new and heretofore unsuspected source of danger." Donora had revealed "the almost completely unknown effects on health of many types of air pollution existing today," he added.

It was a rank Washington smokescreen. Alcoa had spent much of World War II and its aftermath grappling with massive lawsuits and citizen protests over fluoride air pollution from aluminum plants.[9] Oscar Ewing's legal colleague Frank B. Ingersoll was a partner in the Pittsburgh law firm of Smith, Buchanan, Ingersoll, Rodewald, and Eckert that had fought many of those lawsuits on behalf of Alcoa; Frank L. Seamans of the same firm would coordinate a national corporate legal defense strategy in the 1950s as chairman of the Fluorine Lawyers Committee.

The PHS report itself, "Air Pollution in Donora, Pa—Epidemiology of the Unusual Smog Episode of October 1948," was written by the Manhattan Project's wartime fluoride consultant, Helmuth Schrenk. He was particularly adamant in his efforts to disqualify fluoride as the killer agent. "The possibility is slight that toxic concentrations of fluoride accumulated during the October 1948 episode," Schrenk wrote.[10]

The PHS report, however, made no mention of the high fluoride levels in Donora vegetation that Kettering researcher Edward Largent had gathered during a cloak-and-dagger trip to Donora in the summer of 1949. Kettering's Dr. William Ashe had written a letter of introduction for Largent on July 11, to the Director of Industrial Relations at the Donora Works, Mr. E. Soles: "Largent . . . will be around Donora for a day or two, looking into the problem of the effects of particulate fluorides upon foliage and crops. There is no direct relationship between this matter and the smog disaster, but there may be an additional problem which could cause the company considerable embarrassment. . . . I suggest that the purpose of his mission be kept entirely to yourself."[11]

Philip Sadtler had blamed fluoride for defoliating Donora's trees and grass. Largent confirmed high fluoride levels in local vegetation.[12] Why the need for Largent's secrecy?

"It sounds like there was a problem with fluorine emissions and it was clandestine because Kettering did not want other people to know about it—clear as that," believes Lynn Snyder. "The clandestine part fits in with the rest of their activities. If they told people like a plant manager, word would get out, and Phil Sadtler's theory would get more credence."

Schrenk's PHS report also dismissed the numerous medical accounts of long-term health problems caused by air pollution in Donora and the common experience of the residents who invariably became sicker when the smelter fumes were trapped in the valley. And critics found the government report to be laden with mathematical errors, especially when it came to determining fluoride emissions. The report guessed that 210 tons of coal burned in homes emitted 30 pounds of fluoride, but 213 tons burned in the mills gave off only 4 pounds. "No possible reason for the difference is offered," said the physician and researcher Dr. Frederick B. Exner. On page 104 of the report, Exner pointed out, waste gas from the blast furnace contains 4.6 mg of fluoride per cubic meter; on page 108 it contains one-tenth as much. "An elaborate piece of hocus-pocus," concluded Exner. "Incompetent, irrelevant and immaterial to prove anything except how easily people—and I mean those who call themselves scientists—can be duped."[13]

The report made no effort to explain why Donora residents were so terribly injured that weekend while the nearby town of Monessen, which had a steel works and the same bad weather, had been relatively unscathed. But Monessen had no zinc works, residents noted. A local newspaper editorialized that the relationship between the Donora Zinc Works and the smog was "something that no investigation is necessary to prove. All you need is a reasonably good pair of eyes."[14]

Allen Kline agreed. "We thought it was common sense that it was the zinc works. That is what was different in Donora."

Sadtler knew he could not compete with the Pubic Health Service. "When the US government says that something is sulfur dioxide and not fluorine," he said, "then people are taking their word and not my word."

Scientist Kathleen Thiessen is an expert on risk analysis and has written about the health effects of fluoride for the U.S. Environmental Protection Agency. For this book she reviewed many of the confidential and unpublished Kettering documents and compared them with the official published conclusion by the Public Health Service on the Donora disaster.[15] Unlike the PHS report, Thiessen concluded that, judging from the information included in the Kettering documents, fatal quantities of fluoride could have certainly have been present in the valley during the disaster weekend, posing a lethal risk to the elderly and the infirm.

To come to this conclusion, Thiessen first made a rough estimate of how much air blanketed Donora that weekend. If the Donora valley was about 2.5 miles long, between 0.5 and 1.5 miles wide, and some 340 feet deep, then between 320 and 960 million cubic meters of air lay over the town, trapped by a temperature inversion. The Donora steel plant had a daily production capacity of 1,450 tons of steel. Thiessen then calculated that, if each ton of steel requires 2 kg of fluoride, then as much as 2,900 kg (6,380 pounds) of fluoride could have been released per day without emission controls. Trapped by the stagnant weather conditions and suspended over Donora, these airborne fluoride concentrations could have soared well above the concentrations set as industry standard for an 8 hour day. (Additionally, of course, the zinc plant was belching out fluoride. But without surviving data on that plant's daily production capacity, Thiessen was not able to make an equivalent calculation for how much fluoride it may also have contributed during the disaster.)

It is not possible, with just the existing documents, to know with certainty whether fluoride killed Donora's citizens, concluded Thiessen. Nevertheless, she indicated, her series of calculations show that "there is the potential that routine releases of fluorine or fluoride, under conditions of little or no air dispersion, could result in air concentrations high enough to be dangerous to some individuals in the general public."

Thiessen was unimpressed with the science behind the official PHS report. She likened it to similar reports written today, where the intent is to obscure the truth, not reveal it. "My take was that they did a very fine job of writing lots of words in the hopes that nobody would see through to the fact that there was not much information there," she said.

Thiessen was especially skeptical of the government's scientific methodology in exonerating fluoride. Months after the disaster the PHS investigators measured urine samples in Donora children. The fluoride levels were low, and the investigators concluded that fluoride had therefore not been a problem during the disaster. It was a ludicrous argument, Thiessen explained. "They made a point in their report to say there is clearly no evidence of chronic fluoride exposure, but you cannot from that say there was no acute exposure on a given weekend six months ago. But they tried to do that. You can't."

Today investigators who want to examine how the PHS reached its conclusions are stymied. The raw data and records of the government's Donora investigation are missing from the U.S. National Archives and cannot be found. It is a shameful omission and a shocking breach of public trust, particularly as the Donora study was the first federal investigation of air pollution. "They may have been thrown out," suggested Snyder, who spent five years looking for these federal records. "Someone may have decided they were too hot to handle and got rid of them. You have to suspect the worst."

Philip Sadtler confirms the worst.[16] Six months after the disaster, U.S. Steel and the Public Health Service ran a test in Donora to simulate and measure the air pollutants that had been present in the atmosphere at the time. Sadtler was in town that day as the zinc and steel plants fired up and began billowing their smoke and fumes. He stepped into the mobile laboratory where government scientists were monitoring the "test smog." "I looked in and the chemist said, 'Phil, come on in.' Very friendly," Sadtler remembered. "He says, 'Phil, I know that you are right, but I am not allowed to say so.'"

The government conclusion—that no single pollutant had caused the Donora deaths—helped to checkmate the Donora families who were suing U.S. Steel. A more grotesque spectacle quickly followed.

As soon as the report was published, Helmuth Schrenk, the fluoride expert who had led the government's investigation, switched sides. He literally crossed the street from the U.S. Bureau of Mines in Pittsburgh, joined the private Mellon Institute as a research director, and signed up as an expert courtroom witness for U.S. Steel, ready to testify *against* the very Donora citizens whose devastated city he had just investigated for the U.S. government.

"It still makes me angry," said historian Lynne Snyder. "For the chief of the investigation to immediately make himself available to be an expert witness against the plaintiffs of the town is something I would like to have information about. Did he receive money from U.S. Steel? Did he receive it after he left the employ of PHS?"

Schrenk joined Robert Kehoe and Harvard University air pollution expert Professor Philip Drinker as expert witnesses for U.S. Steel.[17] The one-two punch of a flaccid official investigation and the defection of its chief investigator to the side of industry crippled the victims' court case. In April 1951, on the eve of the first "test case" trial of smog victim Suzanne Gnora, the plaintiffs' lawyer—the former Pennsylvania attorney general, Charles Margiotti—settled with U.S. Steel. Facing 160 victim claims totaling $4.5 million, U.S. Steel settled for a one-time payment of a quarter of a million dollars to be disbursed among families of the dead and injured. One-third of the money went to Margiotti. The biggest, richest steel corporation in the world admitted no guilt nor accepted any obligation to reduce air pollution.

Allen Kline received a check for $500. Families of the dead garnered about $4,000 apiece, less Margiotti's third, Kline remembered. There was much anger at the courtroom deal. "We were furious," Kline said. "We weren't interested in the suits for money, we were interested in the suits to publicize what we considered a very serious health hazard."

After the settlement the Donora disaster slipped from public attention. Philip Sadtler's report of fluoride poisoning was almost forgotten. Even the Society for Better Living grew tired and gave up fighting the zinc works. "The whole thing just seemed to fade away," Kline said. "I was weary of getting nowhere."

Allen Kline never found out what chemical made him sick that weekend nor what killed so many of his fellow townsfolk. Despite the fumes, Allen Kline remained in the Webster home that his grandfather had built. The newspaperman developed a whole raft of illnesses, including a heart problem, diabetes, and a case of arthritis so crippling that he was forced into retirement, where an electric elevator chair carried him on rails each night upstairs to bed. Kline's daughter, born in the same Webster home, died of cancer. When the zinc mill finally closed in 1957 and the air over Webster cleared, to Allen Kline it was an epiphany. "I didn't know life could be that grand," he said.

"It Was Murder"

NINE YEARS AFTER the disaster, two officials from the U.S. Public Health Service, Antonio Ciocco and D. J. Thompson, returned to Donora, to work with an air-pollution consultant from the University of Pittsburgh, John Rumford. Ciocco and Thompson published data showing that Donora citizens who had been sick during the disaster remained at greater risk of illness and early death.[18] But John Rumford's explosive findings—of fluoride poisoning in Donora—were never published. The suppression of the fluoride findings by the government health experts mirrored perfectly the evasions and omissions of their PHS colleagues a decade earlier. Without alerting the public, Rumford had taken soil measurements from eight locations in Donora, including downwind from the steelwork's blast furnace. In six of his readings, he found 200–800 parts per million of fluoride in soil. Downwind from the blast furnace, however, his two readings were 1,600 and 2,500 ppm respectively. Rumford next studied health data from the disaster, gathering more firsthand information on Donora health complaints and inquiring whether reported illnesses were more severe when temperature inversions trapped pollutants in the valley. His conclusions were simple. According to a PHS official who examined his data, Rumford's basic findings were:

1. That there is a relation between month-to-month variation in sickness and month-to-month variation in . . . air pollution.
2. That there is more illness in an area over which fluorides are blown from the factory.

The "suspected fluorosis" occurred in the same five-block radius downwind from the Donora steel works where half of the disaster dead had lived, Rumford reported. His data also showed that cardiovascular problems grew worse when the smog gathered in the Donora Valley and that former open-hearth steel workers who handled raw fluoride were especially affected by arthritis and rheumatism.[19]

At first the new generation of PHS officials seemed excited by Rumford's work. The Donora disaster might have a silver lining, they even suggested. The health data might offer a road map for a nation struggling to chart new policies to combat air pollution and to determine the health effects of the most dangerous poisons in the atmosphere. The grim health effects of fluoride air pollution were very clear in John Rumford's data, the PHS officials saw. "Dr. Ciocco liked this part about the fluoride findings," reported one of the reviewers of Rumford's work, Nicholas E. Manos, who was the Chief Statistician of the PHS's Air Pollution Medical Program. "In the case of suspected fluorosis, that is, cases of arthritis and rheumatism," Dr. Manos explained, "you have a correlation with a specific agent, a correlation with the wind trajectory, and also a correlation with the presence of those whose occupation places them near the open hearth using raw fluoride."

Similar health problems associated with fluoride air pollution had been seen elsewhere in the country, noted Manos. And Dr. Leon O. Emik, the Chief of Laboratory Investigations for the PHS Air Pollution Medical Program, contemplated initiating a bold nationwide study on fluoride's health effects. "Dr. Emik suggested we study mortality from arthritis and rheumatism from various cities for possible relation with the frequency of fluoride air pollution. We must remember in this connection Mrs. Gleeson's findings of an increase in cardiovascular deaths in Florida after the influx of plants using fluoride," Manos wrote.[20] (Philip Sadtler had gone to Donora, of course, at the request of Florida farmers battling the fluoride-polluting phosphate industry.)

Instead of pointing a fresh finger at an especially dangerous air pollutant, however, John Rumford's fluoride findings remained unpublished. And for more than forty years the 1949 Public Health Service report on Donora exonerating fluoride has stood as the established account of the most famous air pollution disaster in U.S. history. Its critics were largely forgotten, and fluoride slipped almost entirely from most public discussion of air pollution. When the fiftieth anniversary of the disaster was marked in 1998, no newspaper even mentioned fluoride. Philip Sadtler had died two years earlier. At a municipal church ceremony in Donora an EPA official mentioned only that the long-ago Halloween disaster had shown that "pollution can kill people." A second EPA official blamed the deaths on "a mix" of sulfur dioxide, carbon monoxide, and metal dust.

The shabby treatment Donora citizens received from their government can be attributed, perhaps, to national-security concerns—a consequence of the urgency seizing the United States as it stared down the barrel of a fast-approaching global confrontation with Soviet Russia. Fluoride was critical to the U.S. economy and military defense, and industry's freedom to use it could not be seriously hampered during the cold war. "Maybe it is because it happened in the late 1940s when the U.S. attention was really turned to other issues. During the Donora investigation the Soviets exploded Little Joe and the cold war got underway. Berlin was blockaded. A lot of big things in foreign policy were going on at that time," says Lynn Snyder. Or maybe this treatment was simply due to the fact that "it affected a working-class community," she added.

Scientist Kathleen Thiessen also gives a cold-war interpretation to the shunning of Philip Sadtler and the government's histrionic disavowal of fluoride as Donora's killer chemical. "There certainly was a vested interest on the part of the government not to get the public upset about fluoride—after all if we are spewing out thousands of pounds a month or a day or whatever at Oak Ridge, and probably Portsmouth and Paducah [two other fluoride gaseous diffusion plants] and some other places, we don't want the public to get concerned. We don't want to suddenly say, 'Hey, twenty people died because of a fluoride release last weekend.' This would not be good. We might get somebody upset. The aluminum industry of course was part of the cold war effort too."

Philip Sadtler held a more basic view. Until his death he remained clear about what had happened at Donora and who was responsible for these events. "It was murder," he said. "I thought that the directors of U.S. Steel should have gone to jail for killing people."

Although the Donora disaster faded from public view, Federal Security administrator Oscar Ewing was soon back in the nation's headlines. Nine months after his Public Health Service exonerated fluoride of the Halloween tragedy in western Pennsylvania, Ewing had a surprise announcement for the nation: the U.S. Public Health Service was reversing a long-held position. The ex-Alcoa lawyer declared that his agency now favored adding fluoride to drinking water supplies across the United States.

11

As Vital to Our National Life As a Spark Plug to a Motor Car

THE RAW MILITARY power that won World War II flowed directly, as molten metal, from blast furnaces and aluminum pot lines and from the American mastery of the atomic bomb. Fluoride was at the chemical core of all these operations. While the American public was told that fluoride was safe and good for children's teeth, U.S. strategic planners stockpiled fluoride during the cold war for a feared global war with the Communists.[1]

Fluoride was declared a "strategic and critical" material by the government after World War II. In 1950, as the Korean war erupted, President Truman asked the head of CBS television, William S. Paley, to chair a task force to study the United States' mineral reserves—and its vulnerabilities to having imports cut off in wartime.[2]

Fluoride was the lifeblood of the modern industrial economy, the Paley Commission reported. "[Fluoride] . . . is an essential component of enormously vital industries whose dollar value is measured in billions and upon which the whole national industrial structure increasingly depends," wrote one Paley analyst in a document marked "RESTRICTED." "Without this little known mineral," the document continued, "such industrial giants as aluminum, steel, and chemicals would be most severely affected. Little or no aluminum could be produced; steel production would be reduced substantially; the output and quality of important chemical products such as refrigerants, propellants for insecticides, and plastics would be significantly cut down."[3]

Fluoride was "as vital to our national life as a spark plug to a motor car," announced C. O. Anderson, the vice president of the nation's largest fluorspar producer, Ozark Mahoning. (Fluorspar is the mineral ore from which most industrial fluoride is produced). "Your car doesn't run if the spark plug is in the control of any foreign country," Anderson warned the Paley Commission. Fluoride's importance would only grow, predicted Miles Haman, Manager of the Crystal Fluorspar Company in Illinois. "General expansion of industrial facilities and building up of war machines all over the world [would necessitate] using much aluminum and steel and consequently more fluorspar."[4]

There was bad news, Paley's team heard. Fluoride stockpiles had fallen below "danger point" levels and domestic supplies were growing short. "The U.S. is vulnerable security-wise were a hot war suddenly to develop," stated Paley analyst Donor M. Lion.[5] While 369,000 tons of fluorspar had been consumed by industry in the United States in 1950, a million tons would be needed by 1975, the team projected. "If the United States were compelled to rely on natural fluorspar alone, serious obstacles to growth and security would emerge," the group reported.

But a magic bullet promised to ensure a continued strong national defense, planners heard. Short on fluorspar reserves, the United States was blessed with one of the world's largest supplies of natural phosphate, a raw mineral that lay in huge geological deposits in Florida. The mineral was the feedstock for the production of superphosphate fertilizer. It contained significant quantities of fluoride—3 or 4 percent—and traces of numerous other chemicals, including uranium.[6] America was sitting on its own virtually inexhaustible supply of fluoride. Could the phosphate industry supply fluoride for the nation, the government asked?

Sure—if the price was right, answered Paul Manning, a vice president of the phosphate-producing International Minerals and Chemical Corporation. If the fluoride that was then being belched as pollution into the orange-perfumed Florida air—some nineteen tons in 1957 alone—could only fetch a better price on the market, then the phosphate industry might just be willing to trap some of their waste as silicofluoride.[7] "The difficulty with this," Manning told the Commission, "is that sodium silico fluoride is a 'drug on the market,' and the price which can be obtained for it is not attractive enough to result in its production."[8]

The Florida phosphate producers *could* supply fluoride, explained Manning, but they had little current incentive. Despite a hornet's nest of lawsuits from farmers and angry local citizens gassed by fluoride fumes, it appeared cheaper for industry to fight the lawsuits and concomitant efforts to regulate pollution than to trap the toxic emissions.[9] "At the present time we have no idea as to the point to which prices would have to rise to justify the current recovery techniques," Manning told the Commission.[10]

The dilemma was clear. The government wanted the Florida fluoride in case of wartime emergency—but the state's phosphate producers needed a carrot before capturing their toxic waste. "The phosphate industry is primarily interested in super-phosphate, and fluorine recovery is a very minor matter. This is the kind of potential shortage that could develop into a full-blown crisis before a move is made to avert it," warned one Paley analyst.[11]

An elegant solution existed, of course. Using the phosphate industry's waste to fluoridate public water supplies meant that the fertilizer producers would now pay far less, if anything, to dispose of their most troublesome toxic waste. They would be guaranteed a source of taxpayer revenue for installing pollution-control devices; and U.S. strategic planners would win a nearly inexhaustible potential supply of domestic fluoride. There was yet another potential cold-war reason for disposing of fluosilicic acid in public water supplies. The Florida phosphate beds were *also* an important source of uranium, harvested for the Atomic Energy Commission. Because uranium is only a trace mineral in the phosphate deposits, enormous quantities had to be processed to glean worthwhile amounts of uranium, so much waste fluoride was also produced. Permitting that fluoride to be dumped in public water supplies—rather than being disposed of as toxic waste—reduced the cost of such uranium extraction and provided a supply of fluoride.[12]

In 1983 the EPA's Rebecca Hamner acknowledged that fluoridating water with phosphate-industry waste was a fix for Florida's environmental pollution. "This Agency regards such use as an ideal environmental solution to a long standing problem," the Deputy Assistant Administrator for Water wrote. "By recovering by-product fluosilicic acid from fertilizer manufacturing, water and air pollution are minimized, and water utilities have a low-cost source of fluoride available to them," she added.[13]

DID COLD-WAR PLANNERS also encourage water fluoridation to guarantee an alternative supply of fluoride for war industries or to reduce the cost of disposing of fluoride waste generated by uranium production? On June 1, 1950, as communist troops prepared for an invasion of South Korea, the Public Health Service abruptly reversed its opposition and declared that it now favored adding fluoride to water supplies.[14] The PHS now smiled upon fluoride, announced Oscar Ewing, whose Federal Security Agency was in charge of the PHS. He attributed this change of opinion to results from the water fluoridation experiment in Newburgh, New York, which showed a 65 percent reduction in dental cavities in local children.[15]

But the origins of the Newburgh study, as we saw in chapter 6, were manifestly suspicious. And irrespective of the dental data (which have been seriously questioned[16]), the Newburgh fluoridation experiment was a *safety* trial—designed to last for ten years to research potential side effects of drinking fluoridated water. When Ewing announced the government's about-face in 1950, *the safety study was only half complete.*

Ewing was well placed to act on ulterior national security concerns or on behalf of industry. His Federal Security Agency was one of the most powerful cold-war government bureaus. He had been Alcoa's legal liaison to Washington during World War II, shaping the massive expansion of the nation's aluminum industry. And the former Wall Street lawyer was a member of an inner circle of Truman confidants known as the Wardman Park group, who ate each Monday night at Ewing's Washington apartment and whose cigar-smoking, steak-dining members included Clark Clifford, who was famously close to the Pentagon and the CIA.[17]

"No Injury Would Occur"— Harold Hodge Turns the Tide

WATER-FLUORIDATION ADVOCATES greeted the government flip-flop with rapture. Two Wisconsin dentists were especially elated.

Dr. John Frisch and Dr. Frank Bull, the state dental officer, had been among the nation's earliest profluoridation activists, lobbying federal officials with an enthusiasm that bordered on the perverse. In 1944 Dr. Frisch began giving his seven-year-old daughter Marylin water from a jug he'd prepared with 1.5 ppm fluoride. (That same year the *Journal of the American Dental Association* had editorialized, "Our knowledge of the subject certainly does not warrant the introduction of fluorine in community water supplies.") Frisch placed "Poison" labels on the unfluoridated kitchen faucets, to remind Marylin to drink his potion instead.

Three years later the father's passion was rewarded, according to historian Donald McNeil as related in his 1957 book, *The Fight for Fluoridation*. Sitting in a Madison restaurant, Dr. Frisch noticed a "flash" on his daughter's teeth. "He could hardly believe his eyes," McNeil wrote. "It looked like a case of mottling. He rushed her outside in the bright sunlight and thought he noticed it again. Next day he excitedly asked Frank Bull over to get his opinion. Bull concurred. . . . It was mottling." (Remember, fluorosis does nothing to strengthen a tooth, may in fact weaken it, and is a visible indicator of systemic fluoride poisoning during the period that the teeth were being formed. No matter how mild the mottling, it is an "external sign of internal distress," according to the scientist H. V. Smith, one of the researchers who in the 1930s discovered that fluoride was mottling teeth.)[18]

Now, as the PHS endorsed water fluoridation for the rest of the United States, a similar thrill ran through the Wisconsin dentists. "Cease firing!" wrote Frisch. "The hard fight is over," added Frank Bull.[19]

But the fight was just beginning. Almost immediately citizens began to learn some disturbing information. The world's leading fluoride authority, Kaj Roholm, had opposed giving fluoride to children. The AMA and the ADA had all editorialized against fluoridation as recently as the early 1940s. And leading scientists, such as M. C. and H. V. Smith, also worried about adding fluoride to water supplies. "Although mottled teeth are somewhat more resistant to the onset of decay, they are structurally weak; when the decay does set in the result is often disastrous," the husband-and-wife team reported.[20]

The Smiths sounded an obvious warning. "If intake of fluoride (through drinking water) can harm the delicate enamel to such an extent that it fails to enamelize the unborn teeth in children, is there any reason to believe that the destructive progress of fluoride ends right there?" "The range between toxic and non-toxic levels of fluoride ingestion is very small," Drs. Smith added. "Any procedure for increasing fluorine consumption to the so-called upper limits of toxicity would be hazardous."[21]

Fluoride was put to the vote for the first time on September 19, 1950. It was a gloriously unruly and democratic spectacle. The Wisconsin town of Steven's Point had been fluoridating its water for five months, but local activists—including a poet, a railroad repairman, and a local businessman—forced the town council to put the issue to the ballot. After a colorful debate in the pages of the local newspapers, and rallies with activists caroling "Good-bye, Fluorine" to the tune of "Good Night, Irene," fluoridation was defeated in Steven's Point by a vote of 3,705 to 2,166.

A wildfire of citizen protest now flashed across the United States. The antifluoride camp found one of their most distinguished voices in a Michigan doctor, George L. Waldbott. The German-born physician was a medical pioneer and allergy specialist who had carried out the first ever pollen survey in Michigan in 1927 and the first national fungus survey in 1937.[22] In 1933 he reported on sudden deaths from local and general anesthetics, and was the first scientist to report on similar fatal allergic reactions to penicillin, drawing the attention of *Time* magazine. He had written a book on skin allergies called *Contact Dermatitis*, and in 1953 he published the first medical report on the emphysema caused by smoking cigarettes.[23]

Waldbott now turned his attention to fluoride. In the spring of 1953, Waldbott's wife, Edith, pointed him to recent medical criticism of water fluoridation at a February 1952 Congressional hearing on the use of chemicals in food.[24] Waldbott, the vice president of the American College of Allergists, began his own investigations and soon found that fluoride was no different from many other drugs and chemicals: some people were uniquely sensitive and suffered acute, painful, and debilitating allergy to small amounts of additional fluoride in their water.

Again and again Waldbott came across patients in his own practice who, when they ceased intake of their fluoridated water supply, were relieved of symptoms ranging from stiffness and pain in the spine to muscle weakness from stomach upsets to visual disturbances and headaches. His first report of such a patient appeared in medical literature in 1955, and by 1958 he had come across many more cases.[25] In these patients, ranging from an eight-year-old girl to a sixty-two-year-old woman, he ran scientific "double blind" tests in which the patients were given water without knowing whether it was fluoridated or not. The symptoms recurred only if they were given fluoridated water, the scientist reported.[26]

Waldbott was not the only doctor to spot that some people were especially sensitive to fluoride. A former University of Rochester researcher, Dr. Reuben Feltman, who was working on a PHS grant at the Passaic General Hospital in New Jersey, also reported that fluoride supplements given to pregnant women caused eczema, neurological problems, and stomach and bowel upsets.[27]

Medical professionals saw that it was impossible to control how much fluoride somebody ingested. Athletes and other active individuals, or people in hot climates, diabetics, or the kidney-injured drink more and therefore consume more fluoride. There are varying amounts of fluoride in food, while hundreds of thousands of workers are exposed to fluorides in their jobs.[28] There seemed to be little or no margin of safety between the amount of fluoride that was associated with fewer cavities and the amount that would cause injury. "Unfortunately the line between mottling and no mottling is an elusive one and the degree of control to be exercised seems to be very fine," concluded Dr. George Rapp, professor of biochemistry and physiology of Loyola University School of Dentistry.[29] (Even at the level of 1 part per million, at which the "optimal" cavity-fighting effect was reported, dental mottling was seen in a portion of the population, according to the PHS expert H. Trendley Dean.[30])

Fluoride promoters had a simple solution. Mottled teeth were described as a "cosmetic" issue, not a health problem. Most importantly, promoters vigorously denied that any injury to bones or organs could ever be produced from drinking water fluoridated at 1 part per million.

To make that safety argument, the government turned to a familiar face, Dr. Harold Hodge from the University of Rochester. In two key papers for the National Research Council (NRC) and the American Association for the Advancement of Science (AAAS), published in 1953 and 1954 respectively, Hodge maintained that "Present knowledge fails to indicate any health hazard associated with the extra deposition of fluoride in the skeleton that will undoubtedly accompany water fluoridation."[31]

For a generation, these papers would be a primary source for the reassurances given to Congress and to millions of citizens in the United States and around the world of the safety of water fluoridation. The small print at the end stated that they were "based on work performed under contract with the U. S. Atomic Energy Project, Rochester, New York."

Hodge's assurances were profoundly helpful to industry and the nation's fledgling nuclear program. The large doses he found to be safe for the public and for nuclear workers became for several generations of establishment health officials the medical template for discussing the dangers of fluoride exposure, and laid a medicolegal foundation for the courtroom defense that worker sickness could not possibly be due to fluoride.[32]

Hodge also wielded his safety assurances in Congress to cut down the citizen protest against water fluoridation that was springing up across the country. By the mid-1950s, unimpressed by the Public Health Service endorsement—and connected by George and Edith Waldbott's bimonthly newspaper called *National Fluoridation News*, which contained reviews on the latest medical information, updates of antifluoride referenda around the country, and cartoons by *New Yorker* contributor Robert Day—an unruly alliance of doctors, dentists, scientists, and community groups were successfully turning back fluoridation at the ballot box. Seattle had experienced a tumultuous debate in 1952, voting almost 2 to 1 in a referendum against fluoride. The following year Cincinnati voters also said no. By the mid-1950s the tide of public opinion appeared to be moving against fluoride, according to the historian Donald McNeil.

"[By December 1955] The U.S. Public Health Service reported that of 231 communities voting on fluoridation 127 had rejected it," McNeil wrote. "Adverse referenda votes in twenty-eight communities had discontinued established projects. Six months later the proponents had won eight more elections campaigns, the anti-fluoridation forces forty-five," he added.[33]

In 1954 national legislation banning fluoridation was proposed in Congress by Rep. Roy Wier of Minnesota. The suggested law, HR 2341, was titled "A Bill to Protect the Public Health from the Dangers of Fluoridation of Water." It forbade any federal state or local authority from adding fluoride to water supplies. Hearings were held at the end of May in Room 1334 of the New House Office Building, with a great array of medical figures testifying against and in favor of the bill.[34]

George Waldbott led the opposition. Symptoms of chronic low-level fluoride poisoning, such as "nausea, general malaise, joint pains, decreased blood clotting, anemia" were "vague and insidious" testified Waldbott, and could therefore easily be blamed on something other than fluoride—which made a correct diagnosis difficult, particularly for doctors who knew little about fluoride's toxic potential. Waldbott repeated his arguments that as a result of the danger of allergic reaction, the varying amounts of water drunk by different people, the risk to kidney patients or diabetics, and the extra fluoride consumed in food, "there can be no such thing as a safe concentration." "Neither the benefit nor the safety of fluoridation water supplies are sufficiently proven to warrant experimentation with human life," Waldbott told Congress.

But once again Harold Hodge stepped into the breach, saving the day for the government. He blunderbussed fluoride opponents with his National Academy of Sciences–approved data. The Rochester scientist was the nation's leading fluoride authority, a member of the Mellon Institute's Industrial Hygiene Association, chairman of the prestigious National Academy of Sciences Committee on Toxicology—and, of course, the former chief toxicologist of the Manhattan Project. It would take a massive dose of fluoride, Hodge testified—between 20 and 80 milligrams consumed daily for 10 to 20 years—to produce injury. Waldbott was mistaken, water fluoridation was "harmless," Hodge insisted. "Even if all the fluoride ingested in the drinking water (1 part per million) in a lifetime were stored in the skeleton," Hodge told Congress, "no injury would occur."[35]

Hodge's sober assurances provided the coup de grâce for the

legislation. The proposed law banning fluoridation expired in committee and never made it to the floor of the Congress for a full vote. And Hodge's safety data were repeated for a generation, mantralike, in countless speeches, official documents, pamphlets, magazine articles, and textbooks. They were widely used by the American Dental Association and the World Health Organization. As recently as 1997 these same numbers were cited by the federal Institute of Medicine.[36]

And no one noticed when, in an obscure paper published in 1979, after all the tumult and shouting had died down, Hodge quietly admitted that his safety figures had been wrong (see chapter 17).

12

Engineering Consent

VISITING THE CAMBRIDGE, Massachusetts, home of Edward L. Bernays was a thrilling and unsettling experience. On the occasion of his hundredth birthday in 1991, I spoke with him for the British Broadcasting Corporation's World Service.[1] The nephew of Sigmund Freud was in good health, briskly walking me to an old-fashioned elevator that rose into his private office.

The elevator seemed like a time machine. Bernays seized the brass control switch, and the lattice cage doors slammed shut. The diminutive old man smiled, his eyes twinkling. His audience was captive, and once again the tiny hands of Mr. Edward L. Bernays—the "father of public relations"—gripped the levers of power. The doors opened. We entered a softly lit photo gallery. Bernays shuffled forward, pointing proudly. There he was, rubbing shoulders with men of power from the twentieth century, like the omnipresent character in the Woody Allen movie *Zelig*: Bernays at the signing of the Treaty of Versailles; Bernays with Henry Ford, with Thomas Edison, with Eleanor Roosevelt, with Eisenhower, with Truman; and Bernays with George Hill, the head of the American Tobacco Company. (Bernays's wife was the leading feminist Doris Fleischman. He was a master of exploiting such modern liberal sentiment. On behalf of his tobacco client Bernays had once persuaded women's suffrage activists to march in the 1929 New York Easter Parade holding cigarettes as "torches of liberty.")[2]

The tiny propagandist counted among his clients the dancer Nijinski, the singer Enrico Caruso, and some of the most powerful corporations in America, including CBS, Procter and Gamble, and Allied Signal. Bernays also had close ties to the U.S. military. As a young man in World War I he had been a foot soldier in the government's Committee on Public Information, creating some of the nation's earliest war propaganda. He volunteered those skills for the U.S. Army in World War II, and during the cold war he was in communication with the CIA. Other resume items included advising the United Fruit Company during the U.S. government's overthrow of the elected government of Guatemala; shaping strategy for the U.S. Information Agency (USIA); and advising the government of South Vietnam.

Bernays also persuaded Americans to add fluoride to water.[3]

"I do recall doing that," he said softly during another interview at his home in 1993. Although Bernays was then 102 years old, his memory was good. Selling fluoride was child's play, Bernays explained. The PR wizard specialized in promoting new ideas and products to the public by stressing a claimed public-health benefit. He understood that citizens had an often unconscious trust in medical authority. "You can get practically any idea accepted," Bernays told me, chuckling. "If doctors are in favor, the public is willing to accept it, because a doctor is an authority to most people, regardless of how much he knows, or doesn't know.... By the law of averages, you can usually find an individual in any field who will be willing to accept new ideas, and the new ideas then infiltrate the others who haven't accepted it."

In 1913, for example, Bernays played on medical and liberal sympathies to boost ticket sales of a Broadway play he had helped to produce. The play, *Damaged Goods*, dealt with the then-controversial subject of venereal disease. Bernays circumvented potential censorship, he said, by creating a politically diverse Sociological Committee of doctors and prominent New York citizens to extol the health benefits of sex education and endorse the new play. This committee, which included John D. Rockefeller and a founder of the ACLU, turned *Damaged Goods* into a Broadway hit. By publicizing the purported health benefits of certain products, Bernays similarly increased sales of bananas for the United Fruit Company, bacon for the Beechnut Packing Company, and Crisco cooking oil for Procter and Gamble.[4]

In his 1928 book, *Propaganda*, Bernays explained his technique more formally. He noted "the psychological relationship of dependence of men on their physicians" and other such "opinion leaders" in society. "Those who manipulate this unseen mechanism of society," he wrote, "constitute an invisible government which is the true ruling power of our country . . . our minds are molded, our tastes formed, our ideas suggested, largely by men we have never heard of."[5]

Before World War II, the diminutive media wizard had been a PR adviser to Alcoa. He operated from the same office building, One Wall Street, where the Alcoa lawyer Oscar Ewing had also worked. In 1950 Ewing had been the top government official to sign off on the endorsement of water fluoridation, as Federal Security Administrator in charge of the US Public Health Service.

"Do you recall working with Oscar Ewing on fluoridation?" I asked Bernays.

"Yes," he replied.

Pressed about his relationship with Ewing, Bernays shifted uncomfortably. A memory that had been crystal clear seconds earlier suddenly clouded. "I had the same relationship that I had to other clients, I treated them the way a lawyer treats a client or a doctor treats a client. We had discussion of the problem at hand and how to meet them. I don't remember him very well," he insisted. Bernays glanced furtively at me: "Obviously I did nothing without their approval, in advance."

Bernays's personal papers detail his involvement in one of the nation's earliest and biggest water fluoridation battles, which took place in New York City. It was a key moment. The fight for fluoride was in full swing around the country, with referenda and public opinion running mostly in favor of the antifluoridationists.[6] Both camps understood the importance of winning in New York. A victory for fluoride in the liberal media metropolis would give fluoride promoters a big boost elsewhere, according to Bernays. "If New York accepts an idea, the other states will accept the idea too," he explained to me.

In one corner of the ring was a vigorous popular movement opposing fluoridation. The protesters were backed by leading doctors, such as Dr. Simon Beisler, a former president of the American Urological Association; Dr. Fred Squier Dunn of the Lenox Hill

Hospital; radiologist Frederick Exner; and Dr. George Waldbott. In the other corner was New York City's Health Department, led by Commissioner Dr. Leona Baumgartner. She was supported by the big guns of the nation's health establishment, including Louis Dublin, formerly of the Metropolitan Life insurance company; Robert Kehoe of the Kettering Laboratory; Detlev Bronk of the Rockefeller Foundation; Nicholas C. Leone of the Public Health Service; and Herman Hilleboe, New York State's Health Commissioner.

During the campaign Bernays secretly advised Health Commissioner Baumgartner on how to sell fluoride to the voters. "All this intrigues me no end," he told Dr. Baumgartner in a December 8, 1960, letter discussing fluoridation, "because it presents challenging situations deeply related to the public's interest which may be solved by the engineering of consent."[7] ("The Engineering of Consent" was a well-known Bernays essay on techniques of media manipulation and public relations.)

Bernays advised the Health Commissioner to write TV network bosses David Sarnoff at NBC and William Paley at CBS, telling them that debating fluoridation "is like presenting two sides for anti-Catholicism or anti-Semitism and therefore not in the public interest."[8] She should approach the TV executives gingerly, he warned, "without necessarily asking them to act in any specific way, but rather generically. . . . This might lead to a revision of the whole policy of what shall and shall not be considered controversial."

Other media strategies included mailing innocuous-sounding letters to influential editors, explaining what fluoridation entailed. "We would put out the definition first to the editors of important newspapers," Bernays said. "Then we would send a letter to publishers of dictionaries and encyclopedias. After six or eight months we would find the word *fluoridation* was published and defined in dictionaries and encyclopedias."

During the battle for New Yorkers' hearts and minds the city's Health Department received support from an influential profluoride "citizens committee"—purporting to be interested in fluoride for public-health reasons. The titular head of the Committee to Protect Our Children's Teeth was the famous pediatrician Benjamin Spock. Also lending their names to the Committee's effort was a long list of celebrities, liberals, and notables including Mrs. Franklin

D. Roosevelt, baseball great Jackie Robinson, and trade union leader A. Philip Randolph. A lavish booklet called *Our Children's Teeth* was published by the Committee and distributed around the country. It was a compendium of reassurances of fluoride's safety and denunciations of critics. Safety problems were "nonexistent," wrote Dr. Robert Kehoe from the Kettering Laboratory, while Dr. Hilleboe tarred opponents as "food faddists, cultists, chiropractors and misguided and misinformed persons who are ignorant of the scientific facts involved."[9]

Sold to New Yorkers as a public-health initiative, the Committee to Protect Our Children's Teeth had powerful links to the U.S. military-industrial complex, and to the efforts of big industrial corporations to escape liability for fluoride pollution. In 1956, for example, the Committee's booklet *Our Children's Teeth* was hot off the press. Before most New York parents had an opportunity to read about fluoride's wonders, lawyers for the Reynolds aluminum company submitted the booklet to a federal appeals court in Portland, Oregon, where the company had been found guilty of injuring the health of a local farming family through fluoride pollution (see chapter 13).

Inside the booklet, the judges were told, "are to be found the statements of one medical and scientific expert after another, all to the effect that fluorides in low concentrations (such as are present around aluminum and other industrial plants) present no hazard to man." (Today such a pseudo grass-roots effort would be known as an "astroturf" organization because of its fake popular character and essentially corporate roots.)

The committee was funded by the W. K. Kellogg Foundation, and its goals were to break the political logjam in New York and to help topple dominoes across the country, according to the committee's program director, Henry Urrows.[10] "That was the working assumption—our justification as far as the Kellogg people were concerned—and it turned out that was quite correct because we broke the back of the anti-fluoridation movement by winning in New York and Chicago," Urrows told me.

Although the Committee's expert composition and broad social representation was a classic Bernays-style propaganda technique, Urrows denied that the campaign had anything to do with Bernays, whom he dismissed in clipped, Harvard tones of barely concealed repugnance: "He was a man who would take credit for anything that would reflect credit on him. He was a professional liar." (Urrows may not have known what Bernays was doing, but Bernays kept tabs on Urrows. Correspondence from Urrows to Health Commissioner Leona Baumgartner is found in the Bernays archive.)

More evidence of the Committee's ties to industry can be seen in its staffing and endorsements. General counsel to the committee was Ford Foundation trustee and leading corporate attorney, Bethuel M. Webster. He had been a wartime associate of Harvard president James Conant and of Vannevar Bush, the two leading science bureaucrats who had shepherded the early development of the atomic bomb.[11] And the booklet includes statements from eight DuPont scientists; three scientists from the nuclear complex at Oak Ridge; a doctor from the Army Chemical Center in Maryland; the president of Union Carbide; the former supervisor of uranium hexafluoride production at Harshaw Chemical; the former director of the AEC's Division of Biology and Medicine; Shields Warren, a member of the AEC's Medical Advisory Committee; Detlev Bronk; and Dr. Herbert Stokinger, who had performed many of the Manhattan Project's fluorine toxicity studies for Harold Hodge at the University of Rochester.[12]

According to Urrows, it was "a coincidence" that so many scientists listed in the booklet were associated with the atomic-weapons industry. Fluoride's use in industry was "pervasive," he said. It was therefore unnecessary to list all those various industrial applications in a dental publication, he added. Urrows knew that Dr. Shields Warren, for example, had been associated with the AEC and that the nuclear industry had an interest in fluoride, but he bristled at any suggestion that his committee misled the public by not informing them of fluoride's military uses. "I think what you are doing is injecting a suspicion as though there were a self-interest beyond the public interest. And I think that you are mistaken," Urrows said.

It was not until 1965 that fluoride finally began spilling from New York City faucets. Foes complained bitterly that, while city residents were given a referendum on off-track betting, the fluoride vote had been turned over to the five-man Board of Estimate. An exclusive cocktail party corralling New York's political leaders at the home of Mary and Albert Lasker had launched the final push for fluoride that summer, according to *National Fluoridation News.* Mary Lasker was a member of the Committee to Protect our Children's Teeth and a prominent public health advocate. Her husband was a wealthy advertising executive, whose money came in part from pushing Lucky Strike cigarettes with Edward Bernays for the American Tobacco Company.[13] Guests at the Lasker party on July 25 included Mayor Robert Wagner, members of the Board of Estimate, twelve out of twenty-five members of the City Council, and Brooklyn's borough president Abe Stark.

"This government by cocktails is really unique," commented a press release from the antifluoride Association for the Protection of our Water Supply. "Here is a private one-sided hearing on a most controversial subject, in a meeting by officials in an ex cathedra session. Where does it leave the masses of citizens opposed to fluoridation? Will they have to pool their meager resources and invite the city fathers to an inexpensive bar to hear their story?"

The Committee to Protect Our Children's Teeth had accomplished its broader national mission, said Urrows.[14] "At the time we began work, there may have been—I'm guessing now—5 percent of the public water supplies [in the United States] being fluoridated, at the time we went out of business we had about two-thirds," Urrows added.

The "father of public relations" helped the U.S. Public Health Service to sell fluoride too, it seems. On Valentines Day of 1961, assistant surgeon general and chief dental officer for the Public Health Service, Dr. John Knutson, wrote to Bernays in New York. Knutson asked Bernays to pay a visit to his office to discuss "new approaches to the promotion of water fluoridation." The letter is on government stationery. Bernays answered by return mail, announcing that he expected to be in Washington shortly "to see some of my friends in Government and when the date is set I will make it a point to clear with you for an appointment."[15]

The federal public-relations effort grew in strength during the 1950s and 1960s. From the beginning the scale of the taxpayer-funded propaganda was driven by the strength of public opposition to fluoridation and had as its hallmark disrespect for open debate and a democratic vote.[16]

Big Brother watched. The Public Health Service, the American Dental Association, the American Medical Association, and the American Water Works Association all operated semicovert investigative offices, compiling McCarthyite dossiers on antifluoride medical professionals and sending often second-hand and derogatory "information" to profluoride groups.[17] The government agency for perpetuating such smear campaigns, which "serves as the CIA and the USIA of the pro-fluoridationists" according to *Science* magazine, was a taxpayer-funded outfit inside the NIH, the National Fluoridation Information Service of the Division of Dental Health of the U.S. Public Health Service. The spying unit, remarked *Science,* "makes it its business to know who stands where in the fluoridation controversy."[18]

Medical professionals critical of fluoride were regularly mauled in the press, while doctors and dentists were expelled from their professional organizations for antifluoride heresy.[19] At least one researcher, Dr. Reuben Feltman, who had found that fluoride supplements produce harmful side effects in pregnant women, had his federal funding withdrawn.[20] And the leading fluoride critic, Dr. George Waldbott from Michigan, soon found himself in the cross hairs of fluoride propagandists.[21] In 1988 *Chemical and Engineering News* reviewed the damage that had been done to Waldbott's scientific standing as a result of such attacks. "Rather than deal scientifically with his work," wrote Bette Hileman, "ADA mounted a campaign of criticism based largely on a letter from a West German health officer, Heinrich Hornung. The letter made a number of untrue statements, including an allegation that Waldbott obtained his information on patients' reactions to fluoride solely from the use of questionnaires. ADA later published Waldbott's response to this letter. But the widely disseminated original news release was not altered or corrected, and continued to be published in many places. As late as 1985, it was still being quoted. Once political attacks effectively portrayed him as 'anti-fluoridation', Waldbott's work was largely ignored by physicians and scientists."[22]

Journalists, too, were seized by the zeitgeist. In the summer of 1956 the writer Donald McNeil served as cover for the AMA's Bureau of Investigation in a failed bid to smear a leading antifluoride scientist. Although he would later write propaganda pamphlets for the ADA,

McNeil was then preparing what was regarded as an objective book on fluoride; he would become perhaps the leading media observer of the nationwide debate over fluoride. On July 2, 1956, McNeil wrote to the distinguished radiologist Frederick B. Exner in Portland, Oregon, requesting reprints of Exner's critical paper "Fluoridation." McNeil wrote under a pseudonym, explaining he was an antifluoride activist planning a "door-to-door" campaign in Wisconsin and asking if Exner could give him some idea on the price of reprints.

Secretly McNeil was responding to a personal request from the AMA's chief gumshoe, Oliver Field, to obtain information in order to show "that people are profiting" from the sale of antifluoride literature. (Dr. Exner had no idea of the subterfuge. He duly charged McNeil a.k.a. "Don Marriott" a dollar for a single copy, a rate that fell on a sliding scale to 55 cents per hundred.)[23]

Scientists with an eye for a successful career read the tea leaves closely. A river of federal dollars from the newly flush National Institutes of Health was cascading into research laboratories and college campuses around the nation, profoundly shaping the nation's scientific research priorities. While millions of these taxpayer dollars were spent promoting fluoridation, little money was given to study the potentially harmful effects from fluoride. Instead, the PHS spent lavishly during the cold war, producing profluoride films and public exhibits, as well as funding pseudoscholarly works.

An example of these expenditures was the 1963 booklet, *The Role of Fluoride in Public Health*, produced by the Kettering Laboratory and funded by the PHS. The Kettering Laboratory was simultaneously being funded by several of the biggest fluoride-polluting industries in the United States. The booklet's censorship of details and the Laboratory's interest in proving fluoride safe in low doses can be seen in its near-complete omission of scientists and articles critical of fluoride—and in the tract's propagandistic subtitle, "The Soundness of Fluoridation of Communal Water Supplies."[24]

The American Dental Association—funded in part by millions of dollars in taxpayer grants from the Public Health Service—joined the propaganda campaign, releasing a torrent of movies, slides, booklets, and exhibits, even suggesting scripts for radio programs.[25] One such script—with fake dialogue for doctors, dentists, and a "member" of the Parent Teacher's Association—dealt with the issue of dental fluorosis with Orwellian doubletalk, stating that "Fluoridated water gives the teeth an added sparkle."[26]

A 1952 ADA pamphlet also advised against democracy. "At no time should the dentist be placed in the position of defending himself, his profession, or the fluoridation process," stated the leaflet *How to Obtain Fluoridation for Your Community Through a Citizen's Committee.* Fluoridation "should not be submitted to the voters, who cannot possibly sift through and comprehend the scientific evidence," the pamphlet advised.

Yet the scale of the public-relations campaign mounted on behalf of water fluoridation appears to have startled even the ADA. In August 1952, for example, a blizzard of identical news stories appeared in papers around the country. They all praised fluoride for reducing dental cavities in Newburgh, New York. Curiously, they all did so in exactly the same language. "Who in hell is feeding newspapers canned pro-fluoridation arguments????????" asks a note found by the historian Donald McNeil in the archives of the American Dental Association.[27] "Two clippings, EXACTLY ALIKE, starting with 'Every time we hear a piece of news like the following from one part of the country we are surprised, and a little dismayed, that we don't get the same news from lots of other places.' Then tells of Newburgh's 47 percent reduction in decay" [emphasis in original]. The mystified author then lists several newspapers in Washington, Idaho, Missouri, Iowa, Arkansas, and South Dakota where the promotional story had appeared.

13

Showdown in the West

Martin *vs.* Reynolds Metals

PAUL MARTIN SHUDDERED. A moment earlier he had reached out to examine one of his Hereford cattle, and the animal's elegant curving horn had broken off in his hand. Startled, the rancher looked more closely. The once strong animal had grown skinny and was limping; its coat was matted and its teeth badly mottled. Martin had recently posted a reward in the local newspapers after several of his cattle had gone missing. Then, when he had found his first dead cow, he speculated that someone was shooting and rustling his herd.

Martin looked up to the horizon, past the wild flowering blackberry bushes that garlanded his property. His cattle had continued to die. And now his family was sick. His young daughter, Paula, complained of soreness when she walked. Her ankles "clicked," she said. All three of the family had pains in their bones, serious digestive problems, bleeding gums, a fearful anxiety that kept them awake at night, and a strange asthmalike exhaustion.

The tall rancher realized that the problem was not rustlers. Martin had been in perfect health in December 1946 when he moved into his beautiful new home on the Troutdale ranch. It was a spectacular property, 1500 acres of rich pasture nestled beneath the mighty Columbia River George, through which the greatest of the western rivers departed the Rocky Mountains. Looking back, however, Martin realized that his health had begun to falter in the months after the move to Troutdale. As he walked home to the farmhouse for a lunch of farm-grown fresh vegetables, he slowly nodded. He stared through a farmhouse window, lost in thought.

The window had become badly etched.

In the distance, bordering his property, lay the giant Reynolds Metals aluminum plant. At night, as Martin lay awake, the factory was bathed in electric light, pouring black smoke into the starry Oregon sky. Paul Martin now believed that poison from the Reynolds factory was, somehow, killing his cattle, scarring his property, and poisoning his family.

Paul and Verla Martin's lawsuit against Reynolds Metals in August and September 1955 in Portland, Oregon, was one of the most exhilarating and significant courtroom clashes in modern American history. It was a David-and-Goliath battle: a solitary American farmer standing his ground against the combined legal and financial might of several of the nation's top industrial corporations. The drama in Judge East's district courtroom was captivating. For three weeks a jury listened as several of the world's top scientists, who had come from London, Chicago, and Cincinnati, slugged it out with conflicting medical testimonies, defending themselves against raking volleys of legal cross examinations. A surprise witness materialized, a top scientist perjured himself, and a pair of Harvard-trained medical experts gave devastating explanations of the health problems the Martin family had endured on their Troutdale Ranch.

"This court makes history," stated the leading medical witness for the Martins, Dr. Donald Hunter.

"This is a case of great national importance," proclaimed the Reynolds Metals attorney Frederic A. Yerke Jr., adding that it was "the first case in the history of the country in which an aluminum company has been alleged to have caused injuries to a human being through the emission of fluorine compounds from its plant."[1]

The Martin case stunned corporate America. Until then, no U.S. court had ever ruled that industrial fluoride emissions had caused harm to humans. Such a precedent would open the door to future lawsuits and even jeopardize the nation's war-making ability, industry claimed. Reynolds Metals was joined in court by six aluminum and chemical companies, including Monsanto and

Alcoa, which filed a "friends of the court" brief during the appeals process, pleading that a victory for Martin would drive a stake through the heart of the modern industrial economy by "rendering it unprofitable to conduct such enterprises near places of human habitation."[2] Their expert medical witness was none other than Dr. Robert Kehoe, Director of the Kettering Laboratory. He arrived in Portland early and would spend two weeks at the trial, coaching the company lawyers.

Martin's attorneys played their cards masterfully. They flew in England's top medical specialist in industrial diseases, Dr. Donald Hunter, to be their expert witness, thus catching Reynolds off guard. Hunter's expert credentials matched anything the industry men could offer. The senior physician of the London Hospital, Hunter had written a book on industrial poisons, studied fluoride pollution at an aluminum plant in Scotland, and researched the toxic effects of lead at Harvard Medical School.[3]

When Dr. Hunter rose to testify in late August 1955, he explained to Judge East's court that he had flown directly from Africa to London and then to Portland for the trial. Hunter's testimony marked the end of an even longer journey for the rancher, Paul Martin. His family's mysterious sickness had taken them to some sixteen doctors across the United States—in Chicago, Cincinnati, Baltimore, and New York—where they were confronted with baffled medical professionals in a seemingly endless search to find out what was hurting them. Finally Hunter and a leading Chicago specialist, Dr. Richard Capps, had recognized that the Martin's symptoms were classic symptoms of what Hunter now described to the jury as "subacute" fluorosis.[4]

Hunter was a member of the prestigious Royal College of Physicians in England. The Portland jurors probably smiled as he explained to Judge East that the Royal College had been "created by King Henry VIII in the year of 1518. I think that is 330 years before the state of Oregon began . . . in this office one has to wear a gown which was devised by Henry VIII."[5]

Hunter told the jury that fluoride had killed Martin's cows and injured the family. "Fluorine compounds are deadly poisons to mammalian tissues, and man is a mammal just as much as a cow or a sheep," he explained.[6] Fluoride was so dangerous, Hunter explained, because it was an "enzyme poison." He described research done by English poison gas specialists that had illustrated how fluorine could disrupt cell biology. So lethal were certain fluoride compounds, Dr. Hunter added, that Hitler had used them in World War II to poison generals he wanted to get rid of: "He simply had a banquet, and he ordered men to take the paper off the champagne cork, and he injected fluorides [into the champagne]."[7]

This was too much for the Reynolds lawyer, Frederic Yerke, who interrupted Hunter's testimony: "Object to this, your Honor. I move to strike this as not being competent, relevant or material."

Judge East agreed that it was "a bit dramatic" and urged the English doctor to move on. But Hunter was serious. He told the jury that the Martin family had been poisoned by a chemical so aggressive that it attacked the biological fabric of life itself. "Enzymes are the chemical substances which help the body to work," Hunter explained. "For example, if we go to lunch and we eat a steak, we have in the stomach pepsin, which is an enzyme. It digests the steak, and therefore we are properly nourished . . . modern chemistry shows that enzymes also exist in individual cells, and as everybody knows the human body is made up of masses of cells: cells of the liver, cells of the kidney, cells of the muscles." By hunting enzymes, fluorine compounds were the natural enemies of humanity, the doctor explained: "The enzymes in the cells help the cell to nourish itself and to keep ticking over, which is the process of life. Now, fluorine compounds are such deadly poisons that they go directly for that property of the cell, and they destroy the enzyme process." (Although Dr. Hunter had no way of knowing it, because Harold Hodge never published the data, in 1944 the Manhattan Project at the University of Rochester had explored using a liver enzyme, esterase, as an ultrasensitive detector for fluorine in the workplace. Liver problems, of course, were a cardinal complaint of the Martin family.)[8]

George Meade, Martin's lead attorney, then held up "Exhibit O-1" for the jury. It was the etched window glass from the Martin ranch. The lawyer told the jury that each day several thousand pounds of fluorides had escaped from the Reynolds plant, by the company's own admission. In March 1950, for example, shortly before the Martins abandoned their farm, the plant was belching 3,988 pounds of fluoride into the air every day.[9] Could these fluorides have etched the Martin window glass, Mead asked Dr. Hunter in front of the jury? And if they etched the glass, was that proof that Reynolds fluoride had hurt the family?

Hunter testified that he had seen exactly the same thing in England after the war, where a window was etched with fluoride and a nearby farming family had been hurt. "This is precisely the etched glass window that I saw in Lincolnshire on an ironworks in England, when in 1946, a family like the Martins was overcome with the same symptoms as the Martins," said Hunter. "The effluent was the same thing, hydrogen fluoride and cryolite dust, aluminum fluoride and even silico fluoride which are probably the worse [*sic*] of the lot."[10] Dr. Hunter concluded: "It is my opinion that all three of the Martin family suffer from subacute fluorosis."[11]

A second doctor also diagnosed the Martins with "subacute" fluorosis. Dr. Richard B. Capps of Northwestern University in Chicago was perhaps America's leading specialist on the liver. He too had trained at Harvard and had battled an epidemic of liver jaundice that had plagued U.S. soldiers in Italy during World War II. Dr. Capps testified that medical tests revealed that the livers of both Paul Martin and his daughter Paula were abnormal. He described the Martins' "bizarre" health symptoms—breathing difficulties, stomach problems, bone pain, excess urination, and anxiety—as having been precisely described in the medical literature by the Danish scientist Kaj Roholm.

Paula had been ten years old when the family moved to the ranch. Her health quickly disintegrated. She told the court that when she urinated, "I would be scalded and burned and would have to use Noxema or cream medicines on myself."[12] She was always "short of breath," she added, and unwilling to play sports with other children in the Troutdale High School. Her mother stayed awake at night massaging her painful feet.

Dr. Capps said that the discomfort and "clicking" in Paula's ankles was likely to be caused by fluoride attacking her tendons and bones. The chemical also caused her exhaustion and enlarged thyroid, he explained to the jury. "Fluorine tends to substitute for iodine in such a way that a person who is exposed to fluorine becomes deficient in iodine, and deficiency in iodine causes a certain type of enlargement of the thyroid which is frequently associated with a low metabolism, a deficiency in thyroid function."[13]

The spectacle of decomposing cattle strewn across the Martins' land, and of a glass window scarred by poisonous gases, had left an indelible impression on the Chicago doctor. "I think that if there is enough fluorine to etch a window, it should be able to etch a lung,"[14] Capps told the jury.

Then Capps noted that all three of the Martins had become healthier when they fled the ranch in 1950 and stopped eating the farm's contaminated produce. Their liver tests improved. Their breathing grew stronger, and the fluorine levels in Paul Martin's urine declined. Capps concluded that there was only one medical explanation possible for what had happened on the Troutdale farm: "You are forced to make the diagnosis of poisoning with fluorine," he said.[15]

The star defense witness, Dr. Robert Kehoe, now took the stand. The Reynolds lawyer lobbed a careful softball for the Kettering medical director. "Are you aware," attorney Frederic Yerke queried him, "of any incident or instance based upon your own experience, Doctor, where a man working with fluorides has become disabled by reason of the fact that he has absorbed more than an ordinary amount of the same?" If aluminum workers in wartime factories—which frequently had no pollution controls—had not been sickened by fluoride, went the logic of Yerke's questioning, how could the Martins, who merely lived near a plant, possibly have been injured by smaller amounts of the chemical?

"In my experience, no," Kehoe told the jury. "I have not."

It was a lie worthy of Joseph Goebbels. Just seven years earlier, in the summer of 1948, Kehoe's investigators from the Kettering Laboratory had found 120 cases of bone fluorosis in aluminum workers at Alcoa's plant in Massena, New York. His scientists told Alcoa that thirty-three of the workers were "severe" cases and showed "evidences of disability ranging in estimated degree up to 100 percent." (The Kettering Laboratory's Edward Largent had also found twisted bones and "fluorine intoxication" in workers at the Pennsylvania Salt Company during the late 1940s—although his published study had claimed the men suffered no disability.)[16] The Kettering Laboratory had worked to refute Kaj Roholm's research, arguing that even when fluoride was visible in X-rays of workers' bones, the men bent and hobbling, the medical effect was more likely the result of "hard work," not fluoride. The damaging data from Alcoa and Pennsylvania Salt were never published by Kettering or made public in any way. Both corporations, of course, were funding Kettering's fluoride science.

Kehoe dismissed the significance of the etched glass in the Martin farmhouse. Human lungs were made of sterner stuff, he insisted. Although thousands of pounds of highly toxic fluoride gases and dust had spilled each day for years from the Reynolds plant, felling Martin's cattle, mostly the wind blew away from the farmhouse and, anyway, Kehoe argued, "Glass . . . is much more subject to injury than the human lung."[17] Living in the shadow of the giant Reynolds Troutdale plant was "an entirely harmless situation for human beings," he concluded.

But Hunter and Capps carried the day. On September 16, 1955, the Portland jury decided in favor of the Martins. They awarded the family $48,000 for illness and for medical expenses.[18]

In corporate boardrooms across America the language now grew apocalyptic. The Martin verdict was a precedent that could cost industry billions. Six weeks later, at a private gathering of top industry officials at the Mayflower Hotel in Washington, DC, Alcoa's medical director, Dudley Irwin, told corporate air pollution experts that the Martin ruling was "significant . . . since it is the first one where the plaintiffs allege damage to their health from the everyday emission of an air pollutant."[19]

Reynolds fought the verdict with the desperation of a drowning man. The Appeals Court risked catastrophe for the U.S. economy if it let the Martin ruling stand, Reynolds lawyers claimed, invoking cold-war fears. "Aluminum is vital to our national security, and it is a metal of rapidly increasing importance to the entire economy," the brief began. "A court should be loath to adopt principles of law which would, in effect, make every aluminum plant liable for the unexplained miscellaneous ailments of the population for miles around." And there was the warning: "There is no practical alternative to release of fluorides except cessation of production altogether."[20]

The aluminum company summarized the medical evidence that justified overturning the guilty verdict. Edward Largent's human experiments at the Kettering Laboratory showed that fluoride was safe in moderate doses, the company asserted. Without mentioning that it had helped to pay for the research, Reynolds argued that, because the Kettering scientist had eaten so much fluoride himself, it therefore proved "the harmlessness of the Martins' exposure." After ingesting some 3,000–4,000 milligrams of fluorine over four years, "Mr. Largent had experienced none of the Martin's symptoms or any other symptoms," claimed Reynolds.[21]

And, perhaps for the first time in an American courtroom, the Fluorine Lawyers unveiled a brand new strategy, pointing to the federal government's endorsement of the safety of water fluoridation—and the fad for adding fluoride to toothpaste—as evidence that industrial fluoride pollution could not possibly have been responsible for the alleged injury.

14

Fluorine Lawyers and Government Dentists

"A Very Worthwhile Contribution"

Although big corporations have long used the U.S. government's medical assurances about fluoride safety to defend themselves in fluoride-pollution cases, no collaboration between industry and the federal promotion of fluoride has ever been acknowledged. However, Robert Kehoe's papers show precisely such collusion, detailing how the fluoride research of the National Institute of Dental Research, ostensibly conducted to prove water fluoridation "safe," was covertly performed in concert with industry, which was aware that the medical data would help their Fluorine Lawyers battle American pollution victims and workers in court.

THE REYNOLDS METALS Company employed a legal strategy during the Martin case that would become a staple in American courtrooms. Five years had elapsed since the Federal Security Agency administrator, Oscar Ewing, the former Alcoa lawyer, had endorsed public water fluoridation on behalf of the Public Health Service, which was under the FSA's jurisdiction. During the 1955 Martin trial in Portland, Reynolds reminded the court of fluoride's "beneficial" effects. Fluoride was being added to toothpaste, and 15,000,000 Americans now consumed more fluoride through their drinking water than the Martins had been exposed to, Reynolds's attorneys said.

"This court has thus found to be 'poisonous' an amount of fluorine which scientific and judicial opinion has unanimously found harmless," Reynolds's lawyers argued. "The only thing not explained is how the Martins could have suffered injury from something harmless to the rest of mankind," they added.[1]

Robert Kehoe also understood how public water fluoridation helped industry. His endorsement was featured in the profluoridation booklet, *Our Children's Teeth*. That booklet was simultaneously distributed to New York parents and to the judges on the Martin Appeals Court.[2] "The question of the public safety of fluoridation is non-existent from the viewpoint of medical science," he assured parents and judges alike.[3]

Privately, however, Kehoe was not so sure. "It is possible that certain insidious and now unknown effects are induced by the absorption of fluorides in comparatively small amounts over long periods of time," he had told industry in 1956.[4] And in 1962 Kehoe told Reynolds's medical director, James McMillan, that there remained "a basic question concerning the non-specific effects of prolonged exposure to apparently harmless quantities of fluoride (this by the way is the thing on which George Waldbott bases his entire campaign against fluoridation)."[5]

Kehoe and his Kettering Laboratory continued to soldier for water fluoridation during the 1950s and 1960s, assailing fluoridation critics as "windbags and windmills." Kettering toxicologists Francis Heyroth and Edward Largent were prominent members of National Academy of Sciences panels that endorsed fluoridation. And in 1963 the Kettering Laboratory published an influential bibliography of the medical literature favoring fluoridation, entitled *The Role of Fluoride in Public Health: The Soundness of Fluoridation in Communal Water Supplies.*

Kehoe saw how fluoride safety studies performed by government dental researchers helped his industry patrons in court. Farmers and workers would have a much harder time successfully suing corporations for fluoride pollution if the U.S. Public Health Service had performed its own studies and then vouched for the chemical's safety. "The results of such [dental] investigations are highly advantageous," Kehoe explained to the corporations sponsoring his fluoride work at Kettering, "in that the problem [of 'proving' fluoride safe] exists outside of industry, thereby involving situations in which the economic factors tend to be of different type and significance than those which are often alleged to be active in the industrial world, and often involving investigators who are not subject to accusations of bias based on industrial associations."[6]

Kehoe approached the Public Health Service in April 1952 on behalf of the industries sponsoring his fluoride research, to ask that the health agency perform additional fluoride safety studies.[7] "I was requested by the group [of industries], for whom I have acted as a spokesperson and chairman, in the consideration of this work, to approach your division of the U.S. Public Health Service, with the idea of determining whether or not an investigation of that type might not be conduced by the Service," Kehoe wrote to Dr. Seward Miller.[8]

Government proved cooperative. The top medical investigator at the National Institute of Dental Research, Dr. Nicholas Leone, was especially helpful. In August 1955, for example, during the Martin trial, the public servant Dr. Leone spoke with a senior attorney for Reynolds, Tobin Lennon, who also was a member of the Fluorine Lawyers Committee, directing Kehoe to a federal safety study on fluoride that Leone had recently concluded in Texas.

No record could be found of anyone from this U.S. government agency ever helping the Martin family.

Leone's study had examined people living in two Texas towns, where there were different amounts of natural fluoride in the water supply. The town of Bartlett had between 6 and 8 parts per million in its water, while nearby Cameron had 0.4 ppm. Dr. Leone had given both towns a clean bill of health, reporting in 1953 that no harmful "significant differences" were seen in the two populations. Although George Waldbott and others had vigorously attacked the study's scientific method and conclusions, the so-called Bartlett Cameron Study became, along with Harold Hodge's Newburgh study, a lynchpin of the government's case for water fluoridation—medical "proof" that adding fluoride to water would be harmless.[9]

As the Martin trial hung in the balance, the government's Dr. Leone burned up the long-distance telephone lines to Oregon, answering questions from Reynolds's attorney Lennon on the findings of the Bartlett Cameron study.[10] Dr. Capps had testified on Paul Martin's behalf that fluoride had injured his client's liver, so Reynolds's lawyer wanted information about fluoride's effects on such "soft tissues."[11]

Although the Bartlett Cameron study had not examined soft tissue, such data would soon be at hand, Leone reassured industry. New government-funded studies were in the pipeline. And they spelled good news for the Fluorine Lawyers. In the spring of 1957, as corporate America awaited the Martin Appeals Court verdict, Alcoa's Dr. Dudley Irwin traveled from Pittsburgh to the sparkling new campus of the National Institutes of Health in Bethesda, Maryland, to meet directly with Dr. Leone and discuss "the current status of the fluoride problem." Irwin was the medical coordinator for the Fluorine Lawyers Committee.

Dr. Leone laid out the plans for the new fluoride safety studies that the federal dental agency was then readying. The two men then discussed how those studies could be presented to "best suit our purpose," according to a March 5, 1957, letter Leone sent to Irwin to thank him for his visit. In particular Leone gave the Alcoa doctor details of a "human autopsy" study then being conducted in Provo, Utah, in which "soft tissues" were being analyzed. Leone was serving as "a consultant" to the Utah study, he wrote, and he had personally designed the autopsy protocol.[12] "The interests of the Provo group," Leone explained in his letter, "relate directly to atmospheric pollution of fluorides and its effects on humans. As you know, it has been proven beyond a question of a doubt that similar conditions have an effect upon animals."

Irwin paid close attention. Provo was the site of perhaps the most serious litigation problem then facing the Fluorine Lawyers. Since World War II, when a mighty steel plant had been located in the Utah County valley, near the city of Provo—far from any threat from Japanese bombers—local farmers had been in an uproar over pollution that they claimed had decimated their cattle. By 1957 the Columbia-Geneva Division of U.S. Steel had settled 880 damage claims totaling $4,450,234 with farmers in Utah County. An additional 305 claims for a further $25,000,000 were filed against the company.[13]

Nicholas Leone's researchers, working on a PHS grant, were studying the soft tissues and bones of Utah County residents, he told Irwin. "Inmates of a mental institution close by comprise the study material," he added.

The Bethesda meeting between Alcoa's fluoride doctor and the government scientist went well. They made plans for a future rendezvous. "In view of the vast amount of material soon to be available for publication," Leone concluded in his letter to Irwin, "we are all very enthused about a group presentation at some carefully selected meeting in the near future. I believe we discussed that briefly while you were here and I hope that you have had opportunity to give further thought to the type of meeting that would best suit our purpose. A one-shot presentation and publication in a single issue or monograph should be of more value than publication in a number of publications. . . . Again, it was a pleasure seeing you and I hope we have the opportunity for further discussions in the near future. Best personal regards. Sincerely Nicholas C. Leone, M.D. Chief, Medical Investigations, National Institute of Dental Research."[14] The government scientist enclosed a gift, a copy of a science-fiction novel called *The Pallid Giant* by Pierrepont B. Noyes.

Alcoa's doctor was jubilant at the letter. The government studies would show no harm from fluoride, he had learned (almost certainly from Leone himself). Irwin immediately contacted the Fluorine Lawyers Committee boss, Frank Seamans in Pittsburgh, forwarding to him a copy of the letter he had received from the NIDR's Dr. Leone. Thrilled at the news from Washington, Alcoa's top doctor explained to Seamans, in a letter dated March 13, 1957, exactly how the nation's water fluoridation program, and accompanying health studies, might help American industry: "These clinical investigations pertain to basic studies of individuals residing in areas where the fluoride content of the drinking water varies from 0.04 ppm F. to 8.0 ppm F. You will appreciate that this range of fluoride exposure brackets the range in which a number of us are interested. I have reason to believe the results of these investigations will show no evidence of deleterious effects due to fluoride absorption. The publication of these results will be a very worthwhile contribution," wrote Irwin.[15]

The obliging government dental researcher, Dr. Leone, wanted to share the good news—with a restricted group of industry friends, Dr. Irwin told Alcoa's lawyer. "Dr. Leone has given me his permission to supply copies of this letter to you for distribution to your group of 'fluorine lawyers' on a confidential basis," he wrote Seamans.

Any jubilation in the ranks of the Fluorine Lawyers Committee, however, quickly turned to fresh panic. The following month, on

April 24, 1957, Judges Denman, Pope, and Chambers of the Federal Appeals Court for the 9th Circuit in San Francisco upheld the lower District court ruling in the Martin trial. Their verdict was that Reynolds Metals was guilty of negligence and of poisoning the Martins with fluoride. In a gesture even more frightening for industry, Judge Denman cited a legal principle known as "absolute liability." It meant that industry was responsible for resulting injury, whether accidentally caused or not. "The manufacturer must learn of dangers that lurk in his processes and his products," wrote Judge Denman in his opinion upholding the District Court. "It was the duty of the one in the position of the defendant to know of the dangers incident to the aluminum reduction process," the opinion added.[16]

Frantic, industry scrambled back to the courts. The entire 9th Circuit Appeals court "en banc" (all judges on the Appeals bench, not just a three-person panel) now confronted a stunning spectacle, as six top U.S. corporations paraded before them, pleading for relief. Alcoa, Monsanto, Kaiser Aluminum, Harvey Aluminum, the Olin Mathieson Chemical Corporation, and a division of the Food Machinery and Chemical Corporation, each joined Reynolds in attacking the Martin verdict, filing an impassioned amicus curiae "friends of the court" brief.

One hundred thousand workers were employed in the U.S. aluminum industry alone, industry told the court, while seven aluminum smelters in the Northwest were located in inhabited areas similar to the Troutdale plant. Judge Denman's ruling had impossibly tightened the screws on the economy and jeopardized the nation's cold war military strength, the corporations argued. "The necessity of a strong aluminum industry to national defense is known to all," their appeal noted. Judge Denman had "reached out and with one swift stroke branded the aluminum industry and all industries involving the use of fluorides, ultrahazardous," it added. The industry attorneys warned of "the tremendous impact of this decision," and argued that "if Judge Denman's opinion placing absolute liability on these industries is to stand, the financial handicap so imposed may well impair their financial stability."[17]

As industry's lawyers scrambled back to the courts, the government's Dr. Nicholas Leone (from the National Institutes of Health) hurried for an emergency meeting with industry officials at the Kettering Laboratory. He was accompanied by no less than the Director of the National Institute of Dental Research, Dr. Francis Arnold.

On May 20, 1957—a month after the Martin verdict—the public servants met for "a relatively confidential discussion of the issues" with Dr. Kehoe and a Medical Advisory Committee of officials from the industrial corporations sponsoring the Kettering Laboratory's fluoride research. (This Medical Advisory Committee had been established on behalf of the Fluorine Lawyers Committee by Alcoa attorney Frank Seamans.)[18]

Alcoa's Dudley Irwin opened the meeting. He began by reiterating news of the Martin verdict. The U.S. dental investigators and the company officials reviewed the "weakness" of industry's position, according to notes taken by Dr. Kehoe. The group then discussed "the Martin case in relation to the community problems of air pollution, water pollution and food contamination and the position occupied by industry in creating a new environment characterized by potential hazard to the public health," according to Kehoe.

Industry was vulnerable, Kehoe emphasized to the dental researchers. He summarized the Kettering investigation of the Alcoa plant at Massena, New York, where fluoride had disabled workers. (The results of this study had—and still have—never been published.) There was a need, Kehoe told the NIDR's Nicholas Leone and Francis Arnold, "for research of a basic type to establish the facts on which medical opinion can be based so that irresponsible medical testimony will be discouraged."

The government men took their cue. Drs. Leone and Arnold again laid out for industry the several pending studies the NIDR was conducting. The science was ostensibly being performed to demonstrate to the public the safety of water fluoridation, yet for the anxious men at the Kettering Laboratory that spring day, there was a clear understanding that such medical studies could help polluters.[19] The conference concluded with plans for an ambitious strategy to shape the national scientific debate on fluoride by hosting "conferences, symposia," compiling new medical research "for dissemination or publication," and arranging working sessions "to decide what to do and in what sequence," according to Kehoe's notes.

On June 5, 1958, the full Appeals Court in the Martin case gave some ground. It upheld the verdict that the Martins had been poisoned and that Reynolds was negligent, but it withdrew the earlier opinion that the legal theory of "absolute liability" applied. The final decision was still "distressing," the Fluorine Lawyers' Committee head, Frank Seamans, wrote to Kehoe in a melancholy note eight days later. In an enclosed legal summary of the ruling, dated June 12, 1958, he concluded that the verdict "will doubtless have great significance in the further development of the fluorine problem."[20] There was, however, a solitary ray of sunshine: perhaps the rollback of the "absolute liability" verdict would help industry in the coming decades. "We can now argue that the Martin case went off on its own particular facts and that it is not a ruling that an aluminum smelter is liable for damage regardless of negligence," Seamans wrote.[21]

Seamans believed that it was the visit and testimony of the English expert, Donald Hunter, that had delivered the knockout blow. The English doctor had left a perhaps disfiguring scar upon corporate America. "The court quoted from the testimony of Dr. Hunter," Seamans wrote to Kehoe, "and said that his testimony was worthy of credit and, in fact, outstanding. These quotations from the testimony of Dr. Hunter about the effects of fluorine on human beings are very unfortunate and may serve to give such claims a push."

Kehoe was bitter about the whole affair. "I have rarely found myself in a more embarrassing situation than I was in the Martin case," he told one of Reynolds's other witnesses. And he complained to a friend, Philip Drinker of Harvard's School of Public Health, about the "cleverness" and "histrionic character" of Donald Hunter's testimony.[22] The professor commiserated. "Hunter's exhibition was cheap cockney showmanship at a pretty low point," Drinker said. "Numerous friends in England have told me he was for sale and he certainly sold himself for this," he added.

Dr. Kehoe would have his revenge. In the days after the Martin verdict worried industry its officials turned again to his Kettering Laboratory for help. Industry stood on a precipice. The danger was clear—and so was the solution. A powerful new scientific orthodoxy must be forged to defeat workers and farmers like Paul Martin and remove the threat from independently minded medical experts such as Donald Hunter and Richard Capps.

15

Buried Science, Buried Workers

THE IMPLICATIONS OF the Martin verdict were frightening. Like that Oregon farming family who had been poisoned by Reynolds's fluoride, tens of thousands of U.S. citizens breathed fluoride fumes from nearby steel, aluminum, and coal-fired power plants. A million more would soon live within eight kilometers of eleven hydrogen fluoride manufacturing plants.[1] And hundreds of thousands of American men and women inhaled fluoride dust and gases each day at work.[2]

Industry's response to the Martin verdict was explained by Robert Kehoe to the medical director of the Tennessee Valley Authority (TVA) in January 1956. "You will remember our recent talk about the concern of representatives of the aluminum and steel companies, over the outcome of the Reynolds Metals suit," Kehoe wrote to Dr. O. M. Derryberry on January 9, 1956. That concern included "perturbation of certain of the sponsors over the medico-legal situation," he wrote. "A group was assembled for a meeting with me early in December," Kehoe told Derryberry, "out of which came their request that . . . I advise them concerning a program of investigation which might be enlarged in scope and accelerated in tempo so as to provide them with adequate ammunition for handling similar situations, as well as those that might arise from apprehension among their employees."[3] Kehoe's goal was to "bring to an end the litigation which threatens to occupy the time, attention, and economy of industry without benefit to the health and welfare of their employees or the public," he told his sponsors.[4]

By February the Kettering Laboratory director had drawn up a game plan, focusing on the Achilles heel that had tripped up Reynolds Metals in the Martin trial. The Public Health Service was providing medical information about the health effects of swallowing fluoride, via its water-fluoridation safety studies. But the Martin trial had hinged on the accusation that *air pollution* had hurt the family, and Kehoe saw a clear need for fresh human experiments.[5]

"There seems to be no documentary information on the matter of human safety in relation to such exposure," Kehoe told the TVA's Dr. Derryberry. "In any case, we are about ready to initiate the experiments on animals, and while these are in progress, we can design and construct the facilities for the investigation of human subjects," he added.

Kehoe pointed to another goal: creating an unassailable medical orthodoxy that would block scientists from serving as effective expert witnesses in future court cases. His laboratory's earlier efforts to control scientific information about fluoride had almost borne fruit in the Martin trial, he remarked, but the surprise appearance of the Englishman, Dr. Donald Hunter, had upset the apple cart. "Opposing counsel overcame this obstacle by the importation of an expert who, with some charity, may be judged to have been susceptible to the thrill of participating in a grandstand play or, perhaps, of aiding an aggrieved family," wrote Kehoe.[6]

The only solution was a fresh batch of medical experimentation and scientific data, "so overwhelmingly persuasive, both in itself and its dissemination, as to render futile any efforts to combat it." The new Kettering research would "pile negative evidence upon negative evidence," said Kehoe. "This would result in such difficulty in finding a competent and credible expert witness as to thwart the attempts of counsel to make a case for a potential plaintiff," he added.[7]

The Kettering foot soldiers were given their marching orders at a planning session in the fall of 1956. They were under no illusions about their mandate. "The sponsor group is concerned with the litigation questions that may arise in the future as demonstrated by those that have occurred in the past," noted the scientists who attended the meeting, according to the recorded minutes. "Its purpose is not altruistic," they added. The threat of litigation would be their North Star, guiding research and experiments.

"The sponsors are interested not only in what happens to persons in the plant but also in whether they will be sued or not. They are interested particularly in finding out if the absence of deleterious effects of the absorption of the fluoride ion can be demonstrated," the minutes record. Specifically, what industry needed to learn—sixteen years into the fluoridation of water supplies—was "the physiological effects on the various organ systems of the continued absorption of fluorides." The scientists noted that "something is known about mottled enamel and skeletal changes but [there is] no information concerning effects on other organ systems."[8]

The Martin ruling had exposed the tip of a very dangerous iceberg, Kehoe told an invited audience of government dental researchers and industry lawyers, who had gathered in the Ballroom of the Cincinnati Club for a Fluoride Symposium in Cincinnati in December 1957.[9] The primary threat facing industry, Kehoe explained in his opening remarks, was that workers could use the Martin verdict to buttress lawsuits claiming injury from exposure to airborne fluoride inside factories. "The problem," he went on, was that the court verdict had set the stage for "the greater threat of claims for illness among employees in the industries in which exposure to fluoride is greater than that of any group of persons outside of industry."[10]

In the ballroom sat Harold Hodge from the University of Rochester and Alcoa's Frank Seamans, head of the Fluorine Lawyers Committee. No two people were in a better position to know the risk from airborne fluoride pollution. Twenty-five thousand people worked in aluminum smelting plants in the United States, and tens of thousands toiled in the giant gaseous diffusion plants at Oak Ridge, Paducah, and Portsmouth.[11]

The presentations were biased in favor of industry. Frank Seamans gave a presentation titled "The Medical Aspects of Fluoride Litigation." While the Director of the National Institute of Dental Research, Francis Arnold, discussed the "Present Status of Dental Research in the Study of Fluorides," there were no criticisms of water fluoridation; nor were experts such as Dr. Capps from Chicago or Dr. Hunter from England (both of whom had testified in the Martin trial on the human health consequences of industrial fluoride air pollution) in attendance.[12]

The papers were further culled when it came to their publication. Readers of the American Medical Association's journal *Archives of Industrial Health* (edited by Kehoe's Harvard friend, Philip Drinker), never learned of the symposium remarks on fluoride litigation by Kehoe and Seamans. Nor did they read the paper by D. A. Greenwood from Utah State University, spelling out the stupendous scale of the fluoride lawsuits facing U.S. Steel in Utah.[13]

The Symposium was just one front in industry's campaign to shape a scientific consensus about fluoride. Another was opened that summer of 1957, when industry committed $179,175 to a new fluoride research program at the Kettering Laboratory. It was a down payment on a three-year investigative program that would eventually cost almost half a million dollars. Air pollution would be the major focus of the research. The centerpiece would be an experimental chamber from which forty-two beagle dogs would inhale fine particles of calcium fluoride dust, for six hours a day, five days a week. Alcoa's lawyer, Frank Seamans, handled the money for the new experiment, acting as intermediary between Kehoe, the Fluorine Lawyers, and the Medical Advisory Committee.

On April 16, 1957, Seamans sent a letter to the Fluorine Lawyers, titled "Re: Kettering Research re Human Beings." He laid out how much each corporation would contribute. Checks would be sent on a quarterly basis directly from the companies to the Kettering Laboratory. U.S. Steel, Alcoa, Kaiser Aluminum, Reynolds Metals, and Alcan paid the lion's share, each putting up $30,535 for the first year; Olin Revere Metals, Monsanto Chemical, West Vaco Chemical, TVA, and Tennessee Corporation made smaller commitments. Seamans enclosed a variety of documents. They illustrate the key role the Fluorine Lawyers had in shaping Kettering's medical research, and the importance industry attached to the efforts of the National Institute of Dental Research and other parties on behalf of public water fluoridation.

Enclosures were listed by Seamans as follows:

- Letter from Dr. Irwin under date of March 13, 1957, enclosing a letter from Dr. Leone of the National Institute of Dental Research dated March 5, 1957.

- A publication entitled "Our Children's Teeth." This is the best collection of material dealing with the association between fluorides and human beings that I have seen.
- Lastly, a letter which I am sending to the Medical Advisory Committee, in which an attempt is made to more specifically advise just what the lawyers' group wishes them to do.

> I am sorry that it has taken so long to develop matters to this point. However, I am glad to say that all parties are now in complete agreement and that the work can now go forward. Very truly yours, Frank Seamans.[14]

The crucial inhalation experiments, in which researchers were to "simulate . . . occupational exposure to particulate fluoride," began on October 6, 1958. The forty-two beagles were divided equally into three groups: a control group that received no fluoride; a second group that inhaled a small dose, 3.5 mgs of calcium fluoride per cubic meter of air; and a group that received 35.5 mgs of calcium fluoride per cubic meter.

Kehoe had assembled an expert team of scientists to supervise the dog experiment, according to Eula Bingham, who became head of the Kettering Laboratory in the 1970s and later served as President Jimmy Carter's head of the Occupational Safety and Health Administration (OSHA). They included Robert K. Davis, Klaus L. Stemmer, William P. Jolley, and Edwin E. Larson. "Robert Davis was always the boss," said Bingham. "I really didn't have much contact with him, but he always seemed to be pretty substantial," she added. A pathologist, Klaus Stemmer, "was very well trained in what I would call the old European school of pathology. [He] came over from Germany after the war," said Bingham. "Larson was a very fine person when it came to exposure assessment, and he knew how to put a chamber together so that you could put a dose of whatever the contaminant was in there by inhalation. It was a very substantial training [Larson had], I tell you."[15]

The results of the Kettering beagle experiment were startling—and not at all what the scientists had predicted. "It was anticipated that there would be little or no injury to the lungs of experimental animals," the report noted, "and that the demonstration of the innocuous effects of the respiratory exposure . . . would pave the way for similar experiments with human subjects."

But there could be no human experiments now: the fluoride injured the dogs. Autopsy revealed wounds to their lungs and lymph nodes. The damage had occurred in both groups of animals that were exposed to fluoride, with inflamed lesions on the lung surface and a "fibrosis," or a thickening of the lungs, that was so marked in some cases that the researchers called it "emphysema." "Unexpected," the researchers said, "was the injurious effect exerted by calcium fluoride in the lungs and lymph nodes of the dogs."[16]

The corporate sponsors were quickly informed. "It seems likely that we have produced a dust lung using calcium fluoride as the particulate," Kettering's scientist Albert A. Brust wrote Alcoa's Dudley Irwin in a letter dated February 10, 1960. The fluoride had wreaked havoc with biological tissue, the report explained, when the fluoride ion had attacked the lung's surface. The calcium fluoride had "disassociated" inside the lung, transforming the dust into a corrosive acid deep inside the body, the report stated. "Some degree of solvent action was exerted locally, and the fluoride ion in the resultant solution reacted with the tissue," the report added. The results also showed that fluoride traveled quickly from the lung into the blood stream. "These data appear to confirm beyond all question the efficacy of pulmonary absorption of fluoride," Brust told Irwin.[17]

Frighteningly, long after the dogs had been removed from the inhalation chamber, dust particles remained lodged in their lungs. These particles continued to wreak havoc on the body, dissolving and freeing fluoride ions to mount fresh assaults on the pulmonary tissue, the report recorded. "The results obtained in this experiment are of more than casual interest, especially to investigators in the fields of pulmonary physiology and pathology," the Kettering report noted.

The health effects of airborne fluoride should be studied in workers, the results suggested. "They point to the desirability of conducting systematic investigations of the pulmonary function of representative groups of industrial employees who are being subjected to various types and intensities of exposure to particulate, inorganic fluorides," the authors wrote.

The Fluorine Lawyers understood the frightening legal and health implications of the study. The Kettering data pointed an arrow directly at the heart of key modern industrial enterprises, where the extraordinary incidence of emphysema in workers potentially "dwarfed" even the silicosis crisis of the 1930s.[18] The steel, aluminum, phosphate, gasoline refining, uranium enriching, fluorocarbon, and plastics industries, to name a few, were especially at risk. The general counsel for the TVA, Charles McCarthy, wrote to Kehoe on July 9, 1962, shortly after he received his copy of the report. Its findings were clear, he agreed: workers might be at risk. "The pulmonary findings suggest the need for further investigation of the pulmonary function of exposed workers," noted McCarthy.[19]

Industry's top lawyers received copies of the Kettering dog study—but nobody told America's workers, or their doctors. Instead, the research was buried. Although industry had spent almost half a million dollars on fluoride research at the Kettering Laboratory following the 1955 Martin verdict, the fate of the fluoride-breathing beagles was never made public. The study lay hidden for almost forty years, until, in the course of researching the topic, I found a copy in a basement archive of the old Kettering Laboratory at the University of Cincinnati.

I sent it to the toxicologist Phyllis Mullenix and to an air-pollution expert at the University of California at Irvine, Dr. Robert Phalen.[20] Both suggested that the nonpublication of the study had hurt American workers and misshaped the modern debate over air pollution. Dr. Phalen had written a 1984 book on inhalation experiments and is also a graduate of the University of Rochester. He took his job studying air pollution in Southern California on the recommendation of none other than Harold Hodge. After reading the study, Phalen remarked that he was impressed at the quality of the forty-year-old research.

"It was a very good study," Phalen said. "It was state of the art. I am amazed at how good a job they did." The scientist's conclusions were blunt. It is likely that American workers have inhaled too much fluoride in the workplace for several decades, Phalen told me. "This study is sufficiently strong to cause a reconsideration of the industrial standard," he said.

That's a staggering statement. Many hundreds of thousands of women and men have breathed fluoride in their workplaces since the Kettering study was conducted. Had the threshold for unsafe exposure been set too loosely because the dog research was not published? Occupational standards for workplace exposure to chemicals in the United States are guided by an influential private group known as the American Conference of Government and Industry Hygienists (ACGIH). The group's scientists set what is known as a Threshold Limit Value (TLV) for different chemicals, which is then used by regulatory agencies in setting legal exposure standards, Phalen explained.[21]

"The people who set standards in industry," said Phalen, "review everything they can get their hands on, and then they say, 'What shall we recommend for dusty air in industry for fluoride?' for example." Phalen is baffled at how ACGIH could have left the nation's industrial fluoride standard unchanged since 1946—if it had seen the Kettering beagle study. "As I look at the level that is set today, 2.5 milligrams per cubic meter, it sure looks to me like if [ACGIH] had access to this April 13, 1962 study, they would have recommended a lower level."

Phalen was especially startled to learn that today federal regulatory agencies, such as the U.S. Agency for Toxic Substances and Disease Registry (ATSDR), cannot locate *any* published animal studies on fluoride dust inhalation to cite for the current occupational standard.[22] "I tend to not be a conspiracy-type person," Phalen said, "I was surprised when they said there had been no studies. Why this study wasn't published, I don't know."

Did the standard-setters have access to the Kettering data? I contacted Dr. Lisa Brosseau at the University of Minnesota; she heads ACGIH's standard-setting committee. The beagle study had not been listed as one of the documents ACGIH scientists had consulted in setting the current fluoride TLV.[23] And Dr. Brosseau did not know if past ACGIH review committees had seen the Kettering study. However, she explained, if the 1962 research is not listed on ACGIH's current TLV report for fluoride, then it had not been used in its most recent review. "We will only list those things that we did use," Brosseau said.[24] "It is very possible that we didn't see it," she added.

According to the toxicologist Phyllis Mullenix, the fact that the Kettering data were never published, or made available, is a crime against American workers—with profound health consequences for the rest of the nation. The buried data points at a clear cause-and-effect relationship between an industrial pollutant and an injury widely seen in factories *and* the general population, according to the scientist. "That study is key," said Mullenix, "because it directly links fluoride with emphysema. And that is mind-boggling in terms of public health, because no one has ever made that connection."

Suppressing the 1962 study was a gross dereliction of scientific responsibility, says Mullenix, a medical "cover-up" that has lulled doctors and federal regulators to sleep for forty years. "I regard it as absolutely being hidden," she said. "It was a good study; the results were clear. The memos that went along with it certainly stated that it should be followed up."

Thousands of men and women are stalked by fluoride in the modern workplace yet blinkered to its toxic potential, according to Mullenix. In 1998 she met former aluminum workers from Washington State whose health had been ruined by fluoride. "These men are between thirty and fifty years old and have replaced knees and shoulders; they have leukemias, thyroid problems, and soft tissue diseases. I've never seen such a bunch of young pathetic people with such health problems. I just don't see the outrage. They are just putting them out as old men, and bringing in younger men, over and over again," she said. "Fluoride has impacted the work span of many of our workers, and this is in aluminum factories, petroleum companies, brick, tanneries, steel, glass, plastics, and fluorinated plastics manufacturers. I think that it has had a big impact on our industries that we are not recognizing."[25]

Eating Country Ham

PERHAPS THE FLUORIDE workers most badly treated have been the women and men who won the battle of the cold war, who did our dirty work, laboring in the satanic mills that were America's nuclear bomb factories. Since 1949, an estimated 600,000 worked in government atomic plants, with tens of thousands more employed by private industrial corporations who built the bomb during the early years of the Manhattan Project. But while the U.S. spent an estimated $5.5 trillion to build nuclear weapons, we hid the health risks of working in those factories, denied workers additional hazardous pay, and then fought those very same men and women in court if they became injured or ill and filed for compensation.[26]

"The government told these workers that they had no illnesses," noted Clinton-era Energy Secretary Bill Richardson. "These were heroes and heroines of the Cold War that built our weapons . . . and we turned our backs on them."[27]

"Paducah" Joe Harding was one of those workers, toiling in the Kentucky fluoride gaseous-diffusion plant from 1952 until 1971—when he was fired, without insurance, disability, or benefits.[28] A voice in the wilderness, Harding fought to tell the world that the United States' nuclear-bomb plants were poisoning their workers. In 1950 one of the federal plutonium injectors, Dr. Joseph Hamilton, had worried that proposals to use U.S. prisoners in more human radiation experiments had "a little of the Buchenwald touch." Joe Harding had a similar thought. In a letter written shortly before his death in 1980, and entered into the *Congressional Record* twenty years later, Harding wrote to the Department of Energy about the nation's nuclear weapons program: "It seems that Union Carbide Nuclear Co., all other Corporations that are involved, AEC, Department of Energy, Federal Security, FBI, Justice Department, etc, can do as they please, trample on the public and not be touched," Harding noted. He concluded, "The Germans had a name for this kind of setup. They called it Nazism."[29]

Harding died of cancer the same morning a Swedish TV crew arrived for an interview. At the end weeping sores marched across Joe Harding's body. He struggled to breathe. His stomach and two feet of his intestines had been removed. Bony outgrowths—classic symptoms of extreme fluoride poisoning—sprouted painfully from Harding's palms and joints. The Department of Energy lawyers fought Joe Harding until the end, at one point blaming his sickness on a combination of smoking cigarettes and eating "country ham."[30] After Harding died, the government battled his widow, Clara, in court.[31]

Pressured by union groups and shamed by an ocean of tears, Congress finally enacted legislation in October 2000 that set up a mechanism for compensation of up to $150,000 per injured atomic worker.[32] But the Energy Employees Occupational Illness Compensation Act largely sidestepped the issue of fluoride poisoning. Although a federal study of former bomb-program workers' health found that "respiratory diseases" and "mental disorders" were widespread in the Oak Ridge K-25 gaseous-diffusion plant, there was no mention of a medical link to fluoride, at least for the purposes of worker compensation.[33] (Remember, the buried Kettering dog study had *specifically* linked fluoride to such serious lung problems, while Kaj Roholm and Harold Hodge had each suspected fluoride's role in central-nervous-system disorders, a link confirmed in animals by the laboratory studies of Dr. Phyllis Mullenix at the Forsyth Dental Center in the early 1990s.)[34] "I am not aware of any [nuclear worker] cases that have successfully been compensated for fluoride exposures," said Dr. Ekaterina Mallevskia, a scientist at the Department of Energy-funded Worker Health Protection Program at Queens College in New York, which helps to diagnose the illness of former atomic workers. "We did not pay any particular attention to fluoride; we are concentrating on asbestos, radiation, uranium, plutonium." Fluoride was good for workers, the scientist even suggested, unconsciously mouthing a role written for her a generation earlier by Harold Hodge, Robert Kehoe, and Edward Bernays. "It is more like an insufficient supply than an overexposure. That's why it was initially added to toothpaste," Mallevskia explained.[35]

"No one has ever asked that question"

IT'S NOT JUST workers who are getting hurt by a chemical they never suspected. The Kettering study on beagle dogs is very likely a "smoking gun," linking fluoride to the extraordinary toll taken by air pollution *in the general population*, according to Phyllis Mullenix. Air pollution causes the early deaths of an estimated sixty thousand people in the United States each year—that's 4 percent of all U.S. deaths, and a hundred times the total number of deaths caused by all the other pollutants the EPA regulates.[36] Thirty thousand of these deaths from air pollution are attributed to emissions from electric power plants, which contain fluoride. Countless thousands of additional Americans suffer from other illnesses linked to air pollution, including heart attacks, lung cancer, and breathing disorders such as bronchitis and asthma.[37] Air pollution especially hurts children and inner city residents.[38]

Mullenix once worked as an air-pollution consultant for industry. For eleven years during the 1970s and 1980s she helped the American Petroleum Institute (API)—the oil companies' lobbying group—battle new federal air pollution standards. She had advised corporations such as Monsanto, Amoco, 3-M, Boise Cascade and Mobil Oil, jetting around the country, staying in "fabulous" hotels, all expenses paid. "It was mind-boggling the amount of money that went into it," says Mullenix.

Her specialty was ozone. In the late 1970s the EPA used the Clean Air Act to order a reduction in ozone levels. Industry's lawyers fought back, opposing the new standards and arguing that EPA had the facts wrong. On industry's behalf Mullenix attacked EPA's scientific justification for the proposed ozone policy changes, the so-called criteria document. "It was a shoddy piece of scientific material," she recalls. "Every time EPA came out with another criteria document, I would look for the errors and compare it back to the [scientific] literature. That is what I did for over ten years." Mullenix used her training as a toxicologist to fight what she saw as the EPA's inadequate scientific basis for its attack on ozone pollution.

The efforts to regulate ozone had a fundamental scientific weakness, Mullenix remarked. Laboratory experiments with pure ozone were unable to replicate the many serious injuries and health effects associated with air pollution, she stated. "Study after study, year after year, it was extremely difficult to link ozone with asthma, ozone with emphysema. It just didn't match. That is one of the reasons that I could work for industry."

During her years working for industry, fluoride was never discussed, she told me. "At the time, I didn't know anything about fluoride," she added. "Never, ever was fluoride mentioned as a cause of respiratory distress."

Had the nonpublication of the 1962 Kettering study thrown a generation of scientists off the scent of a key villain, responsible, at least in part, for air pollution's terrible health toll?

"This study, the dog study, I think might have at least triggered some investigators to look at fluorine-containing compounds as a suspect," said Robert Phalen, of the University of California. Instead, most experts today habitually ignore fluoride's role in air pollution. "Whether something like fluoride contributes more than its share, because of an additional irritancy? I would say no one has ever asked that question," he added.

It is a startling oversight, because there is a much greater quantity of fluoride in our air than we once knew. In 1998 the Clinton administration forced several key industries to report the volumes of toxic chemicals they were spilling into the environment. Previously the EPA had allowed industrial sectors, such as the electric utilities and the mining and chemical wholesalers, to avoid reporting that data. The updated information was shocking. Overnight the amount of reported toxic pollution in the United States soared by 300 percent. "Estimate of Toxic Chemicals Is Tripled," headlined the *New York Times*.[39]

Even more dramatic was the increase in the amount of hydrogen fluoride gas that industry now admitted was being spilled into the nation's air. Before the new requirements industry reported that 15 million pounds of HF pollution escaped into the air each year. When the additional industries were added, however, that figure rocketed to almost 78 million pounds, an increase of over 500 percent.[40] Of the almost 63 million pounds of additional HF, 53 million pounds (or 84 percent) came from electric power companies, and most of that came from the burning of coal.

The EPA is studying how the fine particles in air pollution can cause human injury. Does this hydrogen fluoride gas bind with those tiny carbon particles in the atmosphere, contributing to the health damage seen from such particles? What are the synergistic health effects on humans of fluoride *and* sulfur compounds? (Fluoride dramatically increases the toxicity of sulfur compounds on vegetation and animals, according to recent studies in Russia and work performed by the Atomic Energy Commission.)[41]

"You have a good point," said scientist Maria Constantini from the Health Effects Institute (HRI), a shared project of EPA and industry to fund air pollution research. HRI has "never" funded a fluoride study, she said. "Why is it not being measured? People just sometimes look for what they think is there and not for new things."

"HF [hydrogen fluoride] should be looked at," she added. "It could be coating some of the particles and . . . it could be more likely to go down into the deep lung because the particle is carried down in the lung. If it has properties that are toxic properties, depending on the dose, obviously it could be of concern."

The befuddlement of today's air pollution experts is staggering, given the toll of destruction that fluoride has wrought throughout the twentieth century.[42] Fluoride has been the nation's most damaging air pollutant, and almost certainly its most expensive. From 1957 to 1968, fluoride was responsible for more damage claims than all twenty other major air pollutants combined, according to former U.S. National Academy of Sciences fluoride expert Edward Groth.[43] The U.S. Department of Agriculture reported in 1970 that "airborne fluorides have caused more worldwide damage to domestic animals that any other pollutant."[44] And in 1982, L. H. Weinstein of Cornell University's Boyce Thompson Institute reported, "There has been more litigation on alleged damage to agriculture by fluoride than all other pollutants combined . . . of the major airborne pollutants, inorganic fluoride [is] clearly the most toxic," he added.[45]

Weinstein noted fluoride's "extreme" toxicity to vegetation. While ozone or sulfur dioxide hurt plants at a threshold level of 0.05 parts per million, hydrogen fluoride gas produced lesions on some plant leaves at concentrations of one part *per billion*, according to Weinstein.[46] (That suggests fluoride can be up to 50 times more toxic than sulfur dioxide or ozone.)

Despite this manifest chemical danger and extraordinary legal expense—or perhaps *because* of it—federal regulators have long turned their backs on fluoride air pollution. In 1957, the same year Judge Denman issued his devastating legal ruling of human harm in the Martin case, Washington abruptly terminated monitoring of fluoride levels in the nation's air.[47]

That decision came none too soon. Industry's hunger for fluoride grew more voracious in the years following the Martin trial. Hydrogen fluoride use alone more than tripled from 1957 through 1974, from 123 thousand tons to 375 thousand tons.[48] By the end of the 1960s industry was discharging 150 thousand metric tons of fluoride pollution directly into the nation's air.[49]

There is little doubt that the federal decision to end air monitoring helped industry. The feared tsunami wave of fluoride litigation from workers and communities did not break, as industry worried it might, following the Martin verdict.[50] And despite several expensive lawsuits during the 1960s, according to Keith Taylor, an attorney who represented industry in alleged fluoride pollution cases, "We were all comfortable. There were no crises."[51]

Federal aid for fluoride polluters continued. In the early 1970s the EPA elected not to include the chemical on a bad-boy list of so-called criteria air pollutants that are hazardous to human health. Chemicals such as sulfur dioxide, although more voluminous, yet which are only a fraction as toxic as the hydrogen fluoride gas in air pollution, *were* included on the list. Instead, fluoride was categorized in the new Clean Air Act as a "welfare" pollutant, blamed primarily for economic damage, such as injuring crops, rather than human health effects—a chemical favoritism that allowed individual states a permissive flexibility to set emission standards for themselves, instead of adhering to one federal policy.[52] This ruling was based largely on a 1971 National Academy of Sciences report that concluded fluorides presented no direct hazard to human health. According to the logic of the National Academy, cattle were felled, glass was etched, and crops were decimated by a chemical that in similar doses failed to injure people. It was all a grisly farce, of course, a cruel dictate that flew, quite literally, in the face of the sick Americans who lived near fluoride-spewing industrial plants, and of the lessons learned from the Martin trial. Closer to the truth was the observation of top EPA air pollution expert D. F. Walters: fluoride was so toxic a chemical that some form of environmental damage was inevitable, and industries therefore needed the freedom to pollute. Mandating "standards stringent enough to insure complete protection against any welfare effects may require closure of major sources of fluoride emissions."[53]

The Kettering Laboratory's long-ago suppression of the dog study helped to perpetuate a cover-up of fluoride's potential for harm as an air pollutant, says Phyllis Mullenix. "You have a study back in 1962 that says fluoride caused emphysema and there are no studies

Kaj Eli Roholm, pioneering Danish physician who investigated fluoride toxicity in the 1930s. HANS HENDRIK ROHOLM

Francis C. Frary, chief scientist for the Aluminum Company of America (Alcoa) in the 1930s. Concerned about industrial fluoride pollution, he also suggested that fluoride strengthened teeth. REPRODUCED BY PERMISSION OF THE ELECTROCHEMICAL SOCIETY, INC.

Kaj Roholm lecturing.
HANS HENDRIK ROHOLM

The Mellon Institute for Industrial Research in Pittsburgh, founded by leading Alcoa stockholder Andrew W. Mellon, which assisted industry in fighting lawsuits alleging air pollution. CARNEGIE LIBRARY OF PITTSBURGH

Gerald J. Cox, a researcher at the Mellon Institute who had worked on a fellowship from Alcoa and who, in 1939, made the first suggestion that fluoride be added to public water supplies. MELLON INSTITUTE COLLECTION, COURTESY OF THE CARNEGIE MELLON UNIVERSITY ARCHIVES

K-25 Fluoride Gaseous Diffusion Plant at Oak Ridge, Kentucky, which enriched uranium for the Hiroshima atomic bomb. DOE

Manhattan Project–era fluorine protection suit. NATIONAL ARCHIVES

General Leslie R. Groves, in charge of the United States Army Corp of Engineers' WWII–era Manhattan Project to make the atomic bomb. LOS ALAMOS NATIONAL LABORATORY

SECRET THIS DOCUMENT CONSISTS OF 1 PAGE(S) NO. 8 OF 8 COPIES, SERIES A

EIDMM-3b
MD700.N

JLF:mlv
JLF

xxxxxxxx
P. O. Box 287, Crittenden Station
Rochester, 7, N. Y.

29 April 1944

Subject: Request for Animal Experimentation to Determine Central Nervous System Effects.

To: Col. Stafford L. Warren, U. S. Engineer Office, Oak Ridge, Tennessee. (Thru The Area Engineer, Madison Square Area, N.Y.)

1. Inclosed is an outline of a proposed research project to determine central nervous system effects of certain T and F products.

2. Clinical evidence suggests that C616 may have a rather marked central nervous system effect with mental confusion, drowsiness and lassitude as the conspicuous features. It seems most likely that the F component rather than the T is the causative factor.

3. Since work with these compounds is essential, it will be necessary to know in advance what mental effects may occur after exposure, if workmen are to be properly protected. This is important not only to protect a given individual, but also to prevent a confused workman from injuring others by improperly performing his duties.

4. This letter is being routed thru the Area Engineer, Madison Square Area, that approval or disapproval of the information outlined above may be indicated by indorsement.

For the District Engineer:

Mailed on Rochester files Red Class 5/26/44 Rec'd Retd 5/26/44

JOHN L. FERRY,
Captain, Medical Corps,
Assistant.

Incl.:
Outline - proposed research project-nervous effects of T and F products. (2 parts)

DISTRIBUTION
Cpy 1 & 2 - Addressee.
" 3 - [illegible]
" 4 - Capt. Ferry, M.C.
" 5 - Capt. Ferry, RA.
" 6 - Reading File.
" 7 & 8 - Classified File.

SECRET

6723 Int.

Manhattan Project document warning that fluoride (coded "F") rather than uranium (coded "T") likely caused central-nervous-system injury in nuclear workers.
NATIONAL ARCHIVES

The University of Rochester's Strong Memorial Hospital, c. 1946, where plutonium was injected into patients in military experiments that were partly orchestrated by Dr. Harold C. Hodge. EDWARD G. MINER LIBRARY ARCHIVES, SCHOOL OF MEDICINE AND DENTISTRY, UNIVERSITY OF ROCHESTER

Dr. Harold C. Hodge, senior toxicologist for the Manhattan Project, and America's leading scientific promoter of water fluoridation during the cold war. IADR

James B. Conant, president of Harvard University, chemist, and senior government official in the Manhattan Project to make the atomic bomb. G. PAUL BISHOP, COURTESY PAUL BISHOP JR.

Former fluoride workers James Southern, Ralph Deadwyler, and Orfice Coggins, at the Harshaw Chemical Company's secret World War II–era "Area C" plant in Cleveland, OH.
EILEEN BLASS, *USA TODAY*

Harshaw worker Allen Hurt. Once secret Manhattan Project–era medical documents reveal fluoride poisoning of his urine.
EILEEN BLASS, *USA TODAY*

Donora, Pennsylvania, site of the nation's most notorious air pollution disaster, which killed two dozen people and sickened thousands in October 1948.
NATIONAL LIBRARY OF MEDICINE

Public Health Service engineer George Clayton taking measurements in the wake of the Donora disaster. NATIONAL LIBRARY OF MEDICINE

Toxic Fumes Believed Cause of 19 Deaths; Hundreds Stricken

List of 19 dead in Donora smog and pictures, Page 2.

By ASA ATWATER, Pittsburgh Press Staff Writer

DONORA, Nov. 1—The heavy pall of fog which brought mysterious death to 19 elderly persons here this week end has begun to drift away.

Two separate investigations are under way to stalk the "silent killer" which is believed to be a toxic poison in the fog.

The deadly fog struck first Friday night when hundreds of persons—mostly asthma sufferers—experienced difficulty in breathing.

Pittsburg Press
November 2, 1948.

Oct 31. Pittsburgh Press

WEATHER—*Fair and continued mild.*

VOLUME 65, No. 130 ●● PITTS

State of Emergency Declared—

Smog-Born Plague Kills 17 in Donora; Hospitals Overcrowded

Doctors Blame 4 Days of Fog Plus Plant Fumes; Hundreds Leave Town for Safety

DONORA, Oct. 30 (Special)—A state of emergency was declared in Donora today as a mysterious smog-born plague brought death to 17.

Doctors worked without sleep and the Red Cross, American Legion and other groups co-operated to set up an emergency hospital in the town Community building.

Hospitals were jammed to overflowing. Twelve persons

News of the air pollution disaster which took place in Donora, Pennsylvania over Halloween, 1948.

Philip Sadtler, chemical consultant who blamed fluoride pollution for the Donora disaster, and represented New Jersey farmers in WWII–era fluoride pollution claims against the Manhattan Project. TRAUDE SADTLER

Chemist Says Fluorine Gas Caused 19 Smog Deaths

'Several Plants' Blamed for Pollution In Report to Donora Council on Tragedy

Fluorine gas—not sulphur fumes—was the poison in

Philip Sadtler blames fluorine for the Donora deaths

Aluminum Company of America

ALUMINUM RESEARCH LABORATORIES

POST OFFICE BOX 772

ICIS C. FRARY
DIRECTOR

New Kensington, Pa.

December 30, 1948

CONFIDENTIAL

Dr. William F. Ashe
Kettering Laboratory
University of Cincinnati
Cincinnati 19, Ohio

Dear Dr. Ashe:

We have just completed some analytical work which has been discussed with Dr. Dudley A. Irwin, Medical Director of Aluminum Company of America. Dr. Irwin has suggested that I transmit the results of our analysis to you. For your information, the results of our analysis are being transmitted only to you and to Dr. Irwin who is receiving a copy of this letter.

Shortly after the Donora episode, Dr. G. W. Ramsey, Pathologist at the Washington Hospital, Washington, Pennsylvania, sent to me lung tissue and blood from the body of Mike Dorance who died during the period of the trouble at Donora. We made a general examination of the lung tissue in order to determine what elements were present, and the results were more or less of a general nature. The sample showed the presence of a great many elements, including some cadmium in low concentration.

In the course of our examination of the tissue, we made our usual spectrographic test for fluorine which will reveal extremely small amounts, and we were unable to detect any fluorine at all in the lung tissue. It may be of some interest to you to know that when the sample was sent to us, it was immersed in a liquid which may or may not have been excess body fluid. Before ashing the lung, it was removed from this liquid and squeezed several times. This removed as much liquid as possible, but all of the liquid squeezed out, as well as that remaining in the bottle, was carefully ashed and tested. However, the ash of this liquid was so extremely low that we did not have enough sample to make the second spectrographic shot necessary to test for fluorine.

At that particular time, we did not do anything with the sample of blood, because we learned that the body of the man had been embalmed before excision of the tissue.

12/30/48 SHEET NO. 2

At some time later than this, Dr. Ramsey made arrangements with the coroner of Washington County who delivered to me a sample of the embalming fluid which we examined carefully on a qualitative basis. We found no fluorine whatsoever present.

A few days ago, Dr. Irwin suggested that we analyze the sample of blood for fluorine content, and we have just completed that analysis. The sample was received by us and contains 20.3 p.p.m. fluorine.

This figure was obtained by our usual method of analysis which involves ashing the sample at 600°C. in the presence of lime. The ash was subjected to distillation by the Willard-Winter technique, using sulfuric acid, and distilling at a temperature of 165°C. The distillate was concentrated and redistilled from perchloric acid at a temperature of 135°C. Mr. E. J. Largent of Kettering Laboratory is well acquainted with our methods and procedures and, as a matter of fact, he can show you an exact copy of our method. We ran a blank parallel with the blood sample completely through the whole procedure and obtained a blank which amounted to about 10% of the total fluorine found, and the 20.3 p.p.m. fluorine which I am reporting herewith is after the subtraction of the blank.

I trust you will find this information of some use to you.

Very truly yours,

H. V. Churchill

H. V. CHURCHILL, Chief
Analytical Division
ALUMINUM COMPANY OF AMERICA
Aluminum Research Laboratories

HVC/jw

Copy: Dr. Dudley A. Irwin, Pittsburgh

Blood test secretly performed by the Aluminum Company of America on one of the Donora dead, showing high level of fluorine in blood.
MEDICAL HERITAGE CENTER, UNIVERSITY OF CINCINNATI

Dr. George L. Waldbott, internationally renowned allergist and physician who early warned America to the dangers of smoking, and of the potential dangers of even small amounts of fluoride. ELIZABETH RAMSEY

Frank L. Seamans, leading attorney representing the Aluminum Company of America, and head of the industry Fluorine Lawyers Committee. *PITTSBURGH POST-GAZETTE* PHOTO ARCHIVES

Dr. Robert A. Kehoe, Director of the Kettering Laboratory at the University of Cincinnati, and leading defender of industry in fluoride pollution lawsuits. UNIVERSITY OF CINCINNATI, ACADEMIC INFORMATION AND COMMUNICATIONS, CINCINNATI MEDICAL HERITAGE CENTER

Nicholas C. Leone, Chief of Medical Investigations at the National Institute of Dental Research during the 1950s. NATIONAL INSTITUTE OF DENTAL AND CRANIOFACIAL RESEARCH

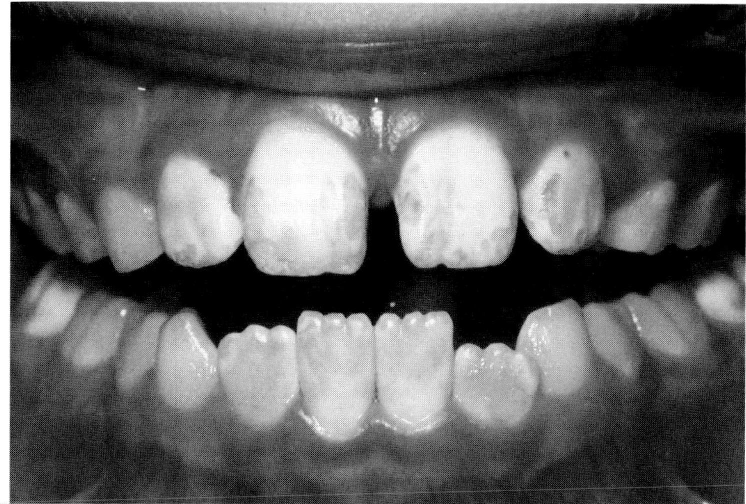

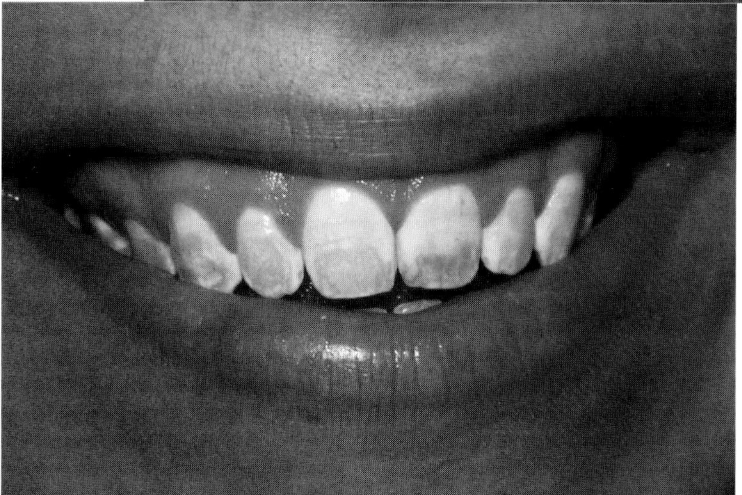

Fluoride poisoned teeth, examples of what is known as "dental fluorosis." DR. HARDY LIMEBACK

The Reynolds Metals Company aluminum reduction plant at Akwesasne, New York.
HENRY LICKERS

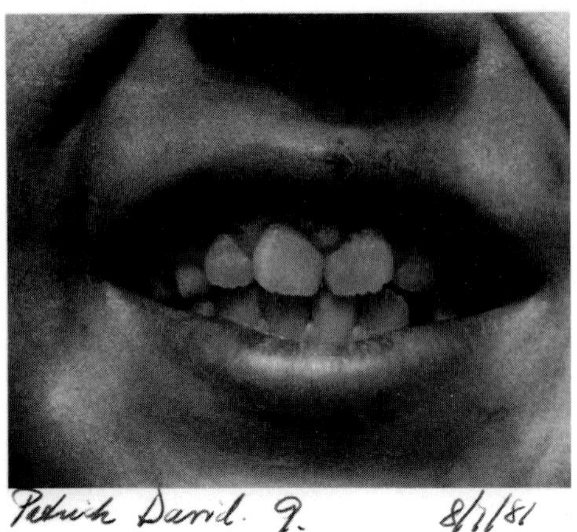

Mohawk child at the Akwesasne reservation in New York, with evidence of fluoride-poisoned teeth.
PROFESSOR LENNART KROOK

National
FLUORIDATION NEWS

Vol. XVII, No. 1 National Fluoridation News — Copyright 1971 — All Rights Reserved January-February, 1971

National Fluoridation News, a newspaper edited by George and Edith Waldbott, which connected the vigorous antifluoridation movement during the 1960s and 1970s.

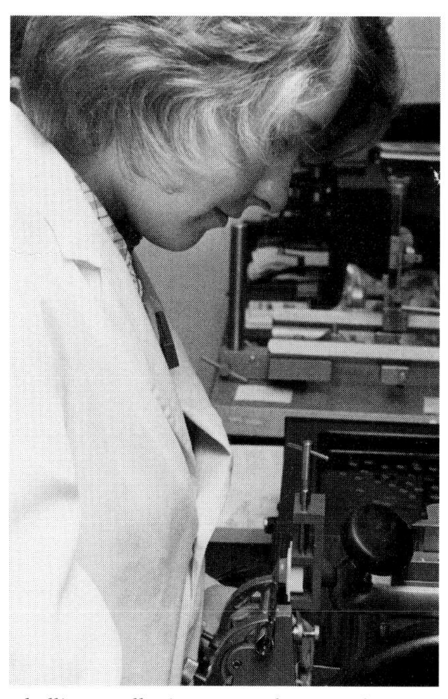

Phyllis J. Mullenix, as a graduate student at the University of Kansas.
PHYLLIS MULLENIX

Rats tested for central-nervous-system effects in the Forsyth Dental Center's RAPID computer system.
PHYLLIS MULLENIX

Phyllis Mullenix's toxicology laboratory at the Forsyth Dental Center, c. 1990. PHYLLIS MULLENIX

Forsyth Dental Center News
RESEARCH INSTITUTE
SCHOOL FOR DENTAL HYGIENISTS
DENTAL INFIRMARY
SPRING 198

New Forsyth Toxicology Dept.

Dr. Phyllis Mullenix has been appointed by Dr. John W. Hein, Director of Forsyth, to head the department of toxicology. In announcing the appointment, Dr. Hein stated:

"Societal concerns are becoming justifiably aroused over the long term implications of traces of toxins in the environment. As a major center of dental science, we at Forsyth believe our institution has a special obligation to answer these concerns by a reexamination and reassessment of the long range toxicity of substances of particular interest to dentistry, as for example, the fluoride ion, mercury (in dental fillings), nitrous oxide (for anesthesia), non precious metal substitutes for gold and many others. But, beyond our interest in the toxicity of specific materials used in dentistry, it is our desire to advance methodology for detecting toxicity. Dr. Mullenix has evolved a new technique which indicates a much more sensitive test than the traditional means of the testing of compounds causing toxic effects on the nervous system. It measures changes in animal behavior rather than changes in structure. Application of this method to nitrous oxide, long considered the safest of general anesthetics, has revealed that this agent can cause damage at certain times during the gestation period in rodents which are only revealed as behavioral changes when adulthood is reached. The far-reaching implications of this research are obvious."

Dr. Mullenix received her Ph.D. from the University of Kansas Medical Center and is a former Fellow in Toxicology of Johns Hopkins School of Hygiene and Public Health. Dr. Mullenix holds many consulting appointments to government and industry and is a faculty member of the Department of Psychiatry of the Harvard Medical School.

Dr. Hein also stated that he had the added pleasure of announcing the appointment of Dr. Harold C. Hodge, internationally known toxicologist, as Research Affiliate in the Department of Toxicology. Dr. Hodge, considered by his colleagues as the dean of modern toxicology, was the founder of the Society of Toxicology and served as its president in 1961. Dr. Hodge has held many important academic and scientific appointments including Professor of Pharmacology and Toxicology, the University of Rochester School of Medicine and Dentistry, Professor of Pharmacology, University of California, San Francisco, and Professor of Environmental Toxicology, University of California, Irvine. While professor at Rochester, Dr. Hodge headed the Division of Pharmacology and Toxicology, Manhatten Project and Atomic Energy Project. Dr. Hodge is also the author of several texts on toxicology and numerous scientific papers have been contributed by him to the pharmacological and toxicological literature.

Dr. Phyllis Mullenix, recently named head of Forsyth's Toxicology Department, with (l) Dr. Harold C. Hodge, Research Affiliate in Toxicology and (r) Forsyth's Director, Dr. John W. Hein.

Forsyth Dental Center News, spring 1984, announcing appointments of Phyllis Mullenix and Harold Hodge. FORSYTH DENTAL CENTER after that?" Mullenix said. "I mean that is a complete dodging of a very important factor that should be looked at. There was no repeat study, no follow-up on fluoride. . . . That is completely the opposite of what happened with ozone," she said. "Everything was blamed on ozone. Everything went into [studying] nitrous oxides, or sulfur oxides." (Unlike the case with fluoride, where the source of the effluent is often obvious and unique, suing a particular factory or industry for use of these more ubiquitous pollutants is much more difficult.)[54]

The Clean Air Act let industry off the hook: federal laws would not protect citizens living near fluoride emitting factories. The aluminum industry was an especially big winner. In 1958 for example, Reynolds Metals—fresh from its defeat in the Martin trial—opened a new aluminum plant near the ancestral Native American farming community of Akwesasne on St. Regis Island in the Gulf of St. Lawrence, which is situated on the border between New York and Canada. *Akwesasne* is a Mohawk Indian word meaning "land where the partridge drums." Those partridges soon fell silent, however, as Reynolds's fluoride filled the air.

By the early 1960s a drumbeat of protest was sounding. Mohawk farmers reported that honeybees and grasshoppers had disappeared from the area, while sick cattle and etched car windows were found downwind from the Reynolds's plant. Although Reynolds was acutely aware of the dangers from fluoride—after all, the company had just received Robert Kehoe's 1962 report on the poisoned beagle dogs—Reynolds did not share the information with the Native Americans, according to the Mohawk biologist Henry Lickers.[55] "For 17 years we allowed Reynolds Metals to come onto the island to look at the problem. And for 17 years they collected data . . . never insinuating there was anything wrong with our cattle," Lickers remarked.[56]

The aluminum industry helped to drive a chemical stake through an ancient culture that had lived in harmony with the earth, said Lickers. "The concept of Peace, the concept of the Great Law—all of those things knit our people together in a strong union. [But] when you poison the environment, the fiber of the community comes apart. Into that void now comes the non-traditional economies—gambling, smuggling—because people no longer can depend upon the old economies."

Evidence that fluoride might be hurting local children at Akwesasne was discovered on a 1978 visit to a Mohawk school by the scientist Bertram Carnow of the University of Illinois School of Public Health. He found a range of health problems on St. Regis Island similar to those that had frequently been linked to fluoride elsewhere. (The complaints echo almost exactly the injuries to Paul Martin's daughter, for example.) "At the school," Carnow's team reported, "teachers stated that . . . the Island children were more irritable and hyperactive and appeared to be suffering from a considerable amount of chronic fatigue. They seemed to be tired all of the time. Additionally, some had complained of aching in the legs, particularly the muscles, and in one case, the son of one of the teachers had so much pain in his feet that he frequently had difficulty in sleeping. Several teachers mentioned poor handwriting as a problem. They felt that in several cases that this might be due to the presence of a tremor. A number of children apparently had rashes, which were noted by one of the teachers. Respiratory infections were frequent and one of the children had developed a goiter."

Among the Akwesasne Mohawks, Carnow concluded, "There would appear to be significant numbers of people with abnormalities of the muscular, skeletal, nervous, and hematologic systems. In addition, there appears to be a large number at high risk because of diabetes and high blood pressure."[57]

In 1980, threatened by Carnow's findings, the Canadian and American governments intervened and arranged for a second team of scientists to visit the tribe for a more in-depth study.[58] Although the report subsequently issued by Dr. Irvine Selikoff of the Mount Sinai School of Medicine in New York was *not* able to conclusively fix the blame on fluoride for local health problems—a determination that eventually helped to undercut the $150 million lawsuit against Reynolds—at least one scientist believes that the Akwesasne verdict has not yet been fully rendered.[59] Phyllis Mullenix is now regularly visiting Akwesasne to advise Mohawk health care providers on the possible relationship between environmental pollution and their sick patients. "A lot of these people have lung problems, asthma, breathing problems—they are all on puffers [inhalers]," she says. Mullenix notes that, while Dr. Selikoff's team found serious breathing difficulties and lung problems in the Mohawks, his scientists were never shown the Kettering Laboratory's fluoride inhalation study, which connects fluoride to lung damage at low doses, and which Reynolds Metals had helped pay for.[60]

Such missing medical evidence has left scientists, doctors, and Native Americans alike in the dark about fluoride's health effects and has shaped an environment where chronic sickness has been blamed, not on fluoride, but on the Indians themselves. "It is bizarre," Mullenix remarked. "This population has been so sick for so long. They said, 'We are Indian—yeah, we are all diabetic, we are all fat, we all have thyroid problems.' They have been told that for so long. A population has accepted illness as a way of life."

What befell the Indians at Akwesasne may have befallen us all. Federal regulators were watching the situation at Akwesasne in early 1980s very closely. A ruling that the Indians had been hurt by fluoride would have increased pressure on the EPA to list fluoride as a hazardous "criteria" air pollutant under the Clean Air Act, and required federal policing of fluoride across the entire country.[61] Instead, the Selikoff team's failure to conclusively link fluoride to Mohawk sickness once again helped what some environmentalists call "the protected pollutant" to wriggle out from under EPA scrutiny.

But *had* Selikoff seen the 1962 Kettering study on the beagles, and the strength of its link between fluoride and lung damage, he might have been forced to rule differently on Akwesasne—and federal regulators might have been forced to look anew at fluoride air pollution across the rest of the country. "The changes that Selikoff was seeing [in the reduced lung capacity of Akwesasne residents] would have made sense," notes Phyllis Mullenix. "His conclusions, in respect to pulmonary function [and its cause-and-effect relationship with inhaled fluoride] would have had to be totally different."

A new focus by the EPA, aggressively targeting fluoride in air pollution, might even make good economic sense, argued the University of California's Robert Phalen, by allowing industry to be more selective in filtering out harmful air poisons. "You can't just turn off all air pollutants, because we will all starve," he said. "You have got to identify the more toxic components and control them in a pin-point fashion. It's like food—do you ban food? No, you say salmonella is a problem and you control it."

16

Hurricane Creek

The People Rule

Scientists have been villains in this story. Robert Kehoe and Harold Hodge buried important research and misled the general public. But scientists have been heroes, too. The pioneering work of Kaj Roholm and George Waldbott in unmasking fluoride's potential for harm was a principled effort to explore fluoride's role in our biology and biosphere. More recently we can see a similar heroic journey in that of Phyllis Mullenix. When her research revealed that fluoride in low doses has effects on the central nervous system, she was fired from her job as the head of the toxicology department at the Forsyth Dental Center in Boston, and her industry funding dried up.

Mullenix has since immersed herself in the medical literature about fluoride and has appeared as an expert witness in several trials in which fluoride was alleged to have injured workers. Although the number of men and women exposed to fluoride in the workplace is enormous—and as we have seen in the data from the Kettering dog study, those workers are likely to have fluoride-induced injuries—nevertheless, fifty years of assurances by the Public Health Service that fluoride in small amounts is good and safe for children make winning damage lawsuits an uphill and often thankless task.

PHYLLIS MULLENIX TOOK her seat on the stand beneath the giant seal of the state of Arkansas mounted high on the courtroom wall. The seal, inscribed in Latin, read *Regnat Populus*—"The

People Rule." The jury leaned forward. Presiding Circuit Court Judge Grisham Phillips peered over his glasses. All eyes were on the female toxicologist and the anticipated confrontation with the tall redheaded attorney Harry M. "Pete" Johnson III, who was now approaching the bench.

Mullenix had changed careers. Since being fired in 1994 she had become perhaps the most prominent scientist in the United States testifying in damage cases about the health risks she saw from low-level fluoride exposure. Mullenix had spoken with sick uranium and aluminum workers in Tennessee and Washington State, met with poisoned Mohawk Indians in New York, and testified in several court cases, helping to win financial settlements for a crippled chemical worker in Georgia and a water-treatment-plant operator in Arkansas. Despite these occasional legal successes, Mullenix believes that doctors and citizens share a blind spot in not viewing fluoride as a chemical poison and an industrial pollutant. "The problem with fluoride is that it is not recognized for what it is. First people think of toothpaste and second they think of drinking water. They have totally ignored the fluoride industry and fluoride workers."

In October 2000, back in Arkansas again, Mullenix found herself in the crosshairs of one of the nation's most powerful corporations, the Reynolds Metals Company of Richmond, Virginia; as we have seen, Reynolds has a long history of fighting fluoride pollution claims and good reason to fear having the chemical more widely recognized as a workplace poison. Reynolds had been one of the principal supporters, and beneficiaries, of Robert Kehoe's fluoride research at the Kettering Laboratory. Now Mullenix was an expert witness for a group of fifty workers suing Reynolds, part of a much larger group of several hundred workers, who also alleged that their health had been damaged while working at the company's Hurricane Creek worksite.[1]

One of the workers was Diane Peebles. The thirty-five-year-old mother of two sat quietly at the back of the Saline County courtroom in Benton throughout the October trial. Since working as a driver at Hurricane Creek in 1995 and 1996, a bizarre spectrum of physical and mental problems had dogged her. Her blood pressure had begun fluctuating wildly, and she had experienced powerful mood swings as well as "lots of headaches and stomach problems," with near constant exhaustion and pain in her joints. "The aching never stops. I wish I had the energy that other people have," she added.

Diane's husband, Scotty Peebles, had taken the witness stand. The stocky, tattooed laborer told the jury that his health had also collapsed in just six months at the Hurricane Creek site. He had been operating heavy machinery in order to bury chemical waste in giant pits. Scotty Peebles shared many symptoms with Diane and the other workers. His lung capacity had been cut almost in half, and his bones had lost mineral density, medical tests showed. His skin had burned bright red and his nose filled with painful blisters at the work site, he testified. Although he had not worked at Hurricane Creek for almost three years, twenty-nine-year-old Scotty Peebles's joints still ached and he, too, was plagued with mysterious headaches and stomach problems, he told the court.

Sitting at the Peebles's kitchen table one October morning during the trial, the soft-spoken Diane suddenly burst into tears. Scotty sat silent, his hand gripping a coffee cup. "It's hard," she blurted out. The strain on the family was sometimes overwhelming, Diane said. Several scientists, including Dr. Mullenix, had testified about the serious and often long-term health risks from fluoride. "The kids want to know, 'Are you sick mom—are you and Dad gonna die?' We tell them we are not going anywhere. I hate having to lie to my children, because I don't know myself. I want to make sure that they are taken care of. That is my biggest fear—because if we are not able to take care of our kids, who is going to?"

Reynolds Metals had hired big-time lawyers to fight the Hurricane Creek workers' claim. Attorney Pete Johnson was from the Virginia-based firm of Hunton &Williams, who since 1910 had defended Standard Oil, Phillip Morris, and a host of banking, electric utility, and railroad companies. Supreme Court Justice Lewis Powell Jr. had once been a partner, and the firm had a reputation in the legal world of having a Southern "old boy" culture. Pete Johnson fitted the mold. The University of Virginia law school graduate was one of Hunton & Williams's younger members, but he had already defended clients in "toxic tort" cases of asbestos and lead poisoning. As Johnson approached the bench and opened his files, Phyllis Mullenix closed her eyes. She smiled, bracing herself, while recalling the words of her husband, Rick, when she had left Boston.

The Reynolds lawyers, he had warned, "are going to chew on your ass a while, but you've got more ass than they've got teeth."

The duel between Mullenix and Johnson over one of the most critical legal issues that had ever faced U.S. industry—fluoride damage to human health—was being fought, fittingly enough, near one of the most historic industrial sites in the United States. Only four miles from the Benton courthouse, Hurricane Creek's red earth once contained some of the nation's richest deposits of bauxite, the raw mineral needed to make aluminum. The Aluminum Company of America had built the nearby company town of Bauxite in the early 1900s to house migrant miners. A National Park Service plaque at the Bauxite museum commemorates the region's vital role in making aluminum for aircraft during World War II.

In October 2000 Benton was ready to make history again. The court case filed by the Hurricane Creek workers was closely linked to what EPA officials call the largest and most environmentally significant waste disposal issue facing aluminum producers in the United States.[2]

The material Scotty Peebles had been burying at Hurricane Creek was a toxic by-product of making aluminum, a waste known as treated spent potliner. The EPA had taken an intense interest in the waste. Each year about 120,000 tons of spent potliner are produced by the aluminum industry in the United States.[3] The waste is impregnated with a witch's brew of fluoride, arsenic, and cyanide. Disposing of it has long been a financial headache for manufacturers—and a flashpoint for conflict with environmental regulators. "There is so much of it and it is somewhat awkward to treat," noted Steve Silverman of the EPA's Office of General Counsel.

Once, ugly black mountains of waste potliner—literally, the waste lining of the steel pots in which the aluminum is smelted—had been stored on site or buried in pits, leaching fluoride and other poisons into groundwater, and winning toxic Superfund status for several aluminum plants across the country, a federal designation that targets the hazardous site for clean-up.[4] But by the early 1990s Reynolds told the EPA that the company had solved the potliner problem. It had invented a process at the Hurricane Creek site to "treat" the waste, heating it with sand and lime in giant furnaces at temperatures of over 1,100 degrees, driving off the cyanide, and then binding the fluoride to the sand and limestone as calcium fluoride.

Hurricane Creek workers Jerry Jones and Alan Williams helped to develop that Reynolds treatment process—becoming, they now believe, two more unwitting victims of industrial fluoride poisoning. In 1988 the two laborers had been part of a work crew of several hundred men that greeted a mighty procession of 100-ton railroad "hopper" cars arriving in Arkansas, hauling potliner waste from aluminum companies in New York, Oregon, and Canada. The experimental treatment plant ran day and night, coiling a plume of black smoke across Saline County. Jerry Jones would climb into the railroad cars, smashing a sledgehammer to loosen the foul-smelling material while wearing only a bandana across his face for protection from the billowing dust. "We knew we were dealing with something awful," he added. "Your sweat would burn, and the stuff smelled just horrendous."

Safety questions drew blunt responses from the Reynolds's contractors, the men recalled. Recession was biting Arkansas hard in the early 1990s, and both Jerry Jones and Alan Williams had young children to feed. "I was told to 'either god-dammed do it, or hit the fucking gate,' because they had over a thousand applications at the office of people waiting to take our jobs," said Jones. "They did not tell us one thing [about safety]."

Alan Williams is a thick-necked former U.S. Marine with a college degree. He became a foreman at Hurricane Creek. He had always been "super physical," he said, but the forty-five-year-old quickly ran into health troubles while at the Reynolds site. "I wasn't sure what the problem was," he said. "My gums had begun to shrink. I quit smoking. I was having chest pains and rashes all over my body. I looked like an alcoholic and I don't hardly drink. It was covering my legs and arms and I was having joint pain. My sex is gone. I'm impotent. It just wasted me away," he said.[5]

By December 1991 the new treatment process was ready. Reynolds assured the EPA the "treated" spent potliner waste would not leach fluoride into ground water at levels the EPA deemed unsafe. That year the treated potliner was removed from the agency's list of toxic materials and "lost its hazardous waste stigma," according to Michelle Peace, an EPA environmental engineer who handled the "delisting" process.

The EPA ruling that the treated potliner was not hazardous was a financial windfall for Reynolds Metals. Instead of paying for the disposal of tens of thousands of tons of highly toxic waste, the company was now permitted to bury up to 300,000 cubic yards of the "treated" material each year in giant unlined pits at Hurricane Creek where, according to Peace, "there was no real associated costs with disposing of that material."[6]

The EPA may have ruled the material safe, but to workers like Scotty Peebles, the acrid dust that filled his truck cab each day was loathsome. Reynolds was experimenting with the "treated" potliner waste as commercial road-grading material, which was called ALROC. It was Peebles's job to haul the ALROC fluoride waste around the site for the test roads. He began to notice changes in the environment after he had begun this process. "It killed all the trees and the grass," Peebles said. "I used to see a lot of deer, then you didn't see too many come around any more."

Reynolds had assured the EPA that fluoride leaching from the "treated" waste would be less than 48 parts per million. But an environmental audit by an EPA contractor found levels at 2,400 parts per million—fluoride levels that "would have impacted human health and the environment," according to Michelle Peace. Nevertheless, the *extraordinary* difference between what Reynolds had promised and what it actually delivered was an honest difference of technique, not a deliberate effort to mislead the government regulators, according to Peace. "[Reynolds] ran the [initial] test as appropriately as they could."

The attorney for the Hurricane Creek workers, Bruce McMath, didn't buy it. He claimed that Reynolds had "hoodwinked" the EPA from the beginning. He showed the Benton jury a Reynolds memo proving, he said, that the company had concealed the truth from the federal regulators.[7] "They knew the treatment process was not going to achieve what they were representing to the EPA it would achieve—or at least, how they knew the EPA was interpreting the data they were giving them." He also noted that Reynolds had hired a former top EPA official to work behind the scenes and help to get the treated potliner delisted. "These corporations have such sustained and long-term working relationships with these agencies," McMath said. "It becomes very difficult for you to overcome that."

Michelle Peace conceded that the EPA had difficulty in evaluating the human-health significance of the revised test data. Her comments are revealing. While the amount of poison leaching into groundwater from the treated potliner was "definitely the greatest for fluoride," nevertheless the agency still saw the cyanide and the arsenic in the waste as the greater health hazard, remarked Peace. Once again industry's historic investment in efforts to spin fluoride's image as good for teeth, and to hide its impact as a pollutant and worker poison, had paid a handsome dividend. "Nobody ever jumped up and down" at the fluoride results, explained Peace. "You need that for your teeth."[8]

The idea that fluoride could be harmful to humans came as no surprise to Reynolds Metals. Nor was the company a stranger to the notion that fluoride's role in dental health could influence the thinking of regulators and jurors. The Reynolds legal team in Arkansas had spent impressive funds in preparation for the October 2000 trial. That fall morning, however, as defense attorney Pete Johnson strode to the bench to begin his cross-examination of toxicologist Phyllis Mullenix, he pointed to the weapon he would use to defend his client. It cost less than four dollars. On the evidence table in full view of Judge Grisham Phillips and the Benton jury, lay a single item—a thin red box of Colgate fluoride toothpaste.

Johnson approached Mullenix and smiled. He held up the toothpaste tube like a trophy. "You wouldn't brush your teeth . . . with Colgate toothpaste, would you, or any toothpaste, for that matter, where they put fluoride in it. Is that right?" Johnson asked Mullenix.

"That's right," the scientist said.

Johnson's legal strategy was familiar. Like the Reynolds lawyer Frederic Yerke in the Martin trial, forty-five years earlier, Johnson now used water fluoridation as a legal defense, pouring scorn on the notion that a chemical added to public water supplies, on behalf of children, could possibly have hurt the Hurricane Creek workers. He raised a polystyrene cup at Mullenix in a theatrical gesture, like a trophy, and slowly sipped the water in front of the jurors.

"You wouldn't drink the water in this courtroom on a regular basis like the folks who work here?" Johnson asked the scientist. "You wouldn't do that, would you?"

"If I could afford to go out and buy the bottled water, I would do so," Mullenix answered.

Hundreds of workers had breathed fluoride dust at the Hurricane Creek plant. The EPA had ordered Reynolds to clean up the site.[9] The federal Occupational Safety and Health Agency (OSHA) had fined a Hurricane Creek contractor for not providing safety equipment and training. Workers alleged serious injury: their bones ached, their lungs gasped for air, weeping sores erupted on flayed red skin, and some employees vomited in the morning before work.[10] But Reynolds's attorney Johnson continued to drill away at the issue of water fluoridation. Mullenix was a loopy dissident, he inferred, out of step with the U.S. Surgeon General, the Public Health Service, and the Centers for Disease Control, all of whom, the jury was reminded, had endorsed fluoridation. It was a legal strategy, trusted and true, that Alcoa's Frank Seamans and his Fluorine Lawyers had understood, a generation before young Pete Johnson had contemplated going to law school.

"In fact," Johnson now said, with a whiff of condescension, "you think that there ought to be a warning sign at the water fountain here outside the courtroom about all the health effects it can cause. Is that right?"

"If someone asked me for my advice, would I drink it, I would say no," Mullenix said. "But I'm not into parading posters or putting labels or warnings up anywhere," she added.

But Johnson soldiered on. Did Mullenix believe, he asked, that fluoridated water was responsible for "thyroid, memory, suicide, depression, neurological [problems], ulcers, stomach problems, eye problems . . . and ear problems?"

It wasn't that simple, Mullenix responded. The scientist explained that most people now received fluoride from multiple sources, not just drinking water. Many foods also frequently contained high levels of fluoride, especially food that was processed and irrigated with fluoridated water. Many agricultural fertilizers contained fluoride. Some popular medications, such as Prozac, were made with fluoride. And workers at numerous industrial sites, such as Hurricane Creek, continued to breathe fluoride at potentially unsafe levels.

"You have to look at the total body burden," Mullenix told the jury.

"The drinking water is a contributor to the total body burden. And then you have to look at the exposure totally and how it adds up."

The jury listened as Johnson continued. "You're against putting fluoride in drinking water because of the medical problems it can cause. . . . Is that right? You are opposed to that?" he asked.

"I really didn't have an opinion about fluoride," Dr. Mullenix answered, "until I had done studies and investigations of it . . . but after doing the studies and considering the health impacts, I would not recommend it as a good practice."

Then Bruce McMath, the workers' attorney for whom Mullenix was an expert witness, walked to the front of the court and handed Judge Grisham Phillips the medical study on the group of beagles, which Reynolds had commissioned at the Kettering Laboratory in 1962. He also gave the judge letters between company officials and the Kettering Laboratory's director, Dr. Robert Kehoe, discussing the research.[11] McMath explained that during the pretrial phase known as discovery, he had asked Reynolds for any company documents about fluoride and its health effects. The company had not given him the Kettering fluoride study, the very study he was now handing to the judge. As Judge Phillips took the long-ago documents from the workers' attorney, Reynolds's past appeared to have caught up with it. The documents linked Reynolds to a medical cover-up, illustrating that scientific information about fluoride's harmful effects had been suppressed for almost half a century.

It was as if fireworks had erupted in the middle of the courtroom. Reynolds's lawyer, Pete Johnson, quickly intervened, walking briskly to the front of the court, huddling with the Judge and hissing at McMath in stage whispers. Reynolds objected to the Kettering study's being admitted as evidence, he said. Johnson was especially outraged at any suggestion that the big aluminum company had "buried the documents or somehow failed to produce documents that it had in its possession," he told Judge Phillips.

McMath fought back. "Your Honor," he insisted. Reynolds had obviously suppressed the health study. "We asked them to produce all their documents," McMath told the Judge, including "studies . . . which pertained to or consisted of human and animal health effects upon exposure to fluoride. And of course, this document is not in there."

McMath wanted the jury to know that Reynolds had commissioned the study in the wake of a fluoride-pollution lawsuit, hidden the results, and then, forty years later, failed to turn the research over to the injured Hurricane Creek workers. "They had hidden it twice," McMath said. To the great distress of McMath and his workers, Judge Phillips banned that argument. Mullenix could discuss the contents of the Kettering study, the Judge ruled, but McMath could not tell the jury about the long-ago Martin lawsuit, or that Reynolds had attempted to keep the study secret.[12] It was a bitter pill for the plaintiffs' attorney. "I thought that they deserved to face that in the court, before the jury," McMath protested.

A moment of truth in the Arkansas trial did come, however, when attorney Johnson questioned Phyllis Mullenix about Scotty Peebles's breathing problems. The waste Peebles had handled at the Hurricane Creek site had contained calcium fluoride, the same chemical that had injured the long-ago Kettering beagle dogs. Peebles said that the foul-smelling dust had filled the cab of his front-end loader, grabbing at his lungs, burning his skin, and inflicting painful headaches. "It was getting in your hair. You would literally breathe this stuff in," he added.

A doctor diagnosed the twenty-eight-year-old with emphysema. Peebles's lung capacity was almost halved, tests showed. Many other Hurricane Creek workers also showed decreased lung capacity.

Mullenix explained how the fluoride dust used in the Kettering experiment had caused lung damage in dogs. Would the same dust hurt workers such as Scotty Peebles at the Hurricane Creek work site, the workers' lawyer McMath asked?

"Yes," replied Mullenix.

Later that day Reynolds's attorney Johnson attempted to shoot down this diagnosis during his cross-examination of Mullenix.

"Are you saying that Mr. Peebles's emphysema was caused by fluoride . . . from Hurricane Creek?" Johnson asked.

"Certainly," said Mullenix.

But Johnson oozed confidence. He had researched the published medical literature thoroughly. "In all those articles you showed us and all these references you gave us," he continued, "do you have any reference that says that fluoride causes or contributes to emphysema?"

Johnson turned, grandly, to the jury. "It is getting late" he reminded them. He swiveled back to Mullenix. "Do you have an article in all of the stuff you have brought and collected that says, 'We did a study and we found that fluoride causes the disease emphysema?'" he asked.

Mullenix played her trump card. "In the Kettering study that we presented earlier, in the pathology reports, the microscopic examination, they use [the term] emphysema lesions," she said. "They use the word 'emphysema,' yes."

Johnson adjusted his glasses. He seemed startled. "You're saying in the Kettering study in the dogs?" His voice trailed off.

"In the dogs," Mullenix repeated.

"In the dogs." Johnson looked at his notes.

"That's correct," said Mullenix.

Judge and jury looked on. The Reynolds lawyer sounded almost incredulous. Animal experiments had connected workers' lung injuries with fluoride? He looked at the bench where his legal support team sat. They stared back.

"They found emphysema, this disease emphysema was caused by fluoride?" Johnson repeated.

"The pathologists' report, in looking at the tissues, said there were emphysematous changes, and that's what was reported," said Mullenix.

"Okay. All right," the Reynolds lawyer finally conceded.

Although Judge Phillips prevented attorney Bruce McMath from telling the Arkansas jury about the long-ago Martin trial—and why the Kettering fluoride research had been commissioned by Reynolds—several former Hurricane Creek workers sitting in the courtroom that Friday afternoon understood what had just taken place.

"I didn't find out 'til yesterday that Reynolds had known anything about [fluoride's inhalation effects]," said Jerry Jones, who had begun working at Hurricane Creek in 1988. "Reynolds had conducted a research test about fluoride in 1962. We should have been told," he said.

"I am angry," said Alan Williams, the former Hurricane Creek shift supervisor. "Reynolds knew in 1962 what fluoride can do to you. They can't say they didn't, because they had their own study."

"Reynolds had a very good idea about fluoride in 1962 based on the testimony I heard here today," said Tommy Ward, a rangy ex-worker who had been in court for most of the trial, watching the jury and listening to the medical experts. Ward had suffered a violent stroke in 1996. He blamed his health problems on his years at Hurricane Creek breathing fluoride potliner. "Mullenix did a superb job," he said. "The jury got enough of that, I could tell. I think the plaintiffs hit a home run today."

Any optimism, however, vanished just four days later. In a decision that left many of the former workers incredulous and angry, Bruce McMath suddenly abandoned the lawsuit against Reynolds. On the Wednesday afternoon of October 25, 2000, Jerry Jones and Alan Williams went to court as usual. "Nobody was around," Jones said.

The trial had been scheduled to end that Friday. Several former workers were hopeful about the outcome. (McMath had seemed confident too, and had even turned down a modest offer from Reynolds to settle the case.) The jurors often passed through a landscaped area outside the courthouse, where smokers and visitors congregated and chatted. Diana Peebles had overheard a juror, she said. "We just have to do something," the juror said, according to Peebles. "They were saying it was just a question of how much they give us," Peebles explained she had overheard.

However, that Wednesday morning, Bruce McMath had told Judge Phillips that he wanted to abandon the trial, in a legal procedure in Arkansas state courts known as non-suiting. McMath believed the jury had turned against him. He feared that the court would rule that the workers had not been injured at the site. It was better, he thought, to withdraw the lawsuit and perhaps allow another legal team to remount it at a later date. "We were going to lose the whole case," he insisted. McMath blamed Judge Phillips for not allowing him to tell the jury that Reynolds Metals had suppressed the Kettering study. And he pointed the finger at state and federal agencies that had let Reynolds bury hundreds of tons of toxic fluoride waste at the Hurricane Creeks site. Reynolds had deceived those agencies, McMath said, by exaggerating how much fluoride its treatment process would remove. But the agencies had backed down, denying that they had been misled, effectively torpedoing his case.

"The EPA basically said in as many words that they did not think they had been deceived or they had acted inappropriately," McMath explained. "Of course they had. To a lawyer or to a sophisticated audience you could see what they had done, but they whitewashed it, and that really took the wind out of our sails in terms of the possibility of punitive damages or indignation with the jury. It became pretty evident to us that we were not going to be successful."

But the Hurricane Creek workers were angry and baffled at the trial's outcome. It seemed bizarre. How could their lawyer first turn down a settlement from Reynolds and then abandon the entire lawsuit? Bruce McMath must have lost a fortune by aborting the Benton lawsuit, said a Little Rock trial attorney, James Swindoll. "You are giving up two hundred grand the minute you do that," he said. "You can't get it back, unless you pursue it a second time."[13]

"We weren't very impressed," said the soft-spoken Diana Peebles. "It just seems very strange to us that this would occur." Several other former workers felt that they had been deceived twice, first by Reynolds Metals and now by their lawyers. After the trial McMath had told Jerry Jones that he was going to "reload" for a second shot at Reynolds and bring a fresh lawsuit against the aluminum company, Jones said. Instead, eleven days later, on November 5, 2000, McMath and his partner, Steve Napper, gathered the workers together for an announcement.

Jerry Jones remembers that day. He had not seen many of his former workmates in years. He was shocked at how their health had deteriorated. Some had developed crooked joints and "big knobs on their knees and fingers," said Jones. Skin sores were visible on many. Others could not lift their arms above their heads. "It was just ugly," said Jones. "It just blew my mind how it is slowly affecting them. You know there is something wrong when they all have the same thing," he added.

The men listened as their lawyers addressed them for the final time. "Boys, we got some bad news," Jerry Jones remembers Steve Napper saying. After three years of representing them, McMath and Napper explained to the gathered men that they were dropping their case. It was too difficult to prove the Hurricane Creek workers had been permanently hurt by their chemical exposures, the lawyers explained. "Find someone else," McMath and Napper told the stunned workers, then shook a few hands and "sidled out" according to the shell-shocked Jerry Jones. "The whole meeting didn't last five minutes," he recalled.

But Bruce McMath is unapologetic for dropping the Hurricane Creek suit. There was little hard data on how much fluoride the men and women had been exposed to, he said. And proving that the chemical had caused so many different injuries, especially in the small sample of workers represented in the Benton case was difficult, he added. Many of the workers also smoked cigarettes. One abused cocaine. "It creates credibility problems," said McMath. "We were looking at a case with thin causation and amorphous damages, so it becomes an impossible proposition."

The fate of the Benton trial was a consequence, perhaps, of fluoride's basic nature. Although fluoride's effects on human health potentially rival or even exceed the injuries caused by any other workplace poison, paradoxically, because fluoride has the potential to cause *so many* kinds of health problems, it is actually harder to fix blame on the chemical. Unlike other chemicals with easy-to-see and unique "signature" effects—such as the mesothelioma cancer caused by asbestos—fluoride is a systemic poison, inflicting different injuries in different people and at different times.

Duking it out with Reynolds Metals also gave Bruce McMath a first-hand look at how water fluoridation has aided industry in the courtroom. Hurricane Creek had been his first fluoride case. Proving fluoride injury to a jury was hard enough; but the federal government's long-ago endorsement of the safety of adding fluoride to public water supplies had placed the entire public-health establishment in fluoride's corner, he said. By waving a toothpaste tube at the Benton jury, Reynolds's Hurricane Creek attorney "was taking advantage of that," McMath pointed out. "Industry has manipulated this public debate to put a smiling face on what is otherwise a toxin, and thereby reduce their cost of doing business in those businesses where fluoride is a waste product," he added. Dumping waste fluoride in reservoirs may help industry, but from a pure public health perspective, McMath said, "This whole thing about putting it in the water is just silly."

After the aborted lawsuit, Jerry Jones and Alan Williams hunted for a new attorney. They met with James Swindoll, who called

McMath's office and received an explanation that McMath had no intention of refiling the workers' case. He remembers the Hurricane Creek workers who visited his office as "some of the most well-informed clients that I ever interviewed." Swindoll declined to represent them, however, and they never found a lawyer willing to refile their claim. "It just looked like a nightmare of a case," he said. "It was going to bankrupt a plaintiff's lawyer."

But because fluoride poisoning isn't easy to prove in a court of law, does it mean that doctors or regulators should abandon the issue? Phyllis Mullenix, for example, continues to take new cases of alleged fluoride poisoning in workers, representing plaintiffs around the country. She is convinced that an epidemic of disease and injury has slipped beneath the radar screen of modern health professionals. It is a sometimes-lonely battle, but the Plains daughter of Olive and Shockey Mullenix cannot walk away from the issue. She remains especially haunted by the anguished phone calls in the middle of the night from crippled former aluminum and chemical workers. They are often suffering obvious central-nervous-system problems, she notes, but they have been cast adrift by today's medical profession. "I get some of the most pathetic individuals calling up," Mullenix says. "They can't get a doctor to listen. The doctors don't know anything about fluoride and think the workers are nuts."

17

The Damage Is Done

BEHIND A CLUTTERED desk at the Newburgh Free Academy, under a portrait of Coretta Scott King, nurse-practitioner Audrey Carey daily performs physical exams on students at the large public school, which has 2,500 children in grades ten through twelve. The former mayor is in a unique position to see some of the health effects from her community's long experiment of adding fluoride to water supplies.

Fifty years earlier, Dr. Harold Hodge had assured local citizens that the Newburgh experiment had proved water fluoridation safe and had urged it upon the entire country. "Health hazards do not justify postponing water fluoridation," he had told Congress in 1954.[1] The Hudson Valley town quickly became the poster child for a global sales effort. Newburgh's smiling youngsters were paraded before scientists from the United Kingdom, New Zealand, and the World Health Organization.[2] And, for six days in 1963 Dr. Hodge sang Newburgh's praises before the Supreme Court in Dublin, prescribing mandatory fluoridation for Ireland.[3]

Ireland, and several other countries, swallowed his story. But today, back in Newburgh, Audrey Carey is no longer certain. The most visible effects from fluoride in Newburgh water are not fewer cavities, but instead the high rates of speckled and mottled teeth. Carey's friends and family, among many others in the community, have this condition, which is known as dental fluorosis. And after fifty years Newburgh children have virtually the same amount of dental decay as their counterparts in the neighboring town of Kingston, which was the "control" city in the original experiment.

Kingston has resisted all efforts in subsequent years to add fluoride to its water supply. But following Newburgh's fluoridation, the rate of fluorosis was always higher there than in Kingston, and during the 1990s it rose again. Fluorosis also occurs more frequently in African American children, according to recent surveys done by the New York Department of Health.[4]

"I see the mottling that occurs, mainly in poor children," Carey told me. She also sees it in her own family: both her grandchildren have dental fluorosis. Although their mother is now "very careful in reading the products she buys, to make sure that there is no fluoride," Carey believes that the damage is done. "Medically, it looks very bad for them," she says. "I am not sure what other physical effects they may have, or defects for that matter."

Newburgh's legacy of mottled teeth is shared by much of the rest of the country. Today, many dentists face a disturbing dilemma. Dental decay is still a serious and painful problem, especially in the inner city and even in fluoridated areas, where children are often trapped in a crossfire of poverty, poor nutrition, and a woeful public provision of dental care.[5] In some American cities as many as 3 out of every 4 children have dental fluorosis, and simply adding fluoride to public water supplies may have reached the end of the road as an easy proposal for fixing bad teeth.[6] The dental researcher Dr. Hardy Limeback, of the University of Toronto in Canada, is so concerned about the dangers of fluorosis that he claims fluoride toothpaste should be a prescription drug—at least until a child can spit, after the age of three. And even spitting is not foolproof; fluoride is absorbed directly into the body through the oral mucosa, notes Limeback. Poor nutrition can also raise the likelihood of dental fluorosis.[7] And if there is fluoride in the water supply, fluoride toothpaste may further increase the jeopardy. "Physicians have to get involved," Limeback insists. Before prescribing fluoride toothpaste, "you have to figure out, is this kid at risk for dental fluorosis?" Better food, regular brushing and flossing, access to a dentist, and using nonfluoride toothpaste may be required. "You can get perfectly healthy teeth with resistant enamel without having any kind of fluoride exposure," notes Limeback. (His son has dental fluorosis, and Limeback no longer keeps fluoride toothpaste in his home.)

Newburgh Mayor Carey's concern that dental fluorosis may signal more serious health problems is also warranted. We are now bathed in fluoride from cradle to grave, from industrial, dental, and a multitude of other and sometimes unexpected sources.[8] But the health implications of such long-term fluoride ingestion remain woefully underexamined. "Dental fluorosis is a bio-marker for systemic fluoride poisoning during early childhood," notes Dr. Limeback. "Teeth are windows to the rest of the body," adds Paul Connett, a chemistry professor and antifluoride campaigner at St. Lawrence University in New York, who likens the symptomatic nature of dental fluorosis to the thin blue gum line that can indicate lead poisoning.[9] Yet when scientists peer behind the polished facade of row upon row of brilliantly shining teeth to explore whether fluoride may be injuring us in other ways, they often get a rude surprise.

In 1992 Dr. Joseph Lyon of the University of Utah coauthored a study published in the *Journal of the American Medical Association* which found that water fluoridation was associated with an increased risk of hip fracture.[10] He was stunned at the lack of interest shown by U.S. public-health agencies in the study's results, and he has since found it difficult to get additional funding to further research this issue, he says. Today the United States has one of the highest rates of hip fracture in the world and is witnessing an epidemic of arthritis in 21 million Americans.[11] Yet doctors are as likely to blame fluoride as flying saucers. "My sense is there has been very little attention paid to toxicity," said Dr. Lyon. "Almost on the grounds that it is an impossibility, and it is a waste of everybody's time and money to even think about it." (Subsequent studies have found similar associations between fluoride in water and bone fractures).[12]

It is not just the elderly who are at risk. Fluoride may be weakening young people's bones as well. In 2001 a study in Mexico reported that dental fluorosis was correlated with a higher incidence of bone fractures in children.[13] In the United States we now pay an annual half-billion-dollar hospital tab as a result of 775,000 childhood sports injuries. Although more young people are now playing sports—particularly girls, who have a high incidence of knee and ankle injuries—Dr. Lyon wonders whether the white, chalky blotches seen on teeth also predict the likelihood of a juvenile sports injury.[14] "Is there some association [between childhood sporting injury and] living in a fluoridated area?" he asked. "There would be a plausible physiologic basis for it."

The assurances that drinking fluoride for a lifetime would be harmless flowed strongest from Dr. Hodge's cold war laboratory at the University of Rochester. In 1954 he had poured oil on the troubled waters of the growing citizens' movement opposing fluoridation—telling Congress that it would require ingesting 20–80 milligrams of fluoride each day for ten to twenty years before injury would occur. After hearing Hodge, Congress rejected the appeals to ban water fluoridation (see chapter 11).

In the late 1980s, however, two antifluoride activists, Martha Bevis and Darlene Sherrell, questioned the data Hodge had given Congress. By then Hodge's numbers had mutated further and were now being draped by fluoride promoters over all possible adverse chronic health effects. The American Dental Association (ADA) stated in a pamphlet that "the daily intake required to produce symptoms of chronic toxicity after years of consumption is 20 to 80 milligrams or more depending on weight."[15]

It was a plain falsehood. Sherrell wrote to the National Academy of Sciences (NAS) asking where the numbers had come from. This dogged researcher spotted that even Hodge had changed his data. Hodge stated in 1979 that 10 mgs of fluoride a day—not 20—would cause "crippling fluorosis."[16] Hodge had given no accompanying explanation for why he had halved his estimate. In any case, the government and the ADA ignored Hodge's correction; they continued to use his higher estimate of the amount of fluoride one could safely consume in a day, even though Hodge himself had repudiated it.[17]

It was only with the help of Florida's Senator Bob Graham that Sherrell won a response in 1990 from the NAS, to whom she pointed out the error. The persistence of the citizen activist paid off. Three years later, in 1993, the NAS National Research Council (NRC) published yet another fluoride report, entitled *Health Effects of Ingested Fluoride*. This time, although there was no accounting or apologizing for the forty years of false reassurances, the numbers were quietly corrected. "Crippling skeletal fluorosis," the NRC stated, "might occur in people who have ingested 10–20 mg of fluoride per day for 10–20 years."[18]

It was an astonishing state of affairs. Two citizen activists, neither of them scientists, had torn away the flimsy garment that had concealed a half-century of scientific deception. The corrected 1993 NRC figures laid bare the facts: countless thousands of Americans have been exposed to dangerous levels of fluoride throughout their lives. In particular, the generation of baby boomers who have ingested a lifetime of fluoridated water and might more accurately be called Hodge's Generation, may be suffering a variety of musculoskeletal and other health ailments that can be traced back to the toxicologist's false promise that fluoride in water was safe.

"The whole thing is bogus," explained the former EPA and U.S. Army scientist Dr. Robert J. Carton. In 1985 he got a close look at what he calls the "dangerous joke" at the heart of the government's fluoride policy and the very real likelihood that fluoride is injuring our bones. That year EPA scientists, including Carton, were asked to set a new and higher national level for the public's permissible exposure to fluoride in drinking water. Until the EPA review Carton had not been aware the subject was controversial. "I was just like everybody else," said Carton, "it was a no-brainer—fluoride is completely safe and effective, all that kind of stuff."

Under Reagan-appointee administrator William Ruckelshaus, EPA senior management had proposed raising the safe permissible level of fluoride in drinking water from 2.3 mg to 4 mg.[19] They had a simple way of justifying this. The blotchy teeth—dental fluorosis—produced by as little as 1 mg of fluoride per liter, which worsened greatly and grew more brittle at 4 mg per liter, were deemed a harmless "cosmetic" side effect. And despite the voluble protests of Carton, fellow EPA scientist Dr. William Hirzy, the Natural Resources Defense Council, and the EPA's employee union—Local 2050 of the National Federation of Federal Employees—the new national standard was approved.

The EPA "got away with it," says Carton—but only at the price of embarrassing its staff as professional scientists and jeopardizing the nation's health. As Carton explains it, even according to the EPA's own figures, 3 percent of the population drink more than five liters of water a day. If that water contains 4 mg of fluoride—the supposedly safe new standard—then those thirsty people will cross the threshold at which even the EPA admitted severe health effects were likely to occur. "You basically have a standard that, based on their *own* information, shows it is going to cause crippling skeletal fluorosis," says Carton. Of course, the 1993 revised estimate by the National Academy of Science for how much fluoride can cause crippling skeletal fluorosis is not 20 mg, *but 10 mg.* That means that the EPA standard is way off and would permit crippling bone injuries in a very great many people. "They are really causing problems," Carton said.

Moreover, the crippling fluorosis estimate specifies a limited time period of ten to twenty years for crippling fluorosis to appear. But fluoride is a poison that accumulates in the body over a lifetime. What happens when you get 10 mg a day for forty or sixty or even eighty years? In that case, you still reach the levels that cause crippling skeletal fluorosis, but at a later age. This simple consideration was not even addressed in the EPA's new exposure standard, says Carton, now retired. "None of it makes sense. All you have to do is look at it for ten seconds and it falls apart," he concludes.

Bone defects possibly linked to fluoride had been noticed at Newburgh back in 1955, after just ten years of water fluoridation. A radiologist, Dr. John Caffey of Columbia University, called the defects "striking" in their "similarity" to bone cancer.[20] They were detected on X-rays and seen more than twice as frequently among boys in Newburgh as among boys in nonfluoridated Kingston. Caffey's cancer suspicions, however, were not discussed in the 1956 Newburgh Final Report. In 1977 a National Academy of Sciences panel took a second look at Dr. Caffey's report, which had been published in 1955. The Newburgh cancer clue had "never been followed up," the experts said. "It would be important to have direct evidence that osteogenic sarcoma [bone cancer] rates in males under 30 have not increased with fluoridation," the panel stated.[21]

Also in 1977 Congress discovered that despite a quarter-century of endorsing water fluoridation, federal health authorities had never cancer-tested fluoride. When cancer tests were finally performed twelve years later, it was found that fluoride caused excess bone cancers in young male rats. The government concluded that the results showed "equivocal" evidence that fluoride was a carcinogen.[22] In truth, fluoride's link to cancer may have been much stronger than authorities conceded. The above-mentioned tests also showed increased liver cancers in rats, but both the bone and liver cancer evidence was systematically "downgraded," according to Dr. William Marcus, chief scientist at the EPA's Division of Water Quality.[23] After Dr. Marcus aired those allegations in an interview on *ABC News*, he was fired (for supposedly unrelated reasons). But a federal judge later ruled that Marcus had been terminated "because he had publicly questioned and opposed EPA's fluoride policy." The toxicologist was reinstated, and the government was ordered to pay damages.[24] Since then additional epidemiological studies have found more cancer in fluoridated areas, especially bone cancer in young men.[25]

Even the verdict of "equivocal" carcinogen is disturbing. Maybe fluoride doesn't cause cancer, but maybe *it does*. Is it worth the risk? "How many cavities would have to be saved to justify the death of one man from osteosarcoma?" asked the late Dr. John Colquhoun, the former chief dental officer of Auckland, New Zealand, and a fluoride promoter turned critic.[26]

Harold Hodge had also reassured American families about fluoride while secretly worrying about the chemical's effects on the central nervous system of nuclear workers. Today central-nervous-system illnesses shadow our young and old alike, with an epidemic of attention deficit and hyperactivity disorder (ADHD) in children, and with 4.5 million elderly citizens who are diagnosed with Alzheimer's dementia. The increase in Alzheimer's in the United States is largely attributed to the aging of the population, but "environmental" causes are also blamed. Does fluoride play a role in causing the disease? Quite possibly: In 1992 the American scientists Robert Isaacson, Julie Varner, and Karl Jensen found that fluoridated water carried aluminum into rat brains, producing Alzheimer's-like changes in brain tissue.[27] Phyllis Mullenix, who gave laboratory mice moderate doses of fluoride and generated symptoms resembling ADHD, fears that the high incidence of both diseases in the general population is direct evidence of fluoride's toxic effects and that both the number and kind of such injuries may worsen in the coming years.

"I think we are going to see a lot more neurological problems that currently have no answers," Mullenix said. "Extremes of behavioral problems are going to start showing up. There will be more children and people with unexplained convulsions, more unexplained cases of Alzheimer's and that kind of thing."

There were other data on Newburgh's health that warranted concern. In the 1956 Newburgh Final Report, researchers noted that young women in Newburgh reached puberty at an earlier age than did girls in nonfluoridated Kingston. Laboratory experiments have recently reproduced similar fluoride effects in gerbils.[28] In other words, fluoride has the ability to impact the female reproductive system and may be lowering the age at which women are reaching puberty. And following the introduction of fluoride into city waters, Newburgh's heart-disease rate was found by researchers to be one of the highest in the United States, another fact missing from the official Final Report.[29] Heart disease also doubled just five years into the nation's other early fluoridation experiment, in Grand Rapids, Michigan. Fluoride concentrates in the arteries, attracting calcium, and "can contribute directly to their hardening," according to scientists.[30] The folly of adding fluoride to water supplies in a nation so burdened by heart disease would seem obvious.

Mayor Carey now sees the 1945 "demonstration project" in her hometown in a very different light. "The more I read and the more I listen, the more I understand that we were subjected to experimentation," Carey stated. The newly uncovered Manhattan Project documents about Newburgh suggest to Carey that her townspeople were not told the truth about the 1945 fluoride experiment. "What happened to all of the samples that they took from me as a child?" she asked. "Where did they end up? What were they taken for? Certainly it wasn't for preventative health care."

Today some dentists are shocked to learn that a classic bait and switch was pulled on the public and on health professionals alike regarding the chemicals used in fluoridation. Pure sodium fluoride was used for the early Newburgh and Grand Rapids experiments, but today 90 percent of fluoridated public water supplies in the United States use not pharmaceutical-grade fluoride but industrial-grade silicofluoride "scrubbed" from the smokestacks of the Florida phosphate industry.[31] Important long-term toxicity tests have never been performed on these silicofluorides, although some studies have associated the chemical with higher levels of blood lead in children who live where they are used for fluoridation. Silicofluorides also frequently contain arsenic at levels that may present a risk of cancer, according to data from the National Academy of Sciences.[32] "You are sticking this poison into the water supply supposedly to prevent dental disease. It is not even doing that—and you are causing cancer just from the arsenic alone. This is totally criminal," argued the University of Toronto's Dr. Limeback.

Even the Paley Commission's long-ago predictions that these silicofluorides produced by the Florida phosphate industry would become an important and valuable source of industrial fluoride have not come to pass (see chapter 11). Today most industrial fluoride used in the United States is the raw mineral fluospar, now mined and imported from China.[33] For now, absent trouble with the Chinese and with a low price of fluorspar on world markets, silicofluoride waste from the Florida phosphate production is not used as an industrial raw material; rather, it is collected, billed to the taxpayer, and dumped into public water supplies around the country—all under the guise of protecting children's teeth.[34] Whether a dentifrice, pollution-control measure, or cold war national security blanket, EPA chemist Dr. William Hirzy put the loony logic of such dumping of industrial silicofluorides this way: "If this stuff gets out into the air, it's a pollutant; if it gets into the river, it's a pollutant; if it gets into the lake, it's a pollutant; but if it goes right straight into your drinking water system, it's not a pollutant. That's amazing!"

While much of the medical profession in the United States remains ignorant about fluoride's potential for harm, there are exceptions. Since 1968, scientists at the International Society for Fluoride Research (ISFR) have catalogued fluoride's impact on human health and the environment. In scores of peer-reviewed papers, their journal *Fluoride* has linked the chemical to multiple human-health effects, including thyroid problems, Down's Syndrome, arthritis, central-nervous-system effects, cardiovascular problems, and breathing difficulties.[35]

George Waldbott—who founded ISFR—believed that fluoride's ability to wreak such biological havoc was a function of its basic nature. Although the exact mechanism of action was then unknown, Waldbott speculated that fluoride buried deep into different organ systems and then disrupted the numerous chemical systems (such as enzymes) that regulate life.

Waldbott may have been right. Enzymes are spectacularly sensitive to fluoride. In files that were only declassified in the mid-1990s it was revealed that in 1944 Harold Hodge's bomb-program researchers at the University of Rochester had experimented with hog liver enzymes to measure fluoride pollution in bomb factories. Fluoride was so much more toxic to the esterase enzyme than uranium that contamination by fluoride and uranium could easily be differentiated.[36] And twenty years after George Waldbott's death scientists may be on the brink of unlocking a crucial cellular mechanism for how fluoride acts on our bodies. That detective story has a disturbing twist. The aluminum industry has spilled a great deal of fluoride into the environment in the last century and has been closely associated with efforts to promote water fluoridation. Ironically, it may be that aluminum *combined* with fluoride is especially responsible for fluoride's toll on health and the environment.

In 1994 the American scientists Alfred G. Gilman and Martin Rodbell won the Nobel Prize for discovering the importance of G-proteins in biology. The protein molecules act as biological amplifiers or relay stations, converting information received at a cell's surface and producing changes inside that cell. For example, when we are angry, the adrenal gland produces the adrenaline hormone. When the hormone reaches the liver or the heart, the G-protein is activated, telling the organ to produce extra energy. The bad news is that G-proteins are easily fooled by aluminum and fluoride, which gang up violently and at a molecular level on our bodies, double-teaming for extra effect, according to the Czech scientist Anna Strunecka, a researcher at King Charles University in Prague. In an abstract titled "Fluoride and Aluminum: Messengers of False Information," Strunecka reports: "It appears probable that we will not find any physiological process which is not potentially influenced by [alumino-fluorides]."[37] She added, "The synergistic action of fluoride and aluminum in the environment, water and food can thus evoke multiple pathological symptoms."

The dangers of pumping fluoride and aluminum into our environment, and our duties to future generations are clear, according to the scientist. "An awareness of the health risks of this new eco-toxicological phenomena . . . would undoubtedly contribute significantly to reducing the risk of a decrease in intelligence of adults and children, and many other disorders of the twenty-first century," noted Strunecka.

The Strange Case of the Missing Debaters

THE POTENTIAL NUMBER of fluoride-linked health issues may be enormous.[38] But the willingness of scientists to confront them is not. Fifty years of state propaganda have left too many scars and phobias. In the spring of 2001 scientist Tom Webster attempted to organize a debate about water fluoridation—and was unable to find anyone willing to speak in defense of the chemical. The Boston University environmental health professor had first grown curious about fluoride in the early 1990s, when his scientist friend Paul Connett had confided that he was worried about the potential negative health effects from small doses of fluoride to which Americans are regularly exposed.

At first Webster himself had been dismissive about the issue of fluoride. "My knee-jerk reaction was, 'Oh man, what are you getting involved in that stuff for? They are all nuts,'" he said. "But then I stopped myself, and I said, 'Well, you know, I actually don't know anything about this.' All I could remember was the Dr. Strangelove image and the John Birch Society. Their two big issues were get the U.S. out of the UN and stop water fluoridation. The more I thought about it, the more I thought, 'Here I am in the public health profession, I teach about this stuff, and I don't know anything about fluoride,'" he remembers. "It turns out there is a huge literature on this which I would never have guessed a couple of years ago."

The professor was baffled. He did not know what to make of the gulf between the nice things the government said about fluoride and the worries of scientists such as Paul Connett.[39] He was especially perturbed by a study he read by a Dr. Phyllis Mullenix showing central-nervous-system effects in rats. "Is this bad?" he said. "My gut reaction was that I don't really like the sound of this."

So Webster scheduled the fluoridation debate. He had joined a new group called the Association for Science in the Public Interest (ASIPI). The members were all professional scientists who had grown concerned that research was too often disconnected from the public interest. Now, as he scrambled to organize a debate at the group's first national conference in May 2001, Webster was scratching his head. Phone call after phone call, letter after letter, he got the same banged-door rejection from profluoridationists. He felt that many of their dismissals had a mechanical, Stepford-wife similarity that almost sounded as if they were reading from a common script. Several respondents had even been quite rude. "I got a couple of really obnoxious replies like, 'How dare you even hold such an event, it is really unprofessional.' One of those was from a guy at the CDC—one of the big fluoridation guys," said Webster. "It reminded me of the kind of stuff that you read about: 'Advice to dentists on why they should never debate antifluoridationists.' It was that kind of thing."

There were even whispers from his own group. A "generation gap" divided scientists, he realized. "One or two people inside the organization said, 'We really shouldn't have a thing on fluoride, it will give us a bad image,'" said Webster. While the younger researchers were willing to host the fluoride debate, Webster found that older members were gun shy as a result of the "painful" experiences many scientists and health professionals had undergone in the 1950s and 1960s. "It is our older colleagues who remember that stuff and how bad it was, and say 'This is just poison for your career,'" said Webster. "This is an old battle from the '50s."

Even liberals in his organization shied from hosting the 2001 debate. "It wasn't about science, it was about the politics," Webster said. "Activist scientists already have a hard enough time in this world. Industry is trying to kill us and it is hard to survive in academia. This is like, why push beans up your nose?"

The May 2001 "debate" in Virginia finally took place and was well attended, despite the lack of any profluoridation speakers, said Webster. His friend Paul Connett spoke. "Most people did not know that there was an issue—fluoride is just not on the radar screen. If people like Connett are crazy, I would have loved to see the CDC people come and squash 'em like a bug. There seems to be almost a taboo about discussing this subject, and that really doesn't seem right in public health."

Tom Webster is not alone in his frustration. That same year, in the fall of 2001, a second scientists' organization, the American College of Toxicology, hosted a "Great Debate" on water fluoridation at its annual Washington, DC, conference. Phyllis Mullenix was a speaker. Again, no one from the profluoride side would speak. The president of the organization, Robert E. Osterberg, had given the debate organizer many names and telephone numbers of scientists at leading drug companies; he was astonished that none of them showed up. "I find it extremely difficult to believe," said Dr. Osterberg, "that companies that make hundreds of thousands of dollars a year by putting fluoride into kids' multiple vitamins wouldn't stand up there and justify why they are doing it, and answer any concerns that people may have."

Epilogue

Blind to the Truth?

Fifty years after Dr. Harold Hodge signed off on fluoride safety at Newburgh, we learned of a potentially disastrous biological threat posed by another class of fluorine chemicals known as perfluorochemicals (PFCs). PFCs are different from the fluorides discussed throughout the rest of the book, both in their chemical composition and in their toxicity.[1] *But just like the fluorides in our toothpaste, PFCs—which include such brand names as Teflon, Gore-Tex, and Stainmaster—are an almost ubiquitous presence in our lives, found in numerous household products and employed in hundreds of industrial applications. And, once again, like fluoride, the story of how the toxicity of PFCs has been investigated or, more accurately, how that information has been suppressed, includes a disturbing link to the nation's nuclear program.*

ON MAY 16, 2000, the giant Minnesota-based industrial corporation 3M made a startling and historic announcement: it was "voluntarily" withdrawing one of America's best-known household products, Scotchgard, from the market. With no current replacement available for the popular fabric protector and dirt repellant, and associated products, an estimated $320 million worth of 3M sales was being washed away. "Sophisticated testing capabilities," 3M explained in a press release, ". . . show that this persistent compound, like other materials in the environment, can be detected broadly at extremely low levels in the environment and in people. All existing scientific knowledge indicates that the presence of these materials at these very low levels does not pose a human health or environmental risk."

"3M deserves great credit for identifying this problem and coming forward voluntarily," announced the EPA Administrator, Carol Browner, in response.

In truth, 3M had come forward about as "voluntarily" as a cornered tomcat in an alley. Behind the crafted public-relations spin of the 3M announcement lies a trail of exposed workers, a potentially profound threat to human health, a global environment once again polluted with a fluorine chemical, decades of corporate delay, and a staggering economic threat to a "fluoropolymer" industry with $2.5 billion dollars in international sales.[2]

It was DuPont that first recognized the commercial potential of organofluorines. By mass-producing refrigerant gases in the 1920s that combined fluorine, carbon, and chlorine (CFCs), the corporation generated a twentieth-century financial windfall.[3] The Manhattan Project quickly commandeered the wizardry of DuPont's fluorine engineers during World War II, using its radical new supersecret PFC oils and seals to lubricate and protect the government machinery in the Oak Ridge gaseous diffusion plant. (The "per" in perfluorinated means that the hydrogen atoms in a normal hydrocarbon chemical bond have been fully replaced with fluorine atoms. The chemical symbol H-C becomes F-C. That lock-tight fluorine-carbon clasp produced ultradurable chemicals that protected the government machinery from even elemental fluorine's corrosive powers.)

After the war a cornucopia of wondrous new household products based on fluorocarbon technology—including plastics, aerosols, pharmaceuticals, waterproofers, pesticides, specialized lubricants, and firefighting foams—soon tumbled from the laboratories and vast research programs that had been assembled by industry and the U.S. military.[4] The ability of the man-made PFC molecules to resist water, oil, and highly corrosive chemicals, made them the unseen servant for a host of modern creature comforts. Today the same types of PFC "polymer" chains that once helped process uranium hexafluoride for the Manhattan Project carry fast-food French fries for McDonalds in greaseproof wrappers and allow spills to be wiped from carpets impregnated with DuPont's Stainmaster fabric protector. "It allows us to do so much which we now take for granted," said British scientist and fluoride historian Eric Banks. He dubbed fluorine "the enabling element," for the bounty it contributes to modern living.

However, just like DuPont's CFC refrigerants—which were once thought safe and inert but then tore a hole in the ozone layer—the manufacture and use of Scotchgard and other PFC chemicals may have very definite *human* health risks. By the end of the twentieth century not only had millions of tons of durable CFCs soared high into the stratosphere, but their PFC cousins had quietly penetrated deep into our bodies and blood.

In 1996 the scientists Theo Colborn and John Peterson Myers and the journalist Dianne Dumanoski published *Our Stolen Future*, examining the ways synthetic chemicals can mimic hormones and disrupt biological growth and development. The book was one of the most important scientific warnings of the modern era and prompted a government review of the "endocrine disrupting" potential of such chemicals. Incredibly, however, it contained not a single reference to PFCs. "We were not aware of them," Dr. Colborn told me. "These did not come on the radar until about six years ago."[5]

How could this have happened, scientists such as Colborn want to know. How could the toxicological significance of an *entire class* of industrial chemicals evade scientists for half a century, slipping under their radar and into our lives and bodies without an alarm bell sounding? "The [PFC] story is a public embarrassment to scientists and regulatory agencies around the world," said a University of Toronto researcher, Scott Mabury. "We know less about organofluorine compounds in the environment in the year 2000 than we knew about chlorinated hydrocarbons when Rachael Carson wrote her book in 1960. That is pathetic. It is pathetic that [such] a compound could reach such high concentrations in human blood tissue and nobody know that it is bio-accumulative and that it is very persistent."

As with fluoride, however, the problem has not been a lack of information on the health effects of PFCs. Instead, the problem is that the research data about PFC toxicity has not been shared with other scientists, federal regulators, or the public. DuPont, for example, has long known that its PFC chemicals pose a potential health risk to workers and consumers. At least two company workers were killed and many others sickened while making Teflon during the war (see chapter 4). Following the wartime deaths, and fearing lawsuits from exposed employees and local citizens, the Manhattan Project's Dr. Harold Hodge from the University of Rochester visited DuPont's Haskell Laboratory in 1944 to discover what DuPont knew about the toxicity of its organofluorines.[6] Following Hodge's visit to DuPont, organofluorines were promptly given a high research priority by the Rochester team. The bomb-program toxicologists were warned that in some cases the toxicity of the organofluorines was worse than that of fluoride.[7] But for years, though Rochester scientists knew that organofluorines were a threat, almost nothing appeared in the medical literature about the toxicity of these important chemicals.

Instead, although health worries continued, the temptation to exploit PFCs for profit proved overwhelming. A 1955 DuPont company document entitled "Teflon—Health Hazards in Heating" notes that if Teflon is "heated above 400 degrees F (204 degrees C) . . . small quantities of harmful compounds are given off. . . . Consequently adequate ventilation must be provided at such temperatures. The concentrations of the volatile products necessary to produce harm have not been precisely established since it has not been possible to duplicate in animal tests the symptoms *observed in humans*" (emphasis added).[8] Nevertheless, on January 23, 1958, a Minneapolis lawyer, Harold D. Field, sought the medical advice of the Kettering Laboratory's Dr. Robert Kehoe. Field had a client who wanted to sell Teflon-lined pans in the United States, he explained. "DuPont has warned our client," Field wrote Kehoe, "that there may be some danger in the use of Teflon for this purpose." And later that year Dr. Albert Henne of Ohio State University contacted Kehoe. A Belgian company, Union Chimique Belge, also wanted to sell Teflon pots and pans in the United States, he told Kehoe. Henne had made some inquiries on the company's behalf.

"You may be interested to learn that . . . DuPont . . . seems to have started a 'rehabilitation' campaign for fluoride in the food business," Henne told Kehoe. He had friends in the legal department at Frigidaire (the unit of General Motors that sold Freon-filled refrigerators), Henne reported. They had assured him that "the sale of coated skillets does not require the formal permission of the Food and Drug Administration." As a precaution, however, would Dr. Kehoe "act as a competent witness in case of a lawsuit?" Henne asked. Kehoe agreed.[9]

Where are the Atomic Energy Commission studies on the toxicity of PFCs? As the Teflon gold rush got under way and nonstick pans became a fixture in our kitchens, it was not until 1968—two decades after the Manhattan Project's Division of Pharmacology had made researching organofluorine toxicity a cold-war priority—that another University of Rochester fluoride scientist, Dr. Donald Taves, published the first data showing that organofluorines were accumulating in human blood.[10] Taves was a colleague of Dr. Harold Hodge, whose scientists at the University of Rochester had warned in 1946 that "organic fluorine compounds appear to be more toxic than the fluoride ion." And although Taves even measured PFCs in his own body, he nevertheless issued a firm reassurance as to the toxicological significance of his disconcerting discovery. "Other chemicals are usually not toxic in blood concentrations similar to those found here for organic fluorides."[11] (At the same time Taves was also collaborating with one of the nuclear industry's big fluorocarbon suppliers, 3M.)

Even today, retired in northern California, those Rochester reflexes remain strong. Dr. Taves agrees with the current safety reassurances from 3M and DuPont: because fluorine and carbon form such a stable bond, their presence in the human body in low doses is of little health concern. "I'm not so sure that they needed to take Scotchgard off the market," Taves said. "That is a very inert chemical."[12]

Similar safety assurances paved the way for the penetration of PFCs into our homes and industry. As a result, while the global PFC industry is now a multi-billion-dollar enterprise, scientists are playing catch-up—filling a fifty-year void in the published data on PFC toxicity. In her 1962 book, *Silent Spring,* scientist Rachel Carson explained how so-called persistent organic pollutants (POPS), such as DDT or PCBs, can pass through the food chain from fish and birds to humans.[13] In the same manner PFCs can accumulate in the human body. The battle over PFCs is shaping up as what may be the *Silent Spring* of the early twenty-first century.[14]

"It is the most important chemical pollutant issue I know of," says former 3M scientist Rich Purdy who, frustrated with 3M's lack of commitment to tackle the PFC issue, resigned in 1999 after nineteen years of work with the company.[15] "PFCs are having an adverse impact on wildlife and possibly humans right now," Purdy adds. "I think they rival the significance of the chemicals that Rachel Carson pointed to," adds a Michigan State scientist, Brad Upham. "I am personally puzzled as to why there is not much more concern about these compounds." (In an interview in September 2002 Upham told me that there had never been a formal request by the National Institutes of Health for scientists to submit proposals to study the toxic effects of PFCs.)

The strength of the carbonfluorine bond in PFCs means that these chemicals can last a very long time. Researchers fear that millions of people may be absorbing the fluorine compounds through treated carpeting, clothing, and furniture and from industrial waste from factories that produce Teflon and similar products. The PFC known as perfluorooctane sulfonate (PFOS), found in Scotchgard, "redefines the meaning of persistence," notes the University of Toronto's Scott A. Mabury. "It doesn't just last a long time; it likely lasts forever."[16]

The global reach of PFCs was revealed in the late 1990s, when 3M measured the level of PFC chemicals in blood samples taken from across the United States and in Europe. The company compared the results with older blood samples taken from Korean war veterans in the 1950s, predating 3M's introduction of Scotchgard. These samples, in comparison, were uncontaminated by the chemical.[17] Researchers from the University of Michigan have also found PFCs in mink, eagles, arctic polar bears, and albatrosses in the Pacific Ocean.[18] "The occurrence of [such chemicals in] albatrosses suggests the widespread distribution of [the chemical] in remote locations," the scientists reported.[19]

Perhaps most disturbingly, the environmental "sink"—or final resting place—of many PFCs is the blood, where they bind to protein and then accumulate in the liver and gallbladder.[20] (Unlike DDT or PCBs, which accumulate in body fat and soil, PFCs are resistant to fat or water. That is what makes them such good waterproofers and fabric protectors.) "It can be like global warming," Rich Purdy told me. "What we produced twenty years ago, we still haven't harvested those effects yet. The peak hasn't hit."

The corporate suppression of information about the human health risks from PFCs was spelled out in internal documents of the DuPont Company only made public in 2002. According to medical studies and memos (reaching as far back as April 1981), DuPont researchers had recorded birth defects in children born to PFC workers at its Teflon plant in Parkersburg, West Virginia. The documents, which were posted on the Internet by the activist Environmental Working Group (EWG) in Washington, DC, revealed that the eyes of some DuPont workers' children were malformed and that there was widespread contamination of the local drinking-water supply by the PFC chemical used to make Teflon, perfluorooctanoic acid [PFOA].[21] Scandalously, and almost certainly illegally, DuPont never reported the birth defects nor the drinking-water contamination to the EPA or the local community.[22]

The EPA has grown increasingly concerned about PFC toxicity.[23] In May 2003 the agency formally asked DuPont to explain why the West Virginia drinking-water and birth-defect data had never been reported to the federal regulators. DuPont's attorney, Andrea V. Malinowski, wrote back, arguing that the birth defects could not "reliably" be linked to PFCs—and therefore did not require that the EPA be informed—and that the levels of PFCs in drinking water were too low to tell the public about.[24] That's a simple falsehood, claimed the EWG, which wants DuPont criminally punished for its actions.[25] The EWG says that DuPont clearly saw the possibility that PFC exposure was linked to the birth defects. Indeed, the company had first examined the health of worker's babies after receiving a 3M laboratory study in March 1981, which showed that PFOA caused eye defects in rats. According to a DuPont document, DuPont's review of children's health had been conducted to answer "a single question"—"does C-8 [PFOA] exposure cause abnormal children?"[26]

"We definitely do have concerns based on the toxicity data that has been submitted," noted Mary Dominiak, the chair of the fluorocarbon work group at EPA's Office of Pollution Prevention and Toxics. "I can't really go further than that because we are currently in the process of updating the hazard assessment."

The willingness of the EPA to review the human health risks from PFCs comes at the same time that federal regulators are also studying the basic issue of fluoride safety, promising to revisit the battlefields of a half-century of pitched conflict over water fluoridation and industrial fluoride pollution. On Tuesday, August 12, 2003, in a cramped room in the National Academy of Sciences building in Washington, DC, a newly formed panel of the National Research Council's (NRC) Committee on Toxicology listened to fluoride safety reassurances from the Centers for Disease Control. They also heard a lengthy criticism of existing safety standards from chemistry professor Paul Connett, a spokesperson for the activist lobbying group Fluoride Action Network. At issue is the EPA's official standard for how much fluoride should be permitted in the public water supply. In 1993, despite a hornet's nest of protest from some of its own scientists, the EPA decided to maintain the maximum contaminant level (MCL) at the level it had set in 1984—4 parts per million. Included in that decision, however, was the caveat that the official standard could be revised if additional scientific studies raised further doubts about fluoride safety. At the public hearing in Washington, Paul Connett pointed out that several new studies *had* been published since 1993, including Phyllis Mullenix's animal experiments at the Forsyth Dental Center, more recent studies from China that have found similar central-nervous-system effects in human beings, and an EPA study that reported that fluoridated water helped to carry aluminum into rats' brains, producing Alzheimer's-like lesions.[27]

According to longtime observers of America's fluoride wars, it is possible that a sea change in federal policy toward water fluoridation may be taking place. Harold Hodge was once the chairman of the NRC's Committee on Toxicology; as recently as 1993 the NRC fluoride panel had rubber-stamped his assurances of fluoride safety. But the new panel includes scientists and academics—Kathy Thiessen and Tom Webster, for example—who have all questioned the wisdom of water fluoridation; another member, Robert Isaacson, was part of the team that linked fluoride and aluminum to the Alzheimer-like lesions in rat brains. Bette Hileman, a reporter for *Chemical and Engineering News* who attended the hearing, stated that Paul Connett's presentation was even greeted with applause from the panel. "This is highly unusual at an NAS/NRC meeting," Hileman remarked. "I would be very surprised if the new NAS report turns out to be a repeat of the one in 1993. The situation has changed."

But the fluoride lobby remains powerful. In the United Kingdom the Labour government of Prime Minister Tony Blair is promoting legislation that would give private water utilities immunity from fluoride-related lawsuits, in a bid to encourage them to fluoridate more communities. For these water companies, such immunity is a key legal requirement if they are to proceed with more fluoridation. In 1996 the toothpaste manufacturer Colgate made a £1000 payment to Sharon and Trevor Isaacs, of Highams Park, Essex, whose son Kevin suffered from dental fluorosis. Colgate acknowledged no liability for the dental damage, although there were hundreds of pending cases of British children with fluorosis-damaged teeth seeking compensation. The *Sunday Telegraph* newspaper reported that "Water companies have fought against fluoride amid fears of litigation."[28]

A great deal is at stake in the NRC review, certainly more than at first meets the eye. The pressure on the EPA to tighten safety standards for water will inevitably bring fresh scrutiny for industrial fluoride users. As Alcoa's Frank Seamans and his band of Fluorine Lawyers knew, the federal government's support of water fluoridation was extraordinarily helpful to corporate America, bolstering industry's legal defense against workers' and citizens' claims of industrial fluoride poisoning. The reverse is also true. If the government admits that fluoride in water is not as safe as they had once reassured us, then industry's fig leaf is jeopardized.

So will the EPA lower the boom on the industrial fluoride polluters? It still doesn't look good. The agency's August 2003 ruling on air pollution, which allows some 17,000 industrial facilities to escape the pollution-control requirements of the Clean Air Act, means that big fluoride polluters, such as coal-burning power stations and aluminum smelters, can continue to vent tens of thousands of tons of hydrogen fluoride gas over our homes and farms.

It is America's industrial workers that most need the protection of regulators. The 1970 Occupational Safety and Health Act guarantees citizens a safe workplace. But eight years before that law was signed, the Kettering dog study showed that inhaled fluoride causes lung and lymph-node damage. The recent unearthing of that long-buried study prompted two leading toxicologists, Robert Phalen and Phyllis Mullenix, to claim that the current standard for occupational exposure to fluoride is almost certainly too high. And with the recent report that emphysema—a key injury in Robert Kehoe's fluoride-breathing dogs—is much more prevalent among industrial workers than once imagined, the inability of federal standard-setters to locate a *single* animal study to justify their current safety standard is especially concerning.[29]

Industry will in all likelihood fight any revision to the water fluoride safety standard. This fierce desire to maintain the existing permissive standards was suggested by the presence of several representatives from the EPA's pesticide division at the NRC public meeting. Dow Chemical is currently using sulfuryl fluoride as a pest fumigant to replace the ozone-depleting methyl bromide. If the fluoride safety standard for water is toughened, Dow's efforts to lobby the EPA to allow increased fluoride residues on our fruit and vegetables will almost certainly be challenged.

As Paul Connett notes, replacing methyl bromide with sulfuryl fluoride is a dubious proposition. "In animal studies it damages the white matter in the brain," Connnet explains. "So Dow is proposing to replace a chemical that causes holes in the ozone layer with one that causes holes in the brain! Some trade-off."[30]

Postscript

ARVID CARLSSON

I AM A pharmacologist and my interest in the fluoridation issue goes back to the sixties, seventies, and eighties when the addition of fluoride to the public water supplies was discussed in Sweden. During that period I studied the scientific literature and the arguments for and against water fluoridation thoroughly. My conclusion was clear: Fluoride is a pharmacologically very active compound with an action on a variety of enzymes and tissues in the body already in low concentrations. In concentrations not far above those recommended it has overt toxic actions. Fluoride added to the drinking water can prevent caries to some extent but it can do so at least as efficiently when applied locally. Moreover, local treatment, preferentially via toothpaste, is more rational, because the caries-preventive action is exerted directly on the erupted teeth. The previous belief that its action is limited to an early period before the eruption of the teeth, is not correct. The systemic action of fluoride via the blood before tooth eruption can lead to damage of the enamel, and mottled teeth. This side effect, as well as other toxic actions of fluoride, is very much reduced when fluoride is applied via toothpaste.

The addition of fluoride to water supplies violates modern pharmacological principles. Recent research has revealed a sometimes enormous individual variation in the reponse to drugs. If a pharmacologically active agent is supplied via the drinking water, the individual variation in response, which is considerable even when the dosage is fixed, will be markedly increased by the individual variation in water consumption. In addition, this measure is ethically questionable and unnecessarily expensive. When the fluoridation issue was debated in Sweden several decades ago I took part in the public debate, and we managed to convince the Swedish Parliament that the addition of fluoride to the water supplies should be rendered illegal. Similar decisions have been taken in most European countries. There is to my knowledge no evidence to suggest that dental health in Europe is worse than in the United States.

During the past two decades water fluoridation has not been debated much in Sweden, and I have not followed the scientific literature in this area closely. I have now read several chapters in Christopher Bryson's book and have found them quite interesting. Christopher Bryson is an excellent narrator, and he reports on recent research previously not known to me. Especially I am intrigued by the story about Phyllis Mullenix and her animal research on the influence of fluoride on behavior and brain development. I am not surprised by the resistance that Phyllis Mullenix so unfortunately experienced. Novel and surprising observations are often met with disbelief by the scientific community, and in this case the prestige of influential people is probably an additional factor.

It is my sincere hope that Christopher Bryson's apparently thorough and comprehensive perusal of the scientfic literature on the biological actions of fluoride and the ensuing debates through the years will receive the attention it deserves and that its implications will be seriously considered.

Dr. Arvid Carlsson, 2000 Nobel Laureate for Physiology or Medicine (for discoveries concerning signal transduction in the nervous system)

Note on Sources

THE FOLLOWING WERE good enough to grant me interviews, and their comments are reproduced throughout:

David Ast, July 16, 1997, July 31, 2002, and August 1, 2002
Eric Banks, April 23, 2001
Edward L. Bernays, December 11, 1993
Eula Bingham, July 15, 2002
George Blackstone, February 25, 2002
Lisa M. Brosseau, July 22, 2002
Georg Brun, March 19, 2001
Audrey Carey, January 2, 2002
Robert J. Carton, September 21, 2002
Theo Colborn, December 9, 2002
Mike Connett, February 7, 2004
Maria Consantini, March 22, 2002
Pamela DenBesten, February 13, 2001
Mary Dominiak, September 12, 2002
John Fedor, May 10, 2001 and October 28, 2001
Hymer Friedell, October 29, 2001
Margaret B. W. Graham, May 14, 2002
Dan Guttman November 8, 2001
John "Jack" Hein, March 21, 2001
William Hirzy, September 16, 2002
John Hoffman, July 27, 2003
Glen Howis, March 25, 1993
Allen Hurt, October 27, 2001
Donald E. Hutchings, June 13, 2002
Jerry Jones, October 20, 2000
Joe Kanapka, November 27, 2002
Kurunthachalam Kannan, September 12, 2002
Allen Kline, March 24, 1993
Arnold Kramish, October 12, 2001, and July 26, 2003
Edward Largent Jr., February 11, 2002

Hardy Limeback, September 26, 2002
Henry Lickers, spring 1993
Joseph L. Lyon, December 4, 2001, and August 8, 2002
James MacGregor, November 19, 2002
Judith MacGregor, June 25, 2002
Arjun Makhijani, May 25, 2001
Ekaterina Mallevskia, August 6, 2002
William J. Marcus, June 14, 2001
Scott Mabury, September 13, 2002
Sal Mazzanobile, November 27, 2001
James Bruce McMath, September 13, 2001, and March 1, 2002
Gabrielle V. Michalek, January 20, 2004
Paul Morrow, November 19, 2003
Phyllis J. Mullenix, multiple occasions including filmed interview February 20, 1999
Olive Mullenix, May 19, 2001
Ralph Nader, spring 1993
Antonio Noronha, summer 1997
Stata Norton, May 19, 2001
Robert E. Osterberg, November 13, 2001
Michelle Peace, June 2, 2002
Diane Peebles, October 22, 2001
Robert Phalen, March 26, 2002
Henry Pointer, October 27, 2001
Gloria Porter, October 28, 2001
Dick Powell, April 23, 2001
Rich Purdy, September 11, 2002
Karin Roholm, May 2001
Philip Sadtler, March 23, 1993
Ted Schettler, June 12, 2002
Bill Schempp, March 24, 1993
Gladys Schempp, March 24, 1993
Steve Silverman, June 18, 2002
John L. Smith, October 27, 2001
George David Smith, May 8, 2002
Karen Snapp, December 1, 2001
Lynne Page Snyder, May 4, 1998
J. Newell Stannard, December 3, 2002

James Swindoll, March 4, 2002
Donald Taves, June 27, 2002
Kathleen M. Thiessen, June 27, 2001, and August 12, 2002
Brad Upham, September 11, 2002
Henry Urrows, June 10, 2002
Sam Vest, June 24, 2001
Tommy Ward, October 20, 2000
Tom Webster, May 31, 2002
Ken Weir, September 17, 2002
Alan Williams, October 20, 2000

Two archives were the main sources of documentary information for this book. The first, the University of Cincinnati's Medical Heritage Center, houses the unpublished medical studies of the Kettering Laboratory of Applied Physiology and the papers of its director, Robert Arthur Kehoe. This archive is cited here as the RAK Collection.

The second, which houses the archives of the Manhattan Project and the Atomic Energy Commission, is the National Archives and Records Administration (NARA). The Atlanta branch of NARA is cited here as the Atlanta Federal Research Center (FRC). Documents from the President's Advisory Committee on Human Radiation Experiments (ACHRE)—a primary source of information on the University of Rochester and Harold Hodge's human experimentation—are also deposited at NARA. The papers of the S-1 Executive Committee of the Office of Science, Research, and Development (OSRD) are located in NARA's Record Group 227.

Additional Manhattan Project and AEC files came from the Oak Ridge Operations Information Office (ORO) and courtesy of the primary research of Pete Eisler of *USA Today.* Joel Griffiths and Clifford Honicker also uncovered documents from the Manhattan Project and the AEC, most notably on the Peach Crop Cases, in the personal papers of General Leslie Groves, on file at NARA. Additional AEC papers were retrieved by Honicker from the University of Rochester. In the text and notes, documents from these researchers and sources are cited as: "via Honicker and Griffiths."

Documents from online search engine-derived archives of the Department of Energy's Human Radiation Experiments Information

Management System are noted here as HREX.

The papers of fluoride historian and ADA pamphlet writer Donald McNeil are at the State Historical Society of Wisconsin in Madison. At that same archive is an important collection of documents from Alcoa on the early history of fluoride research in the United States.

The National Security Archive at George Washington University houses the supporting documents for John Marks's book on CIA drug experimentation, *Search for the Manchurian Candidate* (New York: Times Books, 1979).

The court record of the Martin trial is located in NARA Record Group 276, Boxes 5888 to 5890.

The files of the Buhl Foundation relating to its early funding of dental research at the Mellon Institute are at the Senator John Heinz Pittsburgh Regional History Center in Pittsburgh.

The papers of Ruth Roy Harris on the history of the National Institutes of Dental Research are in the History of Medicine Division at the National Library of Medicine.

The Rockefeller Archive Center in Sleepy Hollow, New York was a source for information on Kaj Roholm's trip to the United States, on early funding of dental studies at the University of Rochester, and on the Committee to Protect our Children's Teeth. The files of the Carnegie Corporation in New York City provided information on the early history of dental research in the United States.

The Truman Presidential Library in Independence, Missouri, houses the personal papers of Oscar Ewing and the papers of the President's Materials Policy Commission, also known as the Paley Commission.

Documents on the history of the Industrial Hygiene Foundation are located at the Mellon Institute in Pittsburgh, with additional papers from the Mellon Institute deposited at Carnegie Mellon Library.

Charles Kettering's personal papers are at the Kettering University in Flint, Michigan, in the Richard P. Scharchburg Collection.

The online archive of the Environmental Working Group was a primary source for documents relating to the history of perfluorinated chemicals, and for the archives of the Chemical Manufactures Assocication (CMA).

Unpublished information on the Donora disaster came courtesy of the late Allen Kline of Webster, Pennsylvania.

An extraordinary resource was the web site of the Fluoride Action Network (www.fluoridealert.org), with its comprehensive and accessible collection of medical studies, news reports and analysis.

Transcripts from the George Bareis, et al vs. Reynolds Metals trial in Arkansas which took place during October 2000 in the Saline County court, were kindly provided by the law offices of James Bruce McMath.

Finally, Martha Bevis of Houston, Texas was able to furnish me with an extraordinary library of information on the history of the fight against fluoridation in the United States.

Notes

Note on Terminology

1. For volatility: "At atmospheric pressure C-216 may combine with almost all known elements, with almost explosive rapidity, giving off extreme heat." Manhattan Project Memo, "Safety and Health Conference on Hazards of C-216 (Code for Fluorine)" To: Safety Section Files. RHTG Classified Doc., 1944-94, Box 166, Building 2714-H, Vault #82761. Such violence also makes fluorine difficult to isolate. Although it is the thirteenth-most abundant element in the earth's crust, it was not until 1886 that a French scientist, Henri Moissan, was finally able to segregate the volatile element. R. E. Banks, "Isolation of Fluorine by Moissan: Setting the Scene," *J. Fluorine Chem.*, vol. 33 (1986), pp. 1–26.
2. J. Emsley et al., "An unexpectedly strong hydrogen bond: Ab initio calculations and spectroscopic studies of amide-fluoride systems," *J. Am. Chemical Soc.*, vol. 103, (1981), pp. 24–28.
3. The National Research Council, for example, "uses the term 'fluoride' as a general term everywhere, where exact differentiation between ionic and molecular forms or between gaseous and particulate forms is uncertain or unnecessary." *Biological Effects of Atmospheric Pollutants: Fluorides* (National Academy of Sciences, 1971), p. 3.

Acknowledgments

1. Said Ralph Nader: "Once the U.S. government fifty years ago decided to push fluoridation, they stopped doing what Alfred North Whitehead once said was the cardinal principle of the scientific method, and that is to leave options open for revisions, and it became a party line, it became a dogma, and they weren't interested in criticism."

Epigraphs

1. "Muskie Hearings": Hearings before a subcommittee on air and water pollution of the committee on public works of the U.S. Senate, 59th Congress, June 7–15, 1966 (Washington, DC: U.S. Government Printing Office), pp. 113–343.

Introduction

1. L. Tye, *The Father of Spin: Edward L. Bernays and the Birth of Public Relations* (New York: Crown, 1998).

2. From 1957 to 1968, fluoride was responsible for more damage claims than all twenty other major air pollutants combined, according to U.S. National Academy of Sciences member Edward Groth. N. Groth, "Air Is Fluoridated," *Peninsula Observer*, January 27–February 3, 1969. See chapter 15 for a list of fluoride damage suits and comparison with other air pollutants.
3. See chapters 7, 9, 10, and 11.
4. For fluoride synergy, see A. S. Rozhkov and T. A. Mikhailova, "The Effect of Fluorine-Containing Emissions on Conifers," The Siberian Institute of Plant Physiology and Biochemistry, Siberian Branch of the Russian Academy of Sciences, trans. L. Kashhenko (Springer-Verlag, 1993), excerpted on the Fluoride Action Network website. Also, Herbert E. Stokinger et al., "The Enhancing Effect of the Inhalation of Hydrogen Fluoride Vapor on Beryllium Sulfate Poisoning in Animals," UR-68, University of Rochester, unclassified; and N. Groth, "Fluoride Pollution," *Environment*, vol. 17, no. 3 (April/May 1975) pp. 22–38. For "Greatest health advance," see *A Century of Public Health: From Fluoridation to Food Safety*, CDC, Division of Media Relations, April 2, 1999. For "Pollution and chemical poisoning of children," see chapters 1, 2, and 16.
5. See chapter 3.
6. See chapters 9 and 3.
7. See chapters 4 through 8.
8. *Wall Street Journal*, September 27, 2001, section A, p. 1.
9. See chapters 9 through 16.
10. The papers of Dr. Harold Hodge of the University of Rochester are closed. Archibald T. Hodge to Mr. J. B. Lloyd, University Archives and Special Collections, Hoskins Library (University of Tennessee), July 7, 1996: "Regarding your letter of June 19, 1996, concerning my father Harold C. Hodge's archives, they will be deposited in total at the University of Rochester Medical Center when a room dedicated to his files is ready." Those of Dr. Ray Weidlein, director of the Mellon Institute, are missing. Gabrielle V. Michalek, the head of archive centers at Carnegie Mellon University, which holds some of the Mellon Institute papers, explained to me that Weidlein had instructed a previous archivist to "throw the papers in the Dumpster." For more on blackballing, see chapter 12.
11. Nile Southern interviewed by Russ Honicker, transcript supplied by Honicker.
12. See chapter 12.
13. Holland discontinued fluoridation in 1976. Water fluoridation was discontinued in West Germany after 1950s. B. Hileman, "Fluoridation of Water," *Chemical and Engineering News*, vol. 66 (August 1, 1988), pp. 26–42. It was also banned in the former East Germany following reunification.
14. "A systematic review of public water fluoridation," *The York Review*, NHS Centre for Reviews and Dissemination, University of York (2000). For the 65 percent reduction in cavities claim, see Oscar Ewing's rationalization for national water fluoridation: Oscar Ewing, "Oral History Interview ," by

J. R. Fuchs of the Truman Library, Chapel Hill, NC, April and May 1969 (interview available online).

15. Interview with Paul Connett, posted on the Fluoride Action Network website.
16. For example, "Recommendations for Using Fluoride to Prevent and Control Dental Caries in the United States," Fluoride Recommendations Work Group, CDC (*MMWR*, vol. 50, no. RR 14, pp. 1–42), August 17, 2001.
17. G. L. Waldbott, A. W. Burgstahler, and Lewis McKinney, *Fluoridation: The Great Dilemma* (Lawrence, KS: Coronado Press, 1978), 149–151.

Chapter 1

1. Jack Hein, author interview, March 21, 2001. Reluctant to give me a formal interview, Hein nevertheless made several comments that have been incorporated here. Mullenix had been teaching at Harvard and doing research in the laboratory of Dr. Herbert Needleman, who was famous for proving that low levels of lead in gasoline would harm children's intelligence.
2. Hein told the British TV journalist Bob Woffinden in 1997 that the compound had been invented by a German chemist, Willy Lange, who was working in Cincinnati. A chemist from the Ozark Mahoning company, Wayne White, had then brought MFP to Rochester. According to Hein, "When Wayne White first came to Rochester with the compound, Harold Hodge looked at it and said, 'Well, I wonder if it's a nerve gas or is it going to prevent tooth decay?'" (tape time code, 04.31.15, 1997). See also the important essay discussing the ability of fluoro-chemicals to inhibit enzyme activity. Willy Lange (The Procter and Gamble Company), "The Chemistry of Fluoro Acids of Fourth, Fifth, and Sixth Group Elements," *Fluorine Chemistry*, vol. 1, ed. J. H. Simons (New York, NY: Academic Press, 1950), 125.
3. Hein was also a luminary in such influential dental organizations as the International Association for Dental Research (IADR). According to Phyllis Mullenix, he had raised funds to build a Washington headquarters for IADR.
4. Hein had been a graduate student under Harold Hodge at the University of Rochester in the 1950s. He told the British TV journalist Bob Woffinden, "We got involved with fluoride because Harold Hodge was interested from his connection over at the Manhattan Project." Interview tape time code 04.26.49, 1997.
5. V. O. Hurme, "An Examination of the Scientific Basis for Fluoridating Populations," *Dent. Items of Interest*, vol. 74 (1952), pp. 518–534.
6. Commemorative plaque at the annex entrance, noting that the industrial donors listed had "insured completion." Also, p. 7 of Forsyth Dental Center brochure, undated: "from 1969 through 1979 . . . federal support for the research programs at Forsyth increased threefold and support from industrial grants increased twofold."
7. *Wall Street Journal*, June 13, 1986, p. 25.
8. Ibid.

9. Letter of recommendation from Mehlman on Agency for Toxic Substances and Disease Registry letterhead, May 31, 1992. "Of the many scientists with whom I have worked, I consider Professor Mullenix to be one of the most talented I have known. I have the highest regard for her scientific ability and integrity," Melman added.
10. In 1994 Phyllis Mullenix sued the dental center alleging, among other things, sexual discrimination. The suit was settled out of court under terms which neither Mullenix not Forsyth are permitted to discuss. Although Mullenix will not discuss her lawsuit, Karen Snapp is blunt about the "dark side" of Forsyth, describing "an old boys club" where chauvinism and bad science mixed freely. She described to this writer several instances of crude sexual harassment at Forsyth and the occasionally sloppy professionalism of some of her colleagues. "I would not describe the atmosphere [at Forsyth] as being highly scientific," she said. "It was very strange, it was very uncomfortable. There were totally incompetent people there who were doing quite well because they played the game. They kind of decided what the results were going to be. If they did not get the result, they would either modify the experiment to give them the result, or just forget about it."

Chapter 2

1. Harold Hodge died on October 8, 1990.
2. The *New York Times*, December 16, 2002, obituary of Florence S. Mahoney.
3. In the 1920s in the United States, for example, between 11 and 16 million out of 22 million school children had defective teeth. Similar conditions were found in the United Kingdom. "In the England of the past the teeth were not as frail or as troublesome as today," Sir James Crichton-Brown told dentists in 1892, after describing the many studies that had found uniformly bad teeth among British children. Dental health in 1920s, estimate of the Joint Committee on Health Problems of the National Educational Association and the AMA, cited in letter from Dr. William Gies to Dr. F. C. Keppel of the Carnegie Corporation, November 18, 1927, Dental Research Program, Box 121, Carnegie Grants IIIa, Carnegie Archive Collection. For United Kingdom, see J. Crichton-Browne, "An address on tooth culture," *Lancet*, vol. II (1892), p. 6.
4. J. S. Lawson, J. H. Brown, J. H. and T. I. Oliver, *Med. J. Aust.*, vol. 1 (1978), pp. 124–125. Cited in M. Diesendorf, "The Mystery of Declining Tooth Decay," *Nature*, vol. 32 (July 1986), pp. 125–129. Falling dental-decay rates presented a dilemma for some in the United States, it seems. A researcher at the Forsyth Dental Center apparently warned, "Recall the European data, for example, which shows declines in caries which are occurring without fluoridation and, indeed, seem to rival the effects obtainable with fluoridation. This could easily become ammunition for the antifluordationists." Cited in e-mail to

Hardy Limeback dated May 15, 2003, from Myron Coplan, of Natick, MA, who explained that he had received the comments directly by mail from the office of Paul DePaola at the Forsyth Center in the early 1980s.

5. See especially J. D. B. Featherstone, "Prevention and Reversal of Dental Caries: Role of Low Level Fluoride," *Community Dent. Oral Epidemiol.*, vol. 27 (1999), pp. 31–40. Also, "Recommendations for Using Fluoride to Prevent and Control Dental Caries in the United States," Fluoride Recommendations Work Group, CDC (August 2001).
6. Linking fluoride to better teeth was not a new idea. As early as 1892 there had been medical speculation that because fluoride was found in dental enamel, it was necessary for strong teeth. In 1925 scientists at Johns Hopkins University tested that theory by feeding rats fluoride. They were disappointed; the fluoride made the teeth weaker, not stronger. They found, "contrary to our expectations, that the ingestion of fluorine in amounts but little above those which have been reported to occur in natural foods, markedly disturbs the structure of the tooth." E. V. McCollum, N. J. E. Simmonds, and R. W. Bunting, "The Effect of Addition of Fluorine to the Diet of the Rat on the Quality of the Teeth," *J. Biol. Chem.*, vol. 63 (1925), p. 553. In 1938 the biochemist Wallace Armstrong of the University of Minnesota may well have contributed to the confusion. He reported that teeth with fewer cavities had more fluoride in them. W. D. Armstrong and P. J. Brekhus, "Chemical Composition of Enamel and Dentin. II. Fluorine Content," *J. Dent. Res.*, vol. 17 (1938), p. 27.

   That data was, in turn, cited by Gerald Cox (whom we will meet in the next chapter) along with Dean's work and his own, permitting him to conclude that "the case for fluoride should be regarded as proved." That was not the conclusion of the editorial writers at the *Journal of the American Medical Association (JAMA)*, who noted after reading Dean's study that "the possibility is not excluded that the composition of the water in other respects may be the principal factor." Dean also said that other differences in the mineral composition of the water in the study cities—especially calcium and phosphorus—were a factor that should not be overlooked. H. T. Dean, "Endemic fluorosis and Its Relation to Dental Caries," *Public Health Reports*, vol. 53 (August 19, 1938), p. 1452. Cited in G. L. Waldbott, *A Struggle with Titans* (New York: Carlton Press, 1965), p. 13. But in 1963 one of the three planks in Cox's argument collapsed when Wallace Armstrong realized that he had gotten it wrong—increased fluoride in the teeth was a function of age and his earlier simple equation of fewer cavities and greater fluoride content was therefore invalid. "Age as a factor in fluoride content was not then (in 1938) appreciated." W. D. Armstrong and L. Singer, "Fluoride Contents of Enamel of Sound and Carious Teeth: A Reinvestigation," *J. Dental Res.*, vol. 42 (1963), p. 133. Cited in Waldbott, *A Struggle with Titans*, p. 119.
7. As we shall see, fluoride's ability to poison enzymes has long been fingered by scientists as a main pathway of its various toxic effects.

8. Fluoridation has been routinely used by bureaucrats to win tax dollars for the NIH and private research institutions. For example, while seeking funding for the entire NIH, Director Dr. Harold Varmus said in 1994 testimony before the Senate Appropriations Subcommittee on Labor, Health and Human Services, Education and Related Agencies, that fluoridation had been the most cost-effective health advance in the history of the NIH. Cited in letter from Gert Quigley of the Forsyth Institute to National Affairs Committee Cohorts, American Association for Dental Research, April 25, 1994. The Quigley memo, presumably reflecting how Varmus's comments had once again endorsed the worth of funding fluoride dental research, is titled "It couldn't have been better if we had written the script." The following month, May 1994, Mullenix was fired from Forsyth.
9. P. M. Mullenix, P. K. DenBesten, A. Schunior, and W. J. Kernan, "Neurotoxicity of Sodium Fluoride in Rats," *Neurotoxicology and Teratology*, vol. 2 (1995), pp. 169–177. (*Teratology* means "the study of malformations.")
10. Letter from Harald Löe, NIDR, to Jack Hein, October 23, 1990.
11. The mixed messages continued. Another official 1996 communication to Mullenix from NIH, rejecting a grant application, nevertheless stated, "The proposal addresses an extremely important question related to public health—whether the officially recommended safe levels of fluoride intake pose risks of adverse health effects, especially impairment of central nervous system function." Cheryl Kitt, PhD, Neurological Disorders and Stroke, to Mullenix, "Clinical Sciences Special Emphasis Panel," August 15, 1996.
12. That was not the impression of Professor Albert Burgstahler. The University of Kansas chemist was a member of the official review committee that examined Mullenix's proposal for NIH funding for further studies. He is also the author of several scientific papers and books on the injurious health effects of small amounts of fluoride and is a past president of the International Society for Fluoride Research. Dr. Burgstahler blamed fear of a "loss of face" at the Public Health Service and among other scientists on the review committee for rejecting her research request. In a letter, July 11, 1996, Burgstahler wrote to Dr. Antonio Noronha of the NIH, "You are well aware of the enormous amount of controversy and sensitivity to loss of face that surrounds the issue of the Mullenix proposal and the very upsetting character of the work she has published on the 50th anniversary of the start of fluoridation in the United States and Canada." He asked, "If any member of the Special Review Committee were to have given a more favorable rating to the proposal, and their names became known to those in funding-decision levels of the USPHS . . . might they not risk jeopardizing further funding from the USPHS for having supported a proposal for research that has already revealed serious errors in USPHS thinking and policy regarding the health hazards of current levels of fluoride exposure in the general population?"

13. M. Hertsgaard and P. Frazer, "Are We Brushing Aside Fluoride's Dangers?" *Salon.com*, February 17, 1999, http://www.salon.com/news/1999/02/17news.html.
14. Tony Volpe and Sal Mazzanobile, who had attended the fluoride toxicity meeting in Jack Hein's office, were installed as Overseers. Forsyth Dental Center brochure, undated, p. 10.
15. Hodge's boss, Manhattan Project Captain John L. Ferry, is the memo's author. Colonel Warren approved the request the same day and allocated a budget of $7,500. Md 3, Md 700, General Essays, Lectures, Medical Report, Box 34, Manhattan Engineer District Accession #4nn 326-85-005, Atlanta FRC, RG 326. (Hodge's two-part research proposal, however, listed as an enclosure "Outline—proposed research project—nervous effects of T and F products," is missing from the files.)
16. At Rochester during the cold war, "The toxicology studies were very comprehensive. They were looking for toxic effects on the bone, the blood, and the nervous system. . . . Without the Manhattan Project and the atomic bomb, we wouldn't know anywhere near as much as we do about the physiological effects of fluoride." Interview with Bob Woffinden and Mark Watts, Channel Four (UK) Transcript, 1997.

Chapter 3

1. Family data from Danish newspaper clippings in Roholm family scrapbook, read in translation by Roholm's daughter-in-law, Karin Roholm. Personal meeting in New York, May 2001.
2. Brun was then ninety-five years old. He published a paper with Roholm on fluoride excretion in workers' urine. *Nordisk Medicin*, vol. 9 (1941), pp. 810–814. Also found at: George C. Brun, H. Buchwald, and Kaj Roholm, "Die Fluorausscheidung im Harn bei chronischer Fluorvergiftung von Kryolitharbeitern," *Acta Medica Scandinavica*, vol. CVI, fasc. III (1941). Citation, photocopy of paper, and several Roholm biographical details provided by Donald Jerne of the Danish Library of Medicine.
3. J. H. Simons, ed., *Fluorine Chemistry*, vol. IV (New York and London: Academic Press, 1965), p. vii.
4. *Fluorine Chemistry*, vol. IV, p. viii. Roholm's memberships included The Society for Health Care, The Younger Doctors' Committee for Continuation Courses in Socialized Medicine, The Danish Association for the Prevention of Venereal Disease, a Committee to Organize a Permanent Hygiene Exhibition, and the Pharmacopeial Revision Committee. Letter to author on January 31, 2002, from Donald Jerne, medical adviser, The Danish National Library of Science and Medicine.
5. Letter from Frank J. McClure (U.S. National Institute of Dental Research) to Lisa Broe Christiansen (Roholm's daughter) on September 19, 1956. (Letter provided to author by daughter-in-law Karin Roholm.)
6. For history of cryolite exploitation, see E. K. Roholm, *Fluorine Intoxication: A Clinical-Hygienic Study, with a Review of the Literature and Some*

Experimental Investigation (London: H. K. Lewis and Co. Ltd., 1937) and R. K. Leavitt, "Prologue to Tomorrow: A History of the First Hundred Years in the Life of the Pennsylvania Salt Manufacturing Company" (The Pennsylvania Salt Company, 1950). The Danish state owned the Greenland cryolite. There were only two buyers, the Øresund Chemical Works of Copenhagen and the Pennsylvania Salt Company of Philadelphia, who held a valuable monopoly for Danish cryolite in the United States and Canada.

7. P. F. Møller and Sk. V. Gudjonsson, "Massive Fluorosis of Bones and Ligaments," *Acta radio,* vol. 13 (1932), p. 269.
8. E. K. Roholm, *Fluorine Intoxication,* pp. 192 and 205.
9. Ibid., pp. 150, 202, 143, and fig 26.
10. Ibid., pp. 142–143, and 178. The U.S. nuclear worker Joe Harding, who suffered from fluoride poisoning, might have recognized this kind of skeletal poisoning; bony outgrowths covered Harding's palms and feet. No American doctor diagnosed these bony outgrowths as a symptom of fluorine intoxication, despite Harding's work in the fluoride gaseous diffusion plant. See chapter 18. See also Joe Harding interview:

> In 1970, I also began noticing and developing something else that was very unusual and new. I had always had perfectly normal and good fingernails and toenails and never any trouble with them. But, along during the summer and fall of 1970, I got some sore places on the balls of my thumb tips and fingertips, where your fingerprints are, that felt like I had maybe stuck a thorn or a splinter real down deep into them. When I would rub my other finger over it, I could feel it way down in there, but yet I couldn't see anything. These kept getting a little more sore, and finally, when the soreness got up near enough to the surface, I kind of dug in. I found something kind of like a piece of fingernail sticking through there. This was very, very painful. I would trim it off back just about as deep as I could reach. It would come back again. It really didn't dawn on me for sure just what this might be at first. But, it didn't take too long till I began to realize that from over on the other side, near the base of my regular fingernails, I was growing fingernails straight through my fingers and coming out on the wrong side. This was pretty painful. I had these on my thumbs and three or four of my fingers. This was the beginning of another very unusual thing for me, which I will talk more about later. . . . In 1971, then, I was still working in the 35 control room, and knee and lungs and hemoglobin in my blood all about the same, skin slowly worse, this fingernail business a little worse, and by this spring, I first noticed that I had something sore under the arch of my right foot. And then I had something getting sore up on the top of the arch bone of my right foot. As time got on, I discovered, I suppose you would call these toenails growing out from under the arch of my right foot, and out under the peak of the arch bone of my

right foot. It was pretty hard for me to keep my shoe tied very tight on that one, and I had to keep digging these things out. (Interview with Dolph Honicker, tape 13.)

11. Roholm, *Fluorine Intoxication*, pp. 138–139. The Dane especially noted an illness called neurasthenia, a condition defined as "an emotional and psychic disorder that is characterized by impaired functioning in interpersonal relationships and often by fatigue, depression, feelings of inadequacy, headaches, hypersensitivity to sensory stimulation (as by light or noise), and psychosomatic symptoms (as disturbances of digestion and circulation)" (ref on pp. 178 and 193). Definition in *Webster's New World Collegiate Dictionary* (New York: Pocket Star Books, 1990).
12. While this field had been "little explored," Roholm added, "it is extremely probable that fluorine acts on the metabolism in various ways and that the symptoms of chronic intoxication have a complicated genesis." Roholm, *Fluorine Intoxication*, p. 286.
13. J. Crichton-Browne, "An Address on Tooth Culture," *Lancet*, vol. 2 (1892), p. 6. Crichton-Browne wrote, "I think it well worthy of consideration whether the reintroduction into our diet, and especially into the diet of childbearing women and of children, of a supplement of fluorine in some natural form . . . might not do something to fortify the teeth of the next generation."
14. E. V. McCollum, N. J. E. Simmonds, and R. W. Bunting, "The Effect of Addition of Fluorine to the Diet of the Rat on the Quality of the Teeth," *J. Biol. Chem.*, vol. 63 (1925), p. 553.
15. For more fluoride in bad teeth, see E. K. Roholm, *Fluorine Intoxication*, p. 150. In mother's milk, ibid., p. 199.

    Earlier speculation from J. Crichton-Browne "An Address on Tooth Culture," was tested experimentally and rejected by McCollum, Simmonds, and Bunting in "The Effect of Addition of Fluorine," *J. Biol. Chem.* Roholm cited both references in his bibliography. The folk notion persisted, however, that fluorine might help teeth. See the suggestions that apparently followed the Alcoa chemist H. V. Churchill's announcement that fluorine caused dental mottling. "At the very meeting where Churchill announced his discovery of large amounts of fluorine in a water supply which caused ugly mottling of teeth a chemist from Hollywood, California, said he felt there must be a threshold point up to which fluorine was desirable. . . . In June 1931, a fellow townsman of Churchill's, a dentist, suggested that fluorine might prevent dental cavities." Donald McNeil, *The Fight for Fluoridation* (New York: Oxford University Press, 1957), p. 37.
16. Roholm, *Fluorine Intoxication*, p. 315.
17. Ibid., p. 321. Further, "Every form of fluorine ingestion is counter-indicated in children when the permanent teeth are calcifying," Roholm wrote on p. 311.
18. Ibid., vi. Also, e-mail, March 8, 2001, to author from Donald Jerne, medical advisor, Danish National Library of Science and Medicine.

19. Volcanic activity in the United States also brings fluoride to the surface. The Old Faithful geyser in Yellowstone National Park shoots forth steam and water poisoned with extraordinarily high levels of fluoride (20 ppm.) See: J. Cholak, "Current Information on the Quantities of Fluoride Found in Air, Food, and Water" (Kettering Symposium, 1957), RAK Collection.
20. In North Africa, scientists blamed fluoride in the soil for crippling local people, Roholm learned. Speder: L'Osteopetrose generalize out "Marmmorskelett" n'est pas une maladie rare. Sa frequence dans l'intoxication fluoree." *J. Radiol. Electrol.*, vol. 20 (1936), p. 1, and *J. Belg. Radiol.*, vol. 140 (1936). In parts of the world today such skeletal fluorosis is endemic. In India, for example, thousands of fresh-water wells drilled by the United Nations during the International Water Decade of the 1980s—to improve local access to clean water and better sanitation—have instead produced a public-health crisis, with many thousands now suffering from skeletal fluorosis. "The problem is enormous, unbelievable," noted Andezhath Susheela, of the Fluorosis Research and Rural Development Foundation in Delhi. Quoted in Fred Pearce, "Wells That Bring Nothing But Ills," *Guardian* (UK), August 2, 1998. See also, Omer Farooq, BBC correspondent in Hyderabad, "Indian Villagers Crippled by Fluoride," *BBCi*, UK Edition, News Front Page News, April 7, 2003.
21. Roholm, *Fluorine Intoxication*, p. 297.
22. H. Ost, "The Fight Against Injurious Industrial Gases," *Ztschr. Agnew. Chem.*, vol. 20 (1907), pp. 1689–1693.
23. K. Roholm, "The Fog Disaster in the Meuse Valley, 1930: A Fluorine Intoxication," *J. Hygiene and Toxicology* (March 1937), p. 131.
24. "In the industrial smoke problem investigators have been interested mostly in the very frequent occurrence of sulfurous waste products . . . but little in fluorine," Roholm remarked. But fluorine compounds were much more toxic that the sulfur compounds, he explained, while "man is more sensitive to fluorine than other mammals." K. Roholm, "The Fog Disaster in the Meuse Valley, 1930: A Fluorine Intoxication," p. 126. Also, G. L. Waldbott, "Fluoride Versus Sulfur Oxides in Air Pollution," *Fluoride*, vol. 7, no. 4 (October 1974), pp. 174–176.
25. "The immense masses of soot and dust emanating from the works have served to promote condensation. Fluorine compounds must have been present in dissolved form in microscopic particles of water and consequently in a very active and easily absorbable form." He added, "It is quite probable that the affection from which these people suffered was an acute intoxication by gaseous fluorine compounds emanating from certain factories in the region." K. Roholm, "The Fog Disaster in the Meuse Valley," p. 126.
26. Ibid., p. 133.
27. H. Christiani and R. Gautier, *Am. Med. Legale*, vol. 94 (1926), p. 821. Cited in F. DeEds, "Chronic Fluorine Intoxication: A Review," *Medicine*, vol. XII, no. 1 (1933). Roholm, *Fluorine Intoxication*, pp. 38–39. P. Bardelli and C. Menzani, "Richerche sulla fluorosis spontanea dei ruminanti," *Ann. D'Igiene*, vol. 45

(1935), p. 399. For worker conditions, see A. W. Frostad, "Fluorforgiftning hos norske aluminiumfabrikkarbejdere," *Tiskr. F. Den norske Laegefor*, vol. 56 (1936), p. 179. Both cited in Roholm.

28. Roholm, *Fluorine Intoxication*, p. 37.
29. Roholm, "The Fog Disaster in the Meuse Valley," p. 136.
30. Roholm, *Fluorine Intoxication*, p. 310. "Physicians should be obliged to notify all diseases acquired while working with fluorine compounds. This is only practiced in USSR and Sweden, where all occupation diseases are notifiable." Roholm notes the Soviet practice approvingly: "In the labour legislation of the USSR great consideration is given to personnel working with fluorine compounds (shorter days, extra holidays, lower pension age, increased pension in the event of invalidity)." See, however, the probable unhappy fate of gaseous diffusion workers in Russia's nuclear program, in David Holloway, *Stalin and the Bomb* (New Haven, CT: Yale University Press, 1994), pp. 189–195.
31. Roholm, *Fluoride Intoxication*, p. 321. Drug reference at p. 311.
32. The Buhl foundation gives grants for education, economics, recreation, and social research. It was established in 1927 by Henry Buhl Jr., owner of Pittsburgh's Boggs and Buhl department store. Weidlein wrote to Charles Lewis, director of the Buhl Foundation, on March 25, 1935: "This investigation was in its origin a part of the Sugar Institute's Industrial Fellowship work but this phase of that problem is no longer related to sugar." Folder 8, Dental Study 1935, Box 32, Buhl Foundation Records, Library and Archives Division, Historical Society of Western Pennsylvania.
33. The estimate of Gauley Bridge deaths is conservative, according to Martin Cherniak's epidemiological study in his *The Hawks Nest Incident* (New Haven, CT: Yale University Press, 1986). For both the scale of the legal threat facing corporations and the key role of the Mellon Institute, see especially D. Rosner and G. Markowitz, *Deadly Dust: Silicosis and the Politics of Occupational Disease in Twentieth-Century America* (Princeton, NJ: Princeton University Press, 1991). For the essential obfuscatory and public relations role of the Mellon Institute in the silicosis debate, see also Rachel Scott, *Muscle and Blood* (New York: E. P. Dutton and Co., Inc., 1974).
34. John F. McMahon to Ray Weidlein, January 16, 1939, Carnegie Mellon Archives, cited in *Deadly Dust*, p. 107.
35. See chapter 4 of *Deadly Dust* for fuller description of Ray Weidlein's key leadership role in forming the Air Hygiene Foundation and shaping its agenda. The Foundation—renamed the Industrial Hygiene Foundation in 1941—would continue to exert a powerful corporate influence in the national debate over air pollution and occupational hazards, including a key early role in the Donora tragedy.
36. E. R. Weidlein, "Plan for Study of Dust Problems," cited in *Deadly Dust*, p. 108.
37. Paul Gross, Lewis J. Cralley, and Robert T. P. DeTreville, "Asbestos Bodies: Their Nonspecificity," *Am. Industrial Hygiene Assoc. J.* (November–December 1967), pp. 541–542.

38. An excellent discussion of the role of Paul Gross and the Mellon Institute in the asbestos story—including the dissent of his fellow scientists—can be found in Rachel Scott, *Muscle and Blood*, pp. 185–189.
39. For scale of asbestos damage awards, see *New York Times*, December 31, 2002, section C, p. 1. Further, recent big asbestos court trials, which have awarded huge sums to plaintiffs, have cited Industrial Hygiene Foundation documents.
40. Alcoa's Francis Frary sat on the membership committee, and the prominent fluoride attorney Theodore C. Waters was a member of the Air Hygiene Foundation's legal committee. An August 30, 1956, letter to Waters from Alcoa's attorney Frank Seamans illustrates their mutual interest in fluoride: "You will recall the occasion of our meeting together in Washington with a group of lawyers who have clients interested in the fluorine problem, at which time we were discussing the U.S. Public Health Service." Waters was also sent information on the 1953 Kettering Fluoride Symposium. See note attached to symposium program, in Kettering files, RAK Collection.
41. Dr. Paul Bovard, "Radiologic Considerations," Symposium on Fluorides, May 13, 1953, paper, p. 2, in Kettering Institute, RAK Collection.
42. G. D. Smith, *From Monopoly to Competition: The Transformation of Alcoa* (New York: Cambridge University Press, 1988), pp. 165 and 175.
43. Russell D. Parker, "Alcoa, Tennessee; the Early Years, 1919–1939," *The East Tennessee Historical Society*, vol. 48 (1946), p. 88. Also, "It was in the hot potrooms of the South Plant—in the smelting or reduction process—that blacks were to be employed on a permanent basis." Smith, *From Monopoly to Competition*, p. 176. Conditions at Massena were so horrendous for workers, and management was so indifferent to their fate, that one young MIT graduate, Arthur Johnson, quit in disgust, he told Smith. Also, in May and June 1948, scientists from the Kettering Laboratory at the University of Cincinnati discovered serious injury and disability in Massena workers. The factory had been producing aluminum since 1912. The investigators confirmed just how dangerous the Alcoa plant had long been. "There can be no doubt that hazardous exposure to fluorides is (and for years has been) present," stated a scientist for Kettering, Dr. William Ashe. He studied 128 men in the "pot" room where the aluminum was smelted: "The most outstanding characteristic of this group," Ashe reported, "is the occurrence of 91 cases of fluorosis of the bone." At least thirty-three of these X-rayed workers "showed evidences of disability ranging in estimated degree up to 100 percent," Ashe concluded. His findings paralleled Kaj Roholm's study of cryolite workers in Denmark. Serious tooth decay, gum disease, and heart problems were common in the Alcoa workers; the scientists added that "an abnormal amount of lung fibrosis among the employees of the pot room was found." Also, "one sees hypertrophic changes in bone along the shafts of the long bones, along the crests of the ilia, the ribs, and the rami of the ischium, in the form of stalagmite-like excrescences which appear similar to changes seen in experimental animals with bone fluorosis. The

interosseous membranes are often ossified. These changes, in no way related to arthritic processes, are believed to be due solely to fluorosis and to indicate that changes about joints may be expected in this disease. Therefore, when one finds, in cases of severe fluorosis of the bone, limitation of motion of the elbow and the X-ray reveals exostoses of unusual density about the elbow, one is probably entirely justified in concluding that the deformity and dysfunction are due to fluorosis, and that disability exists in association with and because of this disease, whether or not the man is aware of it, and whether or not he continues to do his job at the plant." Aluminum Company of America, Niagara Falls Works Health Survey, p. 13, File 4, Box 82, RAK Collection. The Kettering team included the scientist William F. Ashe, who five months later would lead the confidential Kettering investigation of the Donora air pollution disaster. Ashe would receive secret autopsy blood tests from Donora victims, performed by Alcoa, showing high levels of fluoride.

44. The membership of committees of the National Research Council is a guide to some of these relationships: Both Frary and Kettering were members of a Joint Committee, for example, representing the NRC's Science Advisory Board, advising on railway policy. Other members were Frank Jewett, vice president, AT&T; E. K. Bolton, chemical director, DuPont; John Johnston, director of research, U.S. Steel; and Isaiah Bowman, chairman of the NRC and director of the American Geographical Society. Charles Kettering papers, Office Files, Box 96, 87-11.2-296b, and 296f, Scharchburg Archives.
45. Frary was also a poison gas expert, making phosgene poison for the Oldbury Chemical Company in Niagara Falls, before working for the U.S. Army during World War I and then joining Alcoa. See G. D. Smith, op. cit. Also, Margaret B. W. Graham and Bettye H. Pruitt, *R & D for Industry: A Century of Technical Innovation at Alcoa* (New York: Cambridge University Press, 1990).
46. F. DeEds, "Chronic Fluorine Intoxication—A Review," *Medicine*, vol. XII, no. 1 (1933). On industry, F. DeEds: "The possibility of a fluorine hazard should, therefore, be recognized in industry where this element is dealt with or where it is discharged into the air as an apparently worthless by-product. For instance it has been shown by Cristiani and Gautier that the gases evolved at aluminum plants, using cryolite as a raw material, contain sufficient quantities of fluorine to cause an increased fluorine content of the neighboring vegetation, and that cattle feeding on such vegetation develop a cachetic condition," p. 2. His reference is to H. Cristiani and R. Gautier, *Am. Med. Legale*, vol. 6 (1926), p. 336. Further, DeEds calculated that each year 25,000 tons of pure fluorine was "pouring into the atmosphere" from the U.S. superphosphate fertilizer industry alone. He was concerned about where all the fluorine added to soil as phosphate fertilizer ended up. "Assuming an average fluorine content of 4 percent for phosphate rock, and that 75 percent of the fluorine remains in the superphosphate used as fertilizer, it is seen that 90,000 tons of fluorine are being added annually to the top soil. This sizeable quantity gives pause for thought of the potential toxicities concerned therewith." DeEds did not

include the 1933 report of thickened bones in Danish cryolite workers, by P. F. Møller and Sk. V.Gudjonsson, which prompted Roholm's massive study and determination of fluorine intoxication. P. F. Møller and Sk. V. Gudjonsson, "A Study of 78 Workers Exposed to Inhalation of Cryolite Dust," *J. Ind. Hyg.*, vol. 15 (1933), p. 27.

47. One of those studies had been done by Alcoa's H. V. Churchill, who found dental mottling and high levels of fluoride in the well water of Bauxite, Arkansas. Churchill's study was reported in 1931, the same year H. Velu in North Africa and the Smiths in Arizona made the same discovery. (Very curious are the apparently unsuccessful efforts by "Pittsburgh interests" to fund the Smith study in Arizona. That fragmented history is related in McNeil, *The Fight for Fluoridation*, p. 31.) H. Velu, "Le Darmous (oudermes)," *Arch Inst. Pasteur d'Algerie*, vol. 10, no. 41 (1932).
48. "As requested in your letter of June 8th, we have questioned three of our local dentists as to the prevalence of cases of mottled enamel in Massena. All of the dentists stated that they have treated such cases here." Exchange of letters between V. C. Doerschuk, Massena Works, and H. V. Churchill, Aluminum Research Laboratories, June 1931, in Alcoa letters, McNeil Collection, Wisconsin Historical Society.
49. See exchange of letters between H. V. Churchill and C. F. Drake of the City of Pittsburgh Bureau of Water, June 1931. Drake had noted the "Pittsburgh spasmodic fluorine content which appears to have no explanation." He informed Churchill that "an industrial plant not far from New Kensington had been discharging fluorine in the Allegheny River. The officials of that plant discontinued such discharge when requested." Several glass and steel plants were in the vicinity of New Kensington. H. V. Churchill responded, tellingly, "the presence of fluorine in water is apparently not necessarily proof of industrial contamination since it occurs in small amounts in so many water supplies." (In Alcoa letters, McNeil collection, Wisconsin Historical Society.) In 1950, Alcoa was fined for dumping fluoride waste at Vancouver, Washington, into the Columbia River, *Seattle Times*, December 16, 1952. (Cited in Waldbott et al., *Fluoridation: The Great Dilemma* (Lawrence, KS: Coronado Press, 1978), p. 296.)
50. The following decade, an English scientist, Margaret Murray, would call similar dental mottling found near an aluminum smelter in the United Kingdom "neighborhood fluorosis." M. Murray and D. Wilson, "Fluorine Hazards," *Lancet*, December 7, 1946, p. 822. Referring to studies near an aluminum factory in Scotland, they wrote, "In the same part of Inverness-shire we found that the local water supply had a very low fluorine content (0.2 ppm), but we observed "moderate" dental fluorosis in the milk teeth of young children whose homes lay within the district contaminated by vapours from the factory chimneys. Such a condition in the temporary dentition is usually associated with a high maternal intake of fluorine. Children using the same water, whose homes lay outside the affected area, did not show the mottled enamel."

Mottled teeth in children in the factory town of Donora, Pennsylvania, in 1948 was also blamed by Philip Sadtler on fluoride smoke and fumes (author interview), an association that was confirmed around the country by the U.S. Department of Agriculture (USDA) in 1970. The USDA report states: "Where ever domestic animals exhibited fluorosis, several cases of human fluorosis were reported, the symptoms of which were one or more of the following: dental mottling, respiratory distress, stiffness in the knees or elbows or both, a skin lesion, or high levels of F in teeth or urine [six references cited]. Man is much more sensitive that domestic animals to F intoxication." R. J. Lillie, "Air Pollutants Affecting the Performance of Domestic Animals. A Literature Review," Agricultural Research Service, *U.S. Dept. Agric. Handbook,* no. 380 (Washington, DC, August 1970).

Mottling was also seen in children living near DuPont's wartime fluoride operation at Penns Grove, New Jersey. A scientitst active on the Manhattan Project, Harold Hodge, was quick to blame fluoride in water supplies. Roholm reported dental mottling in the children of fluoride workers. Their mothers had transported it from the workplace in breast milk. See *Fluorine Intoxication,* p. 199. The Cornell veterinarian Lennart Krook also sent me photographs of mottled teeth from children on the Akwesasne Mohawk reservation, near the Reynolds aluminum smelter in upstate New York.

51. The notoriously close-knit international aluminum industry could follow accounts of litigation following World War I, which alleged fluoride damage outside an aluminum smelter in Switzerland. They could read the slew of new medical information about chronic health effects, summarized by DeEds. Or they could look inside their own factories. A 1932 study published in English had found "fluorosis" in cryolite workers in Denmark. (P. F. Moller and Sk. V. Gudjonsson, "Massive Fluorosis of Bones and Ligaments," *Acta radiol,* vol. 13 [1932], p. 269.) Sickness was reported in a Norwegian aluminum smelter in 1936: A. W. Frostad, "Fluorforgiftning hos norske aluminiumfabrikkarbejdere," *Tiskr. F. Den norske Legefor,* vol. 56 (1936), p. 179. The following year an investigation at DuPont found "high" fluoride levels in workers' urine. (Letter from Willard Machle, MD, of the University of Cincinnati to Dr. E. E. Evans, Dye Works Hospital, Penns Grove, New Jersey, December 28, 1937, Du Pont file, Kettering Papers, RAK Collection.) And a confidential 1948 study of Alcoa's plant at Massena, New York, confirmed that horribly crippled workers were the result of a fluoride dust hazard that had existed for years. Alcoa may also have faced liability in its flurospar mines. The Franklin Fluorspar Company was an Alcoa subsidiary (see *Mellon's Millions, The Biography of a Fortune: The Life and Times of Andrew W. Mellon,* by Harvey O'Conner [New York: Blue Ribbon Books, Inc., 1933], p. 390). Fluorspar miners in Hardin County, Illinois, wrote to Alice Hamilton about their plight. See *Deadly Dust,* 80, fn 10: D. Rosner and G. Markowitz. The entire issue of how much fluoride contributed to industrial silicosis, or how much fluorosis was misdiagnosed as silicosis, is beyond the scope of

this book. Fluoride was widely used in the foundry place and is found in much mineral ore.

52. By the end of 1935 Gerald Cox's tooth study at the Mellon Institute was not going well. Despite the spring press release trumpeting the imminent discovery of a "factor" preventing decay, Cox's data still "did not reveal any positive effects," he stated in a confidential memo to the Institute's director, Ray Weidlein. On March 24, 1936, almost a year after his Buhl Foundation study had begun, Cox reported to Weidlein that feeding a milk extract, known as XXX liquor, to rats had failed to find the positive results claimed in the previous year's press release. "The data at that time did not reveal any positive effects," Cox told Weidlein, and required therefore "intensive work to re-score all of our sets of teeth. With the new and discriminating system, we have been able to show some positive effects." In April 1936, following Francis Frary's September 1935 suggestion that fluoride had a role in dental health, Cox announced to his Buhl Foundation sponsors that he was proposing to "investigate the effects of dietary fluorine on caries susceptibility." See Mellon Institute Special Report, April 6, 1936, "A study of Tooth Decay," marked *Confidential.* Cox later claimed, somewhat confusingly, that the XXX liquor had contained enough fluorine "to explain the beneficial effects of the early experiments in which it was fed to the mothers." Buhl Foundation Records, Box 33, Folder 7, Dental Study 1936, Library and Archives Division, Historical Society of Western Pennsylvania.
53. The letter linking Alcoa's Francis Frary to Gerald Cox's historic suggestion that fluoride was responsible for good teeth was found in McNeil's personal papers. Cox to author Donald McNeil, August 19, 1956. "The first time I ever gave fluorine a thought was in answer to a question of Dr. Francis C. Frary, who was at that time and until about three or four years ago, Director of Research at Alcoa. He asked if our finding,—I was the speaker in the September 1935 meeting of the Pittsburgh Section of the American Chemical Society—of less caries in rats from mothers on XXX liquor could be due to fluorine." File ADA 53–56, McNeil Papers, Wisconsin Historical Society.

    Whether this is indeed the first time Cox wondered about the usefulness of fluoride in preventing tooth decay is not clear. It is clear, however, that the aluminum industry had been mulling the idea for a while. In the 1931 letter to C. F. Drake, cited above, H. V. Churchill of Alcoa stated that fluorine in low doses "may be positively beneficial."
54. E. R. Weidlein, *Ind. Eng. Chem.*, News Ed., vol. 15 (1937), p. 147. See also G. J. Cox, "Experimental Dental Caries. I. Nutrition in Relation to the Development of Dental Caries," *Dental Rays*, vol. 13 (1937), pp. 8–10, and "Discussion," *JAMA*, vol. 113 (1938), p. 1753.
55. Cox et al., "Resume of the Fluorine-Caries Relationship," *Fluorine and Dental Health*, Publication of the American Association for the Advancement of Science, no. 19 (1942): "The first experimental results, using sodium fluoride were obtained in August 1937."

56. P. C. Lowery to C. F. Kettering, April 25, 1936, filed by letter and year, Office Files, Personal Correspondence, Scharchburg Archive.
57. DuPont had become so wealthy selling munitions during World War I that the company had bought a controlling interest in General Motors. The giant enterprise was only pried apart in the 1950s, following federal antitrust action.
58. D. Rosner and G. E. Markowitz, *Deceit and Denial: The Deadly Politics of Industrial Pollution* (Berkeley: University of California Press, 2002).
59. "Organized Opposition . . . Particularly by the American Standards Association and the New York City Fire Department," Report on Operations of Kinetic Chemicals, Inc., from 1930 through 1943, p. 15. Including "History of Development of Fluorine Chemicals from 1928 through 1930," for presentation to the General Motors Policy Committee, by Donaldson Brown. Prepared by E. F. Johnson and E. R. Godfrey, October 1944. Files of Charles Kettering, Scharchburg Archive.

    Also, "Freon . . . coming in contact with open flames will decompose and you get a certain amount of fluorine and a certain amount of chlorine, and you also, just by happen-stance, get a slight amount of phosgene." Direct examination of DuPont director Willis Harrington, chairman of Kinetic Chemicals. *United States* vs. *DuPont*, Civil Action No. 49 C-1071, p. 3922 (U.S. District Court for the Northern District of Illinois, Eastern Division, 1953).

    There were other concerns, as well. The manufacture of Freon required huge quantities of the extraordinarily corrosive and toxic hydrofluoric acid, and "high" levels of fluoride were soon reported in DuPont workers' urine. Willard Machle, MD, of the University of Cincinnati to Dr. E. E. Evans, Dye Works Hospital, Penns Grove, New Jersey, December 28, 1937, DuPont file, Kettering Papers, RAK Collection.
60. Kehoe et al., "A Study of the Health Hazards Associated with the Distribution and Use of Ethyl Gasoline" (April 1928), from the Eichberg Laboratory of Physiology, University of Cincinnati, Cincinnati, OH, National Archives RG 70, 101869, File 725; cited in Rosner and Markowitz, *Deceit and Denial*, p. 313. Kehoe's essential hypothesis, that low levels of lead in blood were safe and normal, was undercut in the late 1960s by the scientist Clair Patterson of the California Institute of Technology, who examined polar ice and concluded that industrialization had greatly increased lead in the human environment. Kehoe's defense of lead safety was dealt a coup de grâce in the 1970s by Harvard's Herbert Needleman, whose studies with children showed lead to be far more toxic than Kehoe had claimed.

    For Kehoe's contribution to industry profitability, see L. P. Snyder, "'The Death Dealing Smog Over Donora, Pennsylvania': Industrial Air Pollution, Public Health, and Federal Policy, 1915–1963," 1994 PhD thesis available from University Microfilms. See especially chapter 5. Also, J. L. Kitman, "The Secret History of Lead," *The Nation*, March 20, 2000. See also chapter 8 of this book for further discussion of lead.

61. W. F. Ashe, "Robert Arthur Kehoe, M.D.," *Archives of Environmental Health*, vol. 13 (August 1966), p. 139. Cited in Snyder.
62. Ethyl had been established by Standard Oil and General Motors to market TEL.
63. "Studies of the Combination Products of Di-Fluoro-Dichloro Methane" and "Notes on the Toxicity of Decomposition Products from Dichlorodifluoromethane" in Kettering Unpublished Reports, vol. 1.d., RAK Collection. Kehoe dismisses the risk from phosgene, arguing that the presence of irritating HF acid would force prompt evacuation from the danger zone. He does not address the risk to firefighters or to subjects unable to flee the gases. "The only experimental situation which has been found to be responsible for the production of significant proportions of phosgene in the decomposition products of CCl_2F_2 was the result of rapid discharge of the refrigerant in high concentration, through the flame of an oil fire in an enclosed chamber—that is, the conditions were those of a conflagration. Situations which correspond to those which might develop from a leak in a home or building, are uniformly found to produce such relatively low concentrations of phosgene, that no amount of dilution of the decomposition products could eliminate the irritating and warning properties of the acids without eliminating the toxic effects of phosgene."

At a private three-day "Symposium on Fluorides" given for industry at the Kettering Laboratory at the University of Cincinnati in May 1953 Kehoe discussed details of secret human experiments he had performed to test Freon's toxicity for the U.S. government during World War II. He had used himself as one of the gas-chamber test subjects. (See: General Work on Project P.D.R.C. 377 (SECRET) for the Office of Scientific Research and Development, U.S. Government Washington, DC, 7-15-43, unpublished Volumes 1-d, RAK Collection.) Freon produced "unconsciousness after some minutes of exposure to concentrations of the order of magnitude of 11 percent or more," Kehoe recounted. He added, "As the subject of the experiments carried out at the higher concentrations, I was alarmed, fleetingly, at the point of rapid ebb of consciousness, being convinced that the observers outside the chamber were not aware of what was happening to me. Another subject, exposed to much lower concentrations, had considerably less assurance than I and became apprehensive and aggrieved . . . he became quite sure that we were exposing him to a risk which he felt we were concealing from him.

"I describe these as yet unpublished experiments," he told the gathered industry doctors, "since it is something you, as physicians, should know. It is believed, generally, that exposure to Freon 12 is of negligible importance, and that the material is quite harmless. The significance of the matter relates primarily to the repairman, who can get into situations involving the escape of the material from equipment into small enclosures. Such a workman may become unconscious and receive serious physical injury, or even be killed. *It is not true that this is a harmless material.*" Kehoe left unexplained why the repairman himself should not have the information on Freon toxicity. Several of the papers given at the symposium were later published. Kehoe's was not.

64. Kehoe died in November 1992, at the age of ninety-nine. An obituary in the *Cincinnati Enquirer*, November 29, 1992, noted that he had retired from the Laboratory in 1965.
65. W. Langewiesche, "American Ground," *The Atlantic Monthly* (July–August 2002), pp. 44–79. Also published in full as *American Ground: Unbuilding the World Trade Center* (New York: North Point Press, 2002).
66. Numerous and multiple phosgene injuries were reported as a result of chlorofluorocarbon decomposition by the Manhattan Project. Chlorofluorocarbons were used in massive quantities in the K-25 plant at Oak Ridge.

    Freon caused deaths and injuries in the home, too: "Dahlman encountered two [poisoning cases] resulting from heating fluorocarbons above the decomposition temperatures. In the first case, a mechanic operated with an acetylene torch on a refrigerator leaking Freon 12. He developed dyspnea, vomiting, and malaise and required hospital treatment for five days. In the second, an agricultural worker sprayed his bedroom with aerosol Freon fly spray. He then switched on the electric heater and went to bed. During the night he developed vomiting, diarrhea, and malaise and died on the following day." T. Dahlmann; *Nord. Hyg. Tidskr.*, vol. 39 (1958), p. 165. Cited in R. Y. Eagers, *Toxic Properties of Inorganic Fluorine Compounds* (Amsterdam and New York: Elsevier, 1969). (DuPont's New Jersey Chamber Works plant also was blamed for poisoning local farmers and workers with fluoride pollution in the 1940s.) The ozone-depleting gas was scheduled to be phased out by the 1987 Montreal Protocol.
67. One Kettering study monitored fluoride levels in DuPont workers' urine and confirmed that "these results have been high." Letter from Willard Machle, MD, of the University of Cincinnati to Dr. E. E. Evans, Dye Works Hospital, Penns Grove, NJ, December 28, 1937, Report on Operations of Kinetic Chemicals, Inc., from 1930 through 1943, p. 17, RAK Collection. Including "History of Development of Fluorine Chemicals from 1928 Through 1930," for presentation to General Motors Policy Committee, by Donaldson Brown. Scharchburg Archive.

    Freon sales again skyrocketed higher during World War II, with Freon used as a coolant in the K-25 gaseous diffusion plant and as a propellant in DDT antimalaria bug bombs.
68. W. Machle et al., "The Effects of the Inhalation of Hydrogen Fluoride. 1. The Response to High Concentrations. 2. The Response to Low Concentrations," *J. Industrial Hygiene*, vol. 16, no. 2 (1934), p. 129; and vol. 17, no. 5 (1935), p. 221.
69. The Advisory Committee on Research in Dental Caries (Daniel F. Lynch, chairman; Charles F. Kettering, counselor; and William J. Gies, secretary), *Dental Caries: Findings and Conclusions on its Causes and Controls. Stated in 195 Summaries by Observers and Investigators in Twenty-five Countries*, The Research Commission of The American Dental Association (New York, 1939).

70. P. C. Lowery to C. Kettering, Kettering Office Files 1937, "L", 87-11, 1-412, Scharchburg Archive.
71. "Armed with a letter from Dr. Weidlein of Mellon Institute to Mr. A. W. Mellon, he [Friesell] went to Washington to enlist the support of the Public Health Service. Mr. Mellon referred him to Surgeon General Cummings." Letter from H. V. Churchill of Alcoa to Dr. Frederick McKay of the Rockefeller Foundation, May 20, 1931, discussing the role of H. E. Friesell, dean of the University of Pittsburgh's Dental School. Alcoa Documents, Wisconsin Historical Society. Friesell sought to have naturally occurring dental fluorosis studied in Arizona, by University of Arizona scientists H. V. and Margaret Smith (far from the industrial centers of the East).

    See also the letter of August 6, 1930, from C. T. Messner of the Public Health Service to Friesell: "You are probably aware of the fact that the U.S. Public Health Service is a Bureau in the Treasury Department therefore, it might be advisable, especially as our Secretary is from your city, to also urge his endorsement of this program. The slightest interest on his part would influence the Service to a great degree in taking up this problem. I am sure you will hold this statement in strict confidence . . . after your letter is received here I will keep you advised as to how things are going along." File 9, Box 2, McNeil Collection, Wisconsin State Historical Society.

    The following year, in the spring of 1931, the same Captain C. T. Messner at the Public Health Service told H. Trendley Dean he would be studying mottled enamel. Dean stated that he was "assigned" to conduct the epidemiological studies that resulted in the key "fluorine caries hypothesis,"—the scientific basis for U.S. water fluoridation. See Don McNeil interview with Dean, May 3, 1955, in File 13, Box 2, McNeil Collection, Wisconsin State Historical Society.
72. How long Alcoa had known that fluoride produced dental mottling is not clear. (Alcoa was also concerned that the bad teeth in its company town of Bauxite would be linked to aluminum salts and further tarnish the public image of aluminum kitchenware. See McNeil, *The Fight for Fluoridation*, p. 27.) Perhaps it was coincidence that the Alcoa chemist H. V. Churchill's 1931 correlation of bad teeth with fluoride-contaminated well water in the company town of Bauxite appeared in the scientific press just weeks before separate studies confirming fluoride's link to mottled teeth were also published (by Smith and by Velu). What is certain, however, is that as soon as fluoride's links to mottled teeth were public knowledge, Alcoa privately confirmed that dental fluorosis was also found near its aluminum smelter in Massena, New York. See earlier note.
73. H. T. Dean, "Chronic Endemic Dental Fluorosis (Mottled Enamel)," *JAMA*, vol. 107 (1936), pp. 1269–1272.
74. "Ordered" and "hunch" quoted from Don McNeil interview with Dean, May 3, 1955. Dean told McNeil that in 1931, before he began his work, he "had a hunch" there would be fewer cavities in mottled teeth. McNeil Collection, Box 2, File 13. It is not known how Dean arrived at this hunch. Nor

is it known whether Dean had been ordered to "discover" some good news about fluoride. Of interest, however: the man who gave Dean his marching orders, the PHS's C. T. Messner, was the same official who, five years later, met in Detroit with the Freon gas magnate Charles Kettering. This meeting helped to produce the book *Dental Caries*, which also favorably introduced many dentists to fluoride. Indeed, Dean's "hunch" flew in the face of a study done at John Hopkins in 1925 by E. V. McCollum, who was hopeful that fluoride would strengthen teeth but had instead concluded that "the results showed, contrary to our expectations, that the ingestion of fluorine, in amounts but little above those which have been reported to occur in natural foods, markedly disturbs the structure of the teeth." E. V. McCollum, N. Simmons, J. E. Becker, and R. W. Bunting, *J. Biol. Chem.*, vol. 63 (1925), pp. 553–561.

75. H. T. Dean, "Endemic Fluorosis and Its Relation to Dental Caries," *Public Health Rep.*, vol. 53 (1938), pp. 1443–1452. Also H. T. Dean et al., "Domestic Water and Dental Caries," *Pub. Health Rep.*, vol. 56 (April 11, 1941), pp. 756–792. Dean was cross-examined in the 1960 *Schuringa* vs. *Chicago* lawsuit, to enjoin the city from fluoridating water supplies. According to the critic Dr. Richard G. Foulkes, Dean, under cross-examination by Mr. Dilling and aided by F. B. Exner, a radiologist and critic of fluoridation, was forced to admit that his early studies of Galesburg, Quincy, Monmouth, and Macomb and his later studies in twenty-one cities of 7,257 children, did not meet his own criteria of "lifetime exposure" and "unchanged water supply" and were, therefore, worthless. Dr. Exner prepared an "Analytical Commentary" on Dean's testimony. Exner "refers to the transcript and exhibits that show that not only were the basic criteria lacking in Dean's work, but also random variations found in both high and low fluoride areas cancelled out any 'benefits' that appeared in the high fluoride vs. lower fluoride cities,"according to Foulkes. *State of Wisconsin Circuit Court Fond Du Lac County Safe Water Association, Inc., Plaintiff,* vs. *City of Fond Du Lac, Defendant* Case No. 92 CV 579, Affidavit of Dr. Richard G. Foulkes in Support of Motion for Summary Judgment.

76. G. J. Cox, "New Knowledge of Fluorine in Relation to the Development of Dental Caries." *J. Am. Water Works Assoc.*, vol. 31 (1939), pp. 1926–1930. PHS regulations for 1939 stated, for example: "The presence of . . . fluoride in excess of 1 ppm . . . shall constitute grounds for the rejection of water supply." PHS, "Public Health Service Drinking Water Standards," *Public Health Rep.*, vol. 58 (1943), pp. 69–111 (at p. 80). A tenfold margin of safety required that fluoride in water be no higher than 1 part per million, water works engineers agreed. H. E. Babbitt and J. J. Doland, *Quality of Water Supplies in Water Supply Engineering.* 3rd Edition (New York: McGraw Hill, 1939), p. 454. Cited in Waldbott et al., *Fluoridation: The Great Dilemma*, p. 302.

Chapter 4

1. Richard Rhodes, *The Making of the Atomic Bomb* (New York: Touchstone, 1986). On p. 605 Rhodes quotes the French chemist Bertrand Goldschmidt, who wrote that the Manhattan Engineering District was "the astonishing American creation in three years, at a cost of $2 billion, of a formidable array of factories and laboratories—as large as the entire automobile industry of the United States at that date." On congressional secrecy, L. Groves, *Now It Can Be Told* (New York: Da Capo, 1962), p. 362.
2. Lt. Col. E. Marsden to Gen. Groves, December 3, 1943, Memorandum, "Obtaining of Information from C.W.S. on Phosgene, Fluorine, and Fluorine Compounds": "It is requested . . . for the Medical Section of the Manhattan District to be in full possession of all the information on phosgene, fluorine, and fluorine compounds that is presently in possession of the War Department." File EIDM D-2-b. MD 723.13 Memo to the Commanding General, Army Service Forces, Washington, DC, December 3, 1943, from Brigadier General L. R. Groves: "It is requested that Colonel Stafford L. Warren, M.C., be authorized to contact the Chief, Chemical Warfare Service, to obtain all information that may be available in the files of the Chemical Warfare Service . . . on the detection of, and protection against, phosgene, fluorine, and fluorine chemicals." EIDM D-2-a.
3. The enrichment factor was 1.0043. Rhodes, *The Making of the Atomic Bomb*, p. 340. At first, the K-25 plant produced only partially enriched uranium, which was further enriched at Eastman Kodak's Oak Ridge Y-12 plant and then transported as uranium tetrafluoride to Los Alamos. See also Rhodes, 552, 553, and 602.
4. Uranium hexafluoride quantities: "Considerable amounts of special fluorinated chemicals will be supplied to the K-25 plant," including "Uranium hexafluoride 33 tons per month—required by October 1944." See "Functions of Madison Square Area," Md 319.1, Box 26, Report Madison Square Accession #4nn 326-85-005, Atlanta FRC, RG 326. Also memo, "Storage Facilities at the Site For C-616," where Captain L. C. Burman, Corps of Engineers, notes a "2150 lb daily requirement" for hexafluoride. Md 3, Md 700, General Essays, Lectures, Box 34, Manhattan Engineer District Accession #4nn 326-85-005, Atlanta FRC, RG 326. Work force and power consumption: *AEC Handbook on Oak Ridge Operations* (1961), Oak Ridge Public Library.
5. Fresh air: University of Chicago, Metallurgical Laboratory, October 30, 1942, Memorandum to C. M. Cooper from R. S. Apple. Also, memorandum: "Medical Considerations of Work in the Pilot Plant, Philadelphia Naval Yard" from Col. Warren to Rear Admiral Mills, October 25, 1944. C-216 refers to the substance referred to as "fresh air." Md 702.1, Medical Exams Specimens, Box 54, Medical Considerations Accession #4nn 326-85-005, Atlanta FRC, RG 326.

   "Madison Square Area functions as the Materials Section of the Manhattan District to obtain special materials. The principal projects are the location, procurement and refining of uranium ore, preparation of uranium oxide, uranium hexafluoride and uranium metal, and production of fluorinated hydrocarbons." "Functions of Madison Square Area," Md 319.1, Box 26, Report Madison Square Accession #4nn 326-85-005, Atlanta FRC, RG 326. How well the fluoride secrets were kept, at least from foreign governments, is unclear. The Soviet spy Klaus Fuchs had worked on fluorine diffusion at the University of Birmingham in England and spent several crucial months in New York in 1944 with the British Diffusion Mission. He gave the Russians key details of the U.S. fluoride diffusion process, including information about the top-secret sintered nickel barriers through which the gas diffused. See Holloway, *Stalin and the Atomic Bomb*, p. 104.

6. See Rhodes, 494 for K-25 size and complexity. See L. Groves, *Now It Can Be Told* (New York: Da Capo, 1962), pp. 114–115 for corrosion and need to "condition" equipment. Also, at an October 23, 1942, presentation to the S-1 Committee of the OSRD, a precursor to the Manhattan Project, Mr. Z. G. Deutsch of the Standard Oil company, which was building a pilot centrifuge plant to separate uranium at Standard's Bayway refinery in Linden, NJ, stated, "All development work, toward a design of plant for the separation of our isotopes has visualized working with a single material—uranium hexafluoride." He added, "The principal objection to it is its extreme chemical reactivity." See *Manhattan District History*, Book I, vol. 4, chapter 14.
7. On October 19, 1943, top doctors from the Manhattan Project met in Captain John L. Ferry's Madison Square Area offices in New York. Harold Hodge from the University of Rochester was there. So were several doctors from Du Pont, Chrysler, and the Kellex Corporation, as well as the top medical officers for the Manhattan Project, including Col. Stafford Warren. Their secret agenda: "fluorine hazards to workers." Pure fluorine "would consume the skin and flesh," of exposed men, the doctors were warned. Ordinary protective clothing was "not satisfactory." A fluorine explosion would produce a terrifying mix of hydrofluoric acid and "oxygen fluorides." The acid burn might go undetected for twelve hours but would be followed by "extreme pain." Eventually the fluoride "penetrates to the bone, and then will spread along the bone and require amputation," the doctors were told. No one was then certain what the oxygen fluorides might do. Memo: Safety and Health Conference on Hazards of C-216 [code for F] October 19, 1943, Oak Ridge Records Holding Task Group Box 166 Building 2714-H, Vault, #82,761.

   See also, for UF6, Union Carbide Safety Bulletin No S-1, June 16, 1945. UF6 breaks down into HF and uranyl fluoride [UO_2F_2]. The latter, the bulletin notes, "has an action both as a surface irritant and as a poisonous agent acting internally." "When inhaled as a fine dust or fume, it readily goes into solution on the moist linings of the respiratory tract from which it is readily absorbed . . . all of the UO_2F_2 absorbed from any surface is eliminated by the kidneys, which causes kidney damage." "Deep penetrating burns" were produced by surface skin exposure to hydrolysis products, HF and UO_2F_2, Safety Reports, Bulletins, Box 55, Accession #4nn 326-85-005, Atlanta FRC, RG 326.

8. “Prior to the existence of the District, elemental fluorine was a laboratory curiosity.” *The Manhattan District Official History*, p. 3.13, Book 1 General, vol. 7, Medical Program. For most reactive element, R. E. Banks, “Isolation of Fluorine by Moissan: Setting the Scene,” *J. Fluorine Chem.*, vol. 33 (1986), pp. 3–26. For action on steel, above reference, “Memo: Safety and Health Conference on Hazards of C-216” [code for F], October 19, 1943. “Mild steel valves and pipes have been used [to handle fluorine] but it seems that any impurity or foreign substance in the pipe or valve may be the activating agent to start a reaction. Dr. Benning [from Du Pont] exhibited a steel valve . . . which had been consumed by action of C-216. The heat generated by the reaction is tremendous and a considerable flash hazard is present as the reaction is almost instantaneous.”
9. These companies and their roles are described in greater detail in *The Manhattan District Official History*, Book 1, General, vol. 7, Medical Program.
10. The liquid was named after Professor Joseph Simons of Penn State University, who invented a process known as “electro-chemical fluorination,” which used electricity to replace the hydrogen with fluoride in hydrogen-carbon bonds, producing fluorocarbons. (After the war the technology would be licensed to the 3M corporation, which would use it to make, among other things, the fabric protector Scotchgard. See chapter 17.) See J. H. Simons, ed., *Fluorine Chemistry*, vol. 1 (New York: Academic Press, 1950), p. 423.
11. H. Goldwhite, *J. Fluorine Chem.*, vol. 33, p. 113.
12. See “Report on the Fluoro Carbon work” by Harold Urey, September 26, 1942, S-1 files. Further, see Goldwhite. See also *Industrial and Engineering Chem.*, vol. 39, no. 3, p. 292.
13. For example, 35,000 pounds a month of “polytetrafluorethylene” (Teflon); 1,600,000 pounds of “hexafluorxylene”; and 1,400 lbs of “fluorinated lubricating oil.” For delivery schedule of fluorocarbons, see “Functions of Madison Square Area,” Md 319.1, Report Madison Square, Box 26, Accession #4nn 326-85-005, Atlanta FRC, RG 326.
14. Rhodes, *The Making of the Atomic Bomb*; p. 494. Dick Powell author interview; and also Goldwhite, *J. Fluorine Chem.*, above reference.
15. Groves, *Now It Can Be Told*, p. 8.
16. The plant was built in the basement of the Schermerhorn Laboratory in January 1943. Rhodes, *The Making of the Atomic Bomb*, p. 494.
17. “Initiation of Medical Program for Project at Columbia University,” Friedell to the District Engineer, U.S. Engineer Office, Manhattan District, January 20, 1943.
18. Capt. John Ferry to Col. Stafford Warren, November 10, 1943; and Capt. John Ferry to the Area Engineer, Columbia Area, July 14, 1944. “It would be difficult to prove that his illness had not been aggravated by his fume exposure,” Ferry concluded in Spelton’s case. Illness of Mr. Christian Spelton, Md 726.2, Occupational Diseases, Box 55, Accession #4nn 326-85-005, Atlanta FRC, RG 326. For pulmonary fibrosis as symptom, see Roholm, *Fluorine Intoxication*, p. 150.

19. On teeth falling out, see New York Operations Research and Medicine Division, Correspondence 1945–1952, Box 28–47, Box 36 ,"Du Pont File," Atlanta FRC, RG 326. For Priest's fluorine work at Columbia, see *Industrial and Engineering Chem.*, March 1947.
20. Princeton account at Md 319.1, Ferry Report Medical, Box 25, Accession #4nn 326-85-005, Atlanta FRC, RG 326. For Iowa State, case of Max Rankin see Md 702.1, Medical Exams Specimens, Box 54, Accession #4nn 326-85-005, Atlanta FRC, RG 326. For case of Dr. Oscar N. Carlson, report of Allan P. Scoog: Carlson had worked at Ames since 1943. He worked with beryllium fluoride. Multiple hospitalizations were followed with a diagnosis of "diffuse fibronodular pathologic process throughout both lungs . . . occupational fibrosis." *Medicine, Health and Safety, Beryllium*, July 1951–December 1951. NARA II. The gassed Purdue researchers had lung injuries resembling those in soldiers exposed to the World War I poison gas phosgene. Capt. John Ferry to Col. Stafford Warren, May 22, 1944. Also, Capt. John Ferry to Col. Stafford Warren, June 23, 1944, Md 319.1, Report Medical Ferry, Box 25, Accession #4nn 326-85-005, Atlanta FRC, RG 326.
21. Memorandum to Col. Warren from Capt. John Ferry, November 15, 1943, "Visit to DuPont": "The prevailing opinion is that the irritating properties of the HF also formed, will not be a satisfactory guide against the toxicity of the oxyfluoride." "DuPont" File, New York Operations Research and Medicine Division, Correspondence 1945–1952, Box 28–47, Box 36, Atlanta FRC, RG 326; "DuPont," Box 14, Accession #72C2386, Atlanta FRC, RG 326.
22. Memo to Col. E. H. Marsden from Col. Warren. January 6, 1945, "Safety of Operations at S-50," C-616, Box 28, Accession #72C2386, Atlanta FRC, RG 326.
23. "One is impressed," noted Captain John L. Ferry, senior medical officer for the Madison Square Area, "by the similarity between these cases and persons dying from work in beryllium plants." He reminded his boss that "one explanation of the beryllium deaths was that they resulted from exposure to beryllium oxyfluoride." Capt. Ferry to Col. Warren, February 2, 1944. "Fatalities Occurring from a By-Product of T.F.E.," Md 729.3, Safety Program Protection Against Hazards, Book 1, 6/25/42–7/31/44, Box 55, Accession #4nn 326-85-005, Atlanta FRC, RG 326.
24. Dr. G. H. Gehrman, DuPont Medical Director, to Capt. Ferry, May 5, 1944. Md 319.1, General Essays, Lectures, Medical Report, Box 34, #4nn 326-85-005, Atlanta FRC, RG 326.
25. Capt. Ferry to Col. Warren, February 2, 1944. "Fatalities Occurring from a By-Product of T.F.E." Also, DuPont was reluctant to have the government test "their own commercially developed material since several of the components thus far identified give good promise for commercial uses other than that contemplated here." District Engineer Ruhoff to Dr. H. T. Wensel, Clinton Engineer Works, March 30, 1944, Documents 366 and 367, RG 227.3.1.
26. Capt. Ferry to Col. Warren, February 2, 1944. "Fatalities Occurring from a By-Product of T.F.E."

27. Richard Powell, “Fluorine Chemistry: The ICI legacy,” in *Fascinated by Fluorine* (Amsterdam and New York: Elsevier, 2000). He quotes the visiting ICI scientist J. H. Brown on p. 345.
28. Kramish, A., “They Were Heroes Too,” *Washington Post*, December 15, 1991.
29. The secret facility was a pilot version of the massive S-50 “thermal diffusion” factory being readied at Oak Ridge, Tennessee. The plant at Oak Ridge appears also to have presented considerable fluoride health risks to workers, according to the official history. Each time a new hexafluoride cylinder was attached to the S-50 equipment, “the danger of breathing UF6 and of being burned by it in this operation is considerable.” *The Manhattan District Official History*, Book 1, General, vol. 7, Medical Program, p. 3.22.
30. Conant had responded to Col. Warren’s request for information, sending him reports on fluoride from the Chicago Toxicity Laboratory, and OSRD Report #3285 “The Toxicity of Compounds Containing Fluorine.” Conant to H. T. Wensel, October 6, 1943, RG 227.3.1, Document #0398, and Ruth Jenkins (Conant’s secretary) to Wensel, February 15, 1945, RG 227.3.1, Document #0341. Conant sought to keep specialized information about fluoride out of scientific journals during the war. He wrote to the editor of the *Journal of the American Chemical Society*, Arthur Lamb, on December 29, 1943: “I should appreciate it if you would send any papers concerning fluorine compounds to me if they are submitted in the future, and I will try and get in touch with the more conservative reviewers.” Document 0114, RG 227.3.1.
31. TDMR-628 (Technical Division Memorandum Report from Edgewood Arsenal) cited p. 20, OSRD Report 3285. “Among the effects noted were photophobia, headaches . . . as well as difficulty and pain in accommodation.”
32. Low concentrations of the organic compound cited produced “marked weariness, very strong mental depression, reluctance for any physical effort. Quite distinct periods of nervous irritation difficult to control, followed by periods of physical and mental exhaustion, drowsiness and giddiness.” Sporzynski Y.5682 May 5, 1943, cited in OSRD Report 3285, p. 37.
33. E. C. Andrus, D. W. Bronk, G. A. Carden Jr., et al., eds., *Advances in Military Medicine*, 2 vols. (Boston: Little, Brown, 1948), p. 561. See also *Fascinated by Fluorine*, p. 347; and author interview with former Imperial Chemical Industries (ICI) scientist Dick Powell.
34. J. Conant to Dr. H. T. Wensel, Clinton Engineering Works, October 6, 1943. RG 227.3.1 Document 0398.
35. Author interview, July 27, 2003.
36. Capt. Joe Howland, “Studies on Human Exposure to Uranium Compounds,” in Harold Hodge and Carl Voegtlin, eds., *Pharmacology and Toxicology of Uranium Compounds, with a Section on the Pharmacology and Toxicology of Fluorine and Hydrogen Fluoride* (New York: McGraw Hill, 1949), p. 1005. The official history of the Manhattan Project, like Gen. Groves, gives a conflicting account of the disaster. On p. 5.3, Book VI, Section 5, it states only that “several persons were injured.” However, Book I, vol. 6, p. 3.19 notes that,

"Douglas P. Meigs died as a result of burns due to steam. The Aetna, insurance carrier for Fercleve [the contractor], was not permitted to investigate the cause, nor the scene of the accident, but was permitted to make a routine dependency investigation. After complete facts were available to the Insurance Section, the insurance carrier was instructed to make payment as awarded to Meigs's widow by the Bureau of Workmen's Compensation, State of Pennsylvania."

37. A. Kramish, "They Were Heroes Too," *Washington Post,* December 15, 1991. Kramish told me that the Manhattan Project officer, "Dusty" Rhodes was sent to silence the press. The *Philadelphia Record* may have gone to press before he arrived, Kramish thinks. The following morning the newspaper reported that two "specialists" had been killed in an accident. "Gas was released," the newspaper added.
38. Leslie Groves, *Now It Can Be Told* (New York: Da Capo, 1962), p. 121.
39. *Washington Post*, December 15, 1991.
40. At one beryllium factory in Ohio doing secret fluoride work for the Manhattan Project, skin lesions and a crippling lung disease called berylliosis produced an employee turnover rate of 100 percent each month. Captain Mears to Major Ferry, July 30, 1945. He reports on "chemical dermatitis . . . resulting from the fluoride compounds entering through a hair follicle, contaminating a wound, or through a puncture wound by a sharp crystal. In these cases a papule develops slowly with some of the lesions ending in ulceration taking months to heal. Some of the workmen's hands and forearms are covered with inflamed hair follicles, papules, and depressed sharply circumscribed scars." Md 319.1, General Essays, Lectures, Medical Report, Box 34, Accession #4nn 326-85-005, Atlanta FRC, RG 326.
41. "Never before had such quantities of elemental fluorine gas been handled daily," wrote a Manhattan Project doctor, Herbert Stokinger, who saw the daily health risk to American workers. "Continuous exposure to low concentrations from unavoidable losses from the equipment was a source of considerable concern," he added. Hodge and Voegtlin, eds., *Pharmacology and Toxicology of Uranium Compounds*, p. 1024.
42. "Fluorine: Precautions to be Observed in Handling, Shipping and Storage." *Manhattan Project Official History,* Occupational Hazards, Book 1, General.
43. Herbert Stokinger reported that animal deaths were seen in laboratory experiments at 0.3-mg/cu m for fluorine. Hodge and Voegtlin, eds., *Pharmacology and Toxicology of Uranium Compounds*, p. 1033. Also, "The toxicity of oxyfluorides occurring from the liberation of fluorine in the atmosphere" was given a high priority for research. Memo to Col. Warren from Capt. John L. Ferry, November 29, 1943, Md 3, Md 700, General Essays, Lectures, Medical Reports, Box 34, Manhattan Engineer District Accession #4nn 326-85-005, Atlanta FRC, RG 326.

    That toxicity data, only declassified in 1994, is truly spectacular. While exposure of laboratory animals to 0.5 parts per million of pure fluorine for thirty days was considered "safe," a similar, microscopic quantity of oxygen fluoride "was lethal after 14 hours," the scientists reported. See "*Detailed Duties of Harold Hodge*," list of "problems" and "results" encountered by the Rochester Division of Pharmacology and Toxicology. Folder 2, Box SOFO1B219, ACHRE, RG 220; also, C-212 [code for oxygen fluoride]—1 ppm killed all animals (rats and mice), in "Toxicity of C-616, C-212 and C-216" 'Memo to Files' by Capt. B. J. Mears. Medical Crops, Md 3, Md 700, General Essays, Lectures, Medical Reports, Box 34, Manhattan Engineer District Accession #4nn 326-85-005, Atlanta FRC, RG 326. (By comparison, this toxicity appears at least as bad as the World War I poison gas phosgene, which was also found in the bomb plants, as a result of the heating of Freon.) The Chemical Warfare Service had reported to Col. Stafford Warren that, when exposed to phosgene, "mice succumb to chronic exposure of one part per million.") Memorandum for the Files, "Subject: Survey of Phosgene Effects" by Stafford Warren, February 23, 1944, A2, Box 26, Accession #72C2386, Atlanta FRC, RG 326.

Although the Manhattan Project had given a high priority to the experimental investigation of such oxyfluorides, the official and published work does not mention the results—perhaps a worrisome omission, given that the scientists suspected that the compounds might be encountered in the vicinity of bomb plants. The standard text, *Pharmacology and Toxicology of Uranium Compounds*, edited by Carl Voegtlin and Harold Hodge, has no mention of oxygen fluoride. Also, for evidence that scientists suspected oxygen fluoride would be encountered by citizens and workers, after an industrial hygiene survey at Harshaw Chemical in Cleveland in May 1947, Rochester scientists reported that "the results are on the low side, since the efficiency of the sampling procedure we used is not too good for fluorine and oxyfluoride; if considerable quantities of these two gases were present in the air, we probably missed a part of them." See Pharmacology Report #558, The University of Rochester Atomic Energy Project, Box S09F01B227, ACHRE, RG 220.

44. "HF is a protoplasmic poison with great penetrating power and causes deep-seated burns that heal very slowly. . . . When HF comes into contact with the skin, a burn results. If HF is not removed, it tends to keep penetrating with the production of a deep, slow-healing painful ulceration." Capt. John L. Ferry to Dr. Ralph Rosen, Kellex Corp, January 24, 1944, Md 729.3 Safety Program Protection Against Hazards, Book 1, 6/25/42–7/31/44, Box 55, Accession #4nn 326-85-005, Atlanta FRC, RG 326. Also, the chemical was described by Dr. Stokinger as "possibly the greatest single source of minor incapacitation of workers" in the bomb plants. Hodge and Voegtlin, eds., *Pharmacology and Toxicology of Uranium Compounds.*
45. Confirmed in author interview with the ICI fluoride scientist Dick Powell. Such "conditioning" was a massive industrial undertaking. The uranium hexafluoride gas was so corrosive that thousands of pumps, blowers, and piping first had to be treated with either chlorine trifluoride, or elemental

fluorine, leaving a thin film of fluoride on the machinery, protecting it from future corrosion. Joe Harding likened the process to "seasoning the interior of process equipment, like some people have heard of 'burning-in' an old cast iron skillet." Memorandum to Lt. David Goldring from Birchard M. Brundage 1st Lt., Medical Corps, July 13, 1945. "Subject: S-50 Medical Check-ups": Item I. "Definitions 1. Conditioning Area—Building in which parts and apparatus to be used in the Process Area are treated chemically before being placed in operational use." Item II. "Conditioning shop operator—handles the chemical preparation of equipment before it is handled for operations. He also cleans used equipment before its reconditioning and reuse in operations." Item III. "Hazard Classifications. 1. Most serious. A. Transfer room operator. B. Conditioning shop operator." S-50, Box 14. Also, "Conditioning of Equipment," *Manhattan Project Official History*, book VI, p. 5.17.

46. Joe Harding interviewed by Dolph Honicker, undated transcript supplied by his son, Cliff Honicker.
47. Stokinger et al., *The Enhancing Effect of the Inhalation of Hydrogen Fluoride Vapor on Beryllium Sulfate Poisoning in Animals*, UR-68 University of Rochester, unclassified, June 13, 1949. Also, "Fluoride materials are undoubtedly significantly more toxic from the standpoint of acute disease than any other beryllium material now being handled at the Luckey plant." Memo from Merril Eisenbud to W. B. Harris, 2/27/51, Box 3353, MHS 2 Beryllium, Germantown DOE History Archive. Eisenbud also estimated that 50 micrograms of beryllium—inhaled as beryllium fluoride—had "produced acute disease in three individuals" in just twenty minutes, and that "to produce injury by phosgene in a comparable period of time one would have to inhale approximately 50 milligrams!" Health Hazards from Beryllium, Merril Eisenbud, speech presented at a meeting of the American Society for Metals, Boston, March 1954. Document DOE #051094-A-312, ACHRE, RG 220.
48. For deaths: M. Eisenbud, "Origins of the Standards for Control of Beryllium Disease (1947–1949)," *Environmental Research*, vol. 27, no. 1 (February 1982). By June 1949 Robert Hasterlik, the top doctor at Argonne National Laboratory, reported about sixty death from beryllium. *Physics Today* (June 1949), p. 14.

    For sickness: "By far the greatest number of cases occurred in the fluoride handling operations," noted one government report on sickness at the Brush Beryllium Company in Lucky, Pennsylvania. Memo from Merril Eisenbud to W. B. Harris, stamped February 27, 1951: "Acute Beryllium Toxicity—Brush Beryllium Company—Lucky Experience," Div. Biology and Medicine, folder MHS 2 and Beryllium, Box 335, RG 326. At Brush Beryllium Plant in Lorain, Ohio, "In July, 1947, 24 percent of employees in the beryllium metal department were stricken with dermatitis or respiratory disease, compared to 6.4 percent for all other departments. The apparent increase in rates may possibly be explained by the shifts in production to pure metals as the result of AEC contracts." Bob Tumbleson, "Public Relations Problems in Connection with Occupational Diseases in the Beryllium Industry."

The Rochester Atomic Energy Project's Industrial Hygiene Section surveyed the Brush Beryllium Company in Lorain, Ohio, in December 1946 and found up to 64.1 mg/m3 fluoride, with particle sizes below 0.1 micron (a crucial factor in determining toxicity). "The authors conclude that the relatively high fluoride concentrations obtained in the surveyed areas are of particular significance since they may represent a hazard by themselves and also suggest a combined action with beryllium. Further study of this factor is suggested, especially near the beryllium fluoride furnace where the relative fluoride concentration was 1000 times that of beryllium." Bob Tumbleson, Public and Technical Information Service. "Public Relations Problems in Connection with Occupational Diseases in the Beryllium Industry," p. 18, Medicine, Health and Safety—Beryllium (1947–1948), RG 326.

49. Memo from Bob Tumbleson to Morse Salisbury, "Current Status of the Beryllium Problem," January 26, 1948. "Although the four neighborhood cases appeared at Brush in Lorain, the reporter from the Cleveland PRESS interviewed [AEC official] Wyndecker at Clifton, Painesville. . . . Wyndecker tried to quiet him by saying that a large part of their work was being done for AEC and hence was secret." RG 326 Medicine, Health and Safety—Beryllium (1947–1948) National Archive.
50. Turner reported: "Control experiments with electrolytic dust produced with fluorides, but in the absence of Beryllium, caused the same symptoms and mortality. It is evident, therefore, that electrolytic dust owes its toxicity primarily to the halogen radical [fluoride] and not to its content of Beryllium." Robert A. N. Turner, Resident Safety Engineer, Madison Square Area, Manhattan Engineer District, "The Toxicity of Beryllium and Its Salts," p. 2, "Oak Ridge Copy," Box 39, Accession #4nn 326-85-005, Atlanta FRC, RG 326.
51. Robert A. N. Turner, Resident Safety Engineer, Madison Square Area, Manhattan Engineer District, "Poisoning by Vapors of Beryllium Oxyfluorides," p. 1, "Oak Ridge Copy," Box 39, Accession #4nn 326-85-005, Atlanta FRC, RG 326.
52. See Rochester AEP, minutes of "The Second Progress Meeting on Beryllium Toxicity," February 5 and 6, 1947. Also, "The First Progress Meeting on Beryllium Toxicity": 0.5 mg/kg of "5BeO-7BeF2" killed rats, while 0.75 mg/kg of intravenous beryllium fluoride and beryllium oxyfluoride killed rabbits. "Injection of beryllium oxyfluoride . . . caused histologic damage to the kidney probably as a result of the fluoride moiety." (5.0 mg/kg BeSO4, beryllium sulfate, killed rats.) This meeting produced a crucial determination of a permissible limit of 1.5 mg *of beryllium compound* (underlined in original) per 10 m3 of air. By not specifying which compound, public notice was not made of the specific and more toxic nature of the fluoride compounds, it seems. Indeed, just days later, the head of the Rochester AEP, Herbert Stokinger, made a recommendation of 1.5 mg of beryllium per 10 m3 to the AEC for the "Maximal permissible Limit of Exposure to Beryllium." He does not mention nor cite the fluoride toxicity results but rather uses figures from the beryllium sulfate compound, which Rochester had

determined to be ten times less toxic. Stokinger adds, "The suggested level permits an easily attained limit both as regards ventilator and ventilating system." H. E. Stokinger to Fred Bryan, February 18, 1947, Rochester, 400.112 (Pharmacology) Beryllium, Box 48, New York Operations Office, 68F0036, Accession #4kr 326-83-010, Atlanta FRC, RG 326.

Also, researchers at Rochester and at the PHS did not find much toxic effect, chronic or acute, with pure beryllium, which fact allowed industry to deny that there was any great problem from beryllium poisoning. A hint at the agenda of the Rochester group and of Dr. Harold Hodge in particular comes from one of the leading scientists on beryllium toxicity, Dr. Harriet Hardy. "Those responsible for the medico-legal affairs of the AEC should consider the problem of the disability involved in the growing group of individuals with chronic beryllium disease," she writes and adds that "cases of chronic beryllium poisoning are being uncovered daily from a variety of remote and apparently slight beryllium exposures." However, Hardy writes, while "The chronic disease is certainly our most pressing problem, and at present the whole weight of the Rochester work, if I understood Dr. Hodge, is on the acute manifestation. . . . I cannot understand the defeatist attitude about producing chronic changes in animals with beryllium compounds sufficiently approximate to the human pathology." Dr. Hardy to Dr. Warren "Recent trips to Cleveland and Rochester," September 13, 1949, DOE Opennet #1153735.

"Thus, we have a kind of explosive action with the formation of fluorine in status nascendi," Turner stated. "Hence the deeper and most important, more prolonged action of this gas in comparison with that which we see following the inhalation not only of oxides of nitrogen and chlorine but also vapors of fluorine or hydrofluoric acid," p. 6 "The action of the fluorine in such conditions is especially strong and prolonged," Turner adds, "which in fact conditions the specificity of the picture of poisoning by Beryllium oxyfluoride." Robert A. N. Turner, Resident Safety Engineer, Madison Square Area, Manhattan Engineer District "Poisoning by Vapors of Beryllium Oxyfluorides."

53. Although the Maximum Allowable Concentration (MAC) for UO_2F_2 had been officially set by the government at 50 micrograms of uranium per cubic meter, nevertheless, "the lowest concentration of these compounds that will give a uniformly positive response in all animals has not been critically established." Hodge and Voegtlin, eds., *Pharmacology and Toxicology of Uranium Compounds*, p. 2203. Hodge's researchers produced renal injury in a dog at even 50 micrograms/cu m. Dogs were judged to have "unusual susceptibility." Also, Harold C. Hodge and Carl Voegtlin at the University of Rochester to Lt. Col. H. L. Friedell at Oak Ridge, April 26, 1945. Md 3, Md 700, General Essays, Lectures, Box 34, Manhattan Engineer District Accession #4nn 326-85-005, Atlanta FRC, RG 326.

54. "Uranyl fluoride is considered one of the most toxic uranium compounds," wrote Harold Hodge, *Pharmacology and Toxicology of Uranium Compounds*,

p. 33. Also, "It was envisioned that exposures of human beings to this compound would occur mostly by inhalation and almost solely to the fumes, UO_2FO_2 and HF, produced upon its release into the air. Such exposure might take the form of either accidental high concentrations for a relatively short time, possibly repeated several times during a month, or of low level, continuous exposures throughout the period of employment arising from the loss of small amounts of material from systems containing UF_6." *Pharmacology and Toxicology of Uranium Compounds*, p. 1492.

Dangerous levels of fluoride were quickly detected in K25 plant workers' urine. In the early summer of 1952, for example, almost 10 percent of employees tested had too much fluoride in their bodies, doctors reported. And the poisoning was getting worse, "the result of an increase in the magnitude and frequency of individual exposure to fluoride and fluorinated compounds," officials added. Of 535 workers, 58 tested. "Sanitized version of K-25 Plant Quarterly Report for Fourth Fiscal Quarter April 1–June 30, 1952," p. E-9, ORF1000605, Oak Ridge, DOE Public Reading Room.

55. Letter to Ralph Rosen of the Kellex Corporation, which built the K-25 plant. Ferry told Dr. Rosen it was "likely" that the concentrations of gas would be at or "near" the level set for chronic exposure. (However, the MAC for UO_2F_2 was then set at 150 micrograms per cubic meter. That level was reduced to 50 micrograms in 1948, although University of Rochester scientists found kidney damage in dogs at that level too; see note 53 above. No information was found on whether the conditions inside the cold trap chamber changed after 1948 as the MAC was raised.) Captain John L. Ferry to Dr. Ralph Rosen, Kellex Corporation, June 16, 1944. Safety Program Protection Against Hazards, Book 1, 6/25/42–7/31/44, Md 729.3, Box 55, Accession #4nn 326-85-005, Atlanta FRC, RG 326. A similar hazard faced workers at the Harshaw Chemical plant, who made uranium hexafluoride for shipment to Oak Ridge. "Workmen inhale 616 [code for hexafluoride] when disconnecting the receivers from the reactors," noted a report. "A cloud of hydrolyzed 616 escapes during this operation and is not entirely vented," the memo added. Memo from Capt. B. J. Mears, the Madison Square Area, October 11, 1945, to Captain Fred A. Bryan, Medical Section, Manhattan District, Oak Ridge Tennessee. Subject: Urinalysis on Harshaw Chemical Company Workers.
56. Tape-recorded interview with Joe Harding.
57. Several accounts mention the noise and heat inside the gaseous diffusion plant. An early report determining how long men could tolerate working in the "cells" notes temperatures of 118 degrees F and states that "Entrance into a cell which is in operation is a dramatic experience to the uninitiated, apt to be associated with some emotionalism. The noise within the cell might be responsible for part of the light headedness experienced, although the symptom is also recognized as a result of severe heat exposure." "Permissible Work Periods in Cells," Box 9, Accession #72C2386, Atlanta FRC, RG 326.
58. Accidents were frequent at K-25. For example, "On April 1, a release occurred in Building K-1004-A when a cylinder containing 2,559 grams of uranium

became overheated during a material transferring operation, releasing the entire contents of the cylinder." Sanitized version of K-25 Plant Quarterly Report for Fourth Fiscal Quarter April 1–June 30, 1952. In ORF 100605 Oak Ridge DOE Public Reading Room. Also, "On December 30th [1953] . . . a total of 2,506 pounds of uranium hexafluoride . . . was released [when a cylinder failed], contaminating all of Building K-27 . . . the gas was spread widely before the ventilating system could be shut down." K-959—Plant Quarterly Report for Second Fiscal Quarter, October 1–December 31, 1952, p. C-12, in ORF 18729, Oak Ridge, DOE Public Reading Room.

59. Another worker, Sam Ray of Lucasville, Ohio, told Congress in September 2000 that "Compressors would malfunction and process gas (UF6) would leak to the atmosphere. On one occasion, it was so bad that it looked like a fog moving up the mile long building. . . . We have had many small releases that were never reported, as well as documented large releases. Inside of the withdrawal room we had a major release. There were green 'icicles' hanging in the room from crystallize uranium hexafluoride." He also told them that "process gases were routinely vented to the atmosphere" and "fluorine gases from the plant stack area were frequent and resulted in numerous complaints from workers in the area, especially during temperature inversions." Compensation for Illnesses Realized by Department of Energy Workers Due to Exposure to Hazardous Materials. Hearing before the Subcommittee on Immigration and Claims of the Committee on the Judiciary House of Representatives 106th Congress, serial no. 132, p. 210.

60. Col. Stafford Warren to Dr. Fred Bryan, September 24, 1947, DOE stamp 000019, ACHRE, RG 220.

61. Report of Meeting of Classification Board During Week of September 8, 1947, Box SO9FO1B22, ACHRE, RG 220. See also handwritten letter in ACHRE files from an unnamed fluoride worker who worked at Portsmouth Gaseous Diffusion Plant in Piketon, Ohio, in the 1950s. He writes: "In the early years we used to talk about young people dying from cancer and leukemia that worked at the plant and wondered if it was due to working there." DOE document #000010, Box S0901B146, ACHRE, RG 220.

62. An additional five workers were poisoned by "fluorine analogs of phosgene," plant operators at Union Carbide claimed, caused by "pyrolysis of fluorocarbons and fluorolubes." Phosgene can be produced when Freon gas is exposed to very high temperatures. "Summary of K-25 chemical hazards," RHTG 101001, Box 219, RG 326. The document was only declassified in 1997. "Poisoning" was one of the health effects reported at K-25, along with respiratory irritation, burns, and dermatitis.

63. Work Report for June 1944. To: The Chief of the Medical Section, U.S. Engineer Office, Oak Ridge, Tennessee; From: 2nd Lt. Richard Tybout, Corp of Engineers, Medical Section. Document via Pete Eisler, *USA Today*.

64. "A distinct hazard does exist in Area C" that left the Atomic Energy Commission "very vulnerable," Kelly concluded. The government especially feared "pulmonary damage" in workers. While safety levels for uranium

hexafluoride had been set at 40 micrograms per cubic meter, tests showed that on September 30, 1944, dust levels in Area C were as high as 9,130 micrograms per cubic meter—228 times the official tolerance level. March 1, 1945, letter to Harshaw manager Fred Becker from Richard Tybout, 1st Lt. Corps of Engineers Medical Section, via Pete Eisler.

65. Roholm, *Fluorine Intoxication*, p. 26.
66. Analysis of the kidney tissue of one of the victims by the University of Rochester confirmed severe fluoride damage. "The pathological changes in the kidney are accounted for by the overwhelming dose of HF and the acute asphyxia." Capt. B. J. Mears to the District Engineer, Manhattan District, Oak Ridge (Attention: Major J. L. Ferry.) November 1, 1945. Oak Ridge Operations Records Holding Task Group. Classified Documents 1944–1994, RHTG document #38,658, OR0034167, Box 214, Vault, Bldg. 2714-H.
67. Rochester kidney report: "This report is of particular interest because (name redacted) was employed in the C-616 [uranium hexafluoride] plant and his duties required him to remove the receiver from the reactors. It is in this procedure that the employees come in contact with a cloud of PG [process gas] . . . he was exposed to C-616 to the same extent as any other single employee." Capt. B. J. Mears to the District Engineer, Manhattan District, Oak Ridge (Attention: Major J. L. Ferry.) November 1, 1945. Oak Ridge Operations Records Holding Task Group, classified documents 1944–1994, RHTG document #38,658, OR0034167, Box 214, Vault, Bldg. 2714-H.
68. P. Dale and H. B. McCauley, "A Study of Dental Conditions in Workers Exposed to Dilute and Anhydrous Hydrofluoric Acid in Production," December 31, 1943, File G-118, New York Operations Research and Medicine Division, Correspondence 1945–1952, Box 28–47, Atlanta FRC, RG 326.

    Also, on race: "Specimens showing large amount of T [code for uranium] are usually from the colored employees," noted an October 1945 memo from Manhattan Project Capt. B. J. Mears. "Because of their lack of personal responsibility," Mears complained, "this officer recommended that these specimens be collected before the employee starts to work." Of course, if workers gave urine specimens before their shift began, it would have the effect of measuring and recording lower levels of toxic exposure than they were actually receiving. Capt. Mears discriminated between the black workers and "employees who can be trusted." They were allowed to give urine at the end of their shift. Perhaps more importantly, those "trusted" workers, "consistently show T values well below 1 mg per liter." Memo from Capt. B. J. Mears, the Madison Square Area, October 11, 1945, to Captain Fred A. Bryan, Medical Section, Manhattan District, Oak Ridge Tennessee. Subject: Urinalysis on Harshaw Chemical Company Workers, via Pete Eisler.
69. P. Dale and McCauley, *J. Am. Dent. Assoc.*, vol. 37, no. 2 (August 1948), p. 132.
70. Fedor formed a union safety committee, then contacted the Ohio Division of Safety and Health and persuaded that office to do a study of conditions in

the fluoride plant. The state inspectors found fluoride levels as high as 6 and even 18 ppm. State regulation permitted 3 parts per million. In 1949 Fedor submitted the first motion to an American Federation of Labor national convention, seeking greater union involvement in occupational safety issues. Author interview, October 2001.

71. Despite multiple warnings from federal and state government, the industrial accidents, and pressure from John Fedor's safety committee, Harshaw's management seemed strangely unmoved. "Our plant hourly safety committee has been quite concerned about our HF problems, and I believe are exaggerating them, as I believe the hazards in Area C have been exaggerated," Vice President C. S. Parke wrote to the AEC official W. E. Kelly on February 3, 1948. "I speak somewhat as a layman, but we have manufactured HF fluorides for forty years. It is only lately that occupational disease has been suspected. Two of our men are reputed to have fluorosis, but nobody can tell us how this has harmed them. In fact, the inference by some doctors is that they have benefited. Certainly the situation is nothing to get alarmed at." C. S. Parke to W. E. Kelly, February 3, 1948. AEC document via Pete Eisler.

72. Secretly the government was intensely interested in the medical fate of the Area C workers. When the plant finally closed in 1952, AEC doctors proposed covertly "keeping tabs" on former employees—without letting the men and women know why they were being watched. "The ultimate objective is to determine the incidence of lung cancer . . . to justify the current M.A.C.'s [maximum allowable concentrations in the other AEC plants]," Dr. Roy E. Albert, the Assistant Chief of the Division of Biology and Medicine, explained in a 1955 letter to the University of Rochester's Dr. Louis H. Hempleman. "We have racked our brains for any useful subterfuge in carrying out the study but none came to mind which could possibly hold water for any length of time," he added.

The subterfuge they used in the end to examine former workers at the Cleveland City Hospital was explained to a hospital doctor, Dr. Robert R. Stahl. "To put it baldly," Albert wrote Dr. Stahl on August 1, 1955, "I think we are fundamentally interested in the autopsy data, the examination program being a mechanism to keep tabs on the people involved in the survey." Extreme care was needed. If too much medical data were gathered from the workers, "there would be a distinct risk of stimulating lawsuits against the Atomic Energy Commission," Dr. Albert emphasized to Dr. Joseph T. Wearn at the School of Medicine at Western Reserve University in Cleveland, who would supervise the "study."

The plan fell through. Dr. Stahl was appalled when he read the AEC proposal. He pushed the government men away, with an admonition about medical ethics. "The project protocol . . . grossly misrepresents the type of information that AEC is apparently attempting to obtain," Dr. Stahl told Dr. Albert. "Basically," he added, "a health survey is being used as a 'front' for obtaining such autopsy data . . . since this is the basic motive involved neither Dr. Scott nor myself are interested in such a project." The AEC had wanted to keep the men's records secret. "Allow me to remind you," Stahl added, "that a physician has a legal responsibility toward any patient seen to keep this patient's records in his files." File 092694-a, Box S09501B196, ACHRE, RG 220.

One disturbing aspect of this proposed study is the number of people who appear to have known of the gravity of the workers' exposure. For example, a May 7, 1953, memo to the Executive office of the CDC, from Alexander D. Longmuir, chief of the PHS Epidemiology Branch, states, "Thursday morning I received a telephone call from Dr. Roy E. Albert, Medical Officer, New York Operations Office, AEC, 70 Columbus Avenue, New York City. Dr. Albert called me at the suggestion of my personal friend, Dr. David D. Rutstein, Professor of Preventive Medicine Harvard Medical School, because he felt we might be interested in a proposal he had to make. His proposal was the desirability of a follow up of between 400 and 600 employees of the Harshaw Chemical Company in Cleveland, Ohio. These employees were exposed for a period of from one to three years in 1945, 1946, and 1947 to 600 times the tolerance dose of radioactive dust, resulting from the processing of uranium and radon. . . . In view of reports from Europe, that uranium miners suffer an exceptionally high incidence of cancer of the lung, Dr. Albert and his advisory groups recommend that these employees also be studied for the same condition." Memo cc'd Dr. Roy E Albert and Dr. Alexander Gilliam. Medicine Health and Safety, AEC, RG 326.

73. Stafford L. Warren, *The Role of Radiology in the Development of the Atomic Bomb,* p. 856. DOE Opennet accession #NV0729054.
74. In the official review of the material releases from Oak Ridge and the relationship to community health effects, fluoride emissions were not even considered, an omission that concerned at least one top scientist. Letter to Dr. Kowetha A. Davidson, Chair Oak Ridge Reservation Health Effects Subcommittee, Oak Ridge National Laboratory, from Kathleen M. Thiessen, PhD Senior Scientist, SENES, Oak Ridge, January 16, 2001. Re: Oak Ridge Reservation Health Effects Subcommittee and review of the Oak Ridge Dose Reconstruction. Thiessen wrote that "there are a number of contaminants that were never evaluated quantitatively during either the Oak Ridge Dose Reconstruction (1994–2000) or the preceding Phase I Dose Reconstruction Feasibility Study (1992–1993). . . . [I]t is clear . . . that the fluorine and fluoride releases from K-25 alone were very large. . . . It is my professional opinion that the historical fluorine and fluoride releases from the K-25 and Y-12 sites should be assessed quantitatively, both with respect to the amounts of material used and released, and with respect to the potential health implications for off-site individuals." cc: Rear Admiral Robert Williams, ATSDR, Mr. Jack Hanley, ATSDR.
75. Paducah began production in 1954. At Portsmouth, Ohio, which opened in 1954, "the quantity of fluorine to be released was steadily increasing and that this fluorine could not be contained in any holding drum, but must be vented

to keep the cascade in proper operation." Memo to H. L Caterson to K. H. Hart, "Venting of Fluorine from the X-326 Building, October 3, 1955, 1089/120" cited in Arjun Makhijani, Bernd Franke, and Milton Hoenig, *Preliminary Estimates of Emissions of Radioactive Materials and Fluorides to the Air from the Portsmouth Gaseous Diffusion Plant, 1954–1984* (unpublished), p. 19.

76. Several of the Area C workers referred to the bridge damage and to the paint tarnishing on cars. The Sierra Club filed a lawsuit for $9,960,000 in Cuyahoga County Common Pleas Court alleging "fluoride fumes from the plant at 1000 Harvard Ave., SE over the past twenty-two years, have destroyed the nearby Harvard Dennison Bridge." *National Fluoridation News* (January–February, 1971), p. 2.

77. Pharmacology Report #558, Monthly Progress Report for June 1947, Box S09F01B227, ACHRE, RG 220.

78. AEC Monthly Status and Progress Reports, July 1949, via Pete Eisler, *USA Today*. The same document notes the "disastrous Donora episode of last winter."

79. Sadtler told me that after World War II, "I was lecturing to the American Chemical Society in Cleveland . . . [on] 'smoke, dust, fumes and fellow travellers.' . . . And a lawyer came up to me and said the judge wants to do something for the monsignor in a certain section of Cleveland. And we agreed that I would investigate. . . . I did find out that Harshaw Chemical was letting off, I believe, HF."

80. "At the present time, at least 10 percent of the fluorine generated for use in the manufacture of uranium hexafluoride is unavoidably lost in the vent gases from the process. The recovery of this fluorine has become of prime importance since the expansion of the uranium hexafluoride manufacturing facilities to the 48 tons of uranium per day production level. The estimated cost of the vented fluorine will amount to $400,000 per year based on the above percentage lost and a cost per pound of $0.65." Memo, "Recovery of Fluorine from Feed Plant Vent Gases," March 2, 1955, ORF14753, ORF18718 for plant damage. Both in Oak Ridge DOE Public Reading Room.

For dumping, see Capt. Bernard Blum to Lt. Col. Luvern W. Kehe, "Contamination of Water in Poplar Creek," August 10, 1945, Md 319.1, General Essays, Lectures, Box 34, Manhattan Engineer District Accession #4nn 326-85-005, Atlanta FRC, RG 326.

81. See also Chapters 5 and 8. For accounts of pollution in New Jersey and Pennsylvania, see "The Peach Crop Cases." See also, for litigation, E. J. Largent, *Fluorosis: The Health Aspects of Fluorine Compounds* (Columbus, OH: Ohio State University Press, 1961), p. 124. "Claims of damage to plants and animals appeared in almost epidemic numbers along the Delaware River in the Philadelphia area in 1944 and 1945. . . . Since the beginning of this same period of time, a series of claims of fluoride induced damage have appeared in Tennessee."

82. "Fluorine is an extremely toxic and hazardous chemical. There are three potential liabilities associated with its release to the atmosphere. The first

and most significant is the potential effect on agriculture crops and livestock in the surrounding area. . . . The second significant liability is a hazard to personnel in the immediate area, both employees and the general public. The maximum allowable concentration of fluorine in the air as recommended by the national advisory group, the American Conference of Governmental Industrial Hygienists, is 0.1 part per million. Any appreciable release of fluorine to the atmosphere will in all probability result in some concentrations in excess of this level. Although concentrations considerably in excess of this level can be tolerated without permanent injury, a basis for complaint and possible legal action does exist." Letter A. J. Garcia to C. L. Becker, "Fluorine Air Pollution at GAT Plant Site, August 30, 1954, 1089/124." Cited in Arjun Makhijani et al., *Preliminary Estimates.*

83. A. Stern, *Air Pollution* (New York: Academic Press, 1962), p. 391.
84. J. G. Rogers et al., *Environmental Surveillance of the U.S. Department of Energy Portsmouth Gaseous Diffusion Plant and Surrounding Environs during 1987, April 1988*, (ES/ESH-4/V4), 18. See also Makhijani, Franke, and Hoenig, *Preliminary Estimates of Emissions of Radioactive Materials and Fluorides*, p. 20.

    Also, for later HF releases at the Portsmouth, Ohio gaseous diffusion plant, although Ohio had no standards for gaseous fluorides: "As of 1986, Kentucky's seven day ambient air standard was .8 microgram HF/m3 . . . in comparison data recording sheets from 1973 show individual fluoride measurements as high as 5 micrograms/m3. 1982 and 1983 measurements also exceeded the above standard, with the maximum off-site average monthly concentrations of fluorides as HF around the plant varying between 1.94 and 6.09 microgram/m3 in 1982, and between 1.83 and 15.1 microgram/m3 in 1983." Cited in Makhijani et al., *Preliminary Estimates*, p. 21.

Chapter 5

1. Three years later Harold Hodge would look out over another spectacular view as an official observer of the 1946 atomic bomb blast at Bikini Atoll in the South Pacific. H. Hodge, *J. Dental Res.*, vol. 26 (1947), pp. 435–439.
2. Md 600.914, Progress Reports Rochester, Box 47, Accession #4nn 326-85-005, Atlanta FRC, RG 326.
3. Priorities were determined by combining the rating of: "(1) the toxicity and (2) the number of persons who had real or potential exposure to each compound." The top toxicological priorities were (uranium compounds) UO_2F_2 and (nonuranium) F_2 and HF_2. Harold Hodge and Carl Voegtlin, eds., *Pharmacology and Toxicology of Uranium Compounds, with a Section on the Pharmacology and Toxicology of Fluorine and Hydrogen Fluoride* (New York: McGraw Hill, 1949), historical foreword, p. 5.
4. Hein interview with Bob Woffinden, timecode 04.21.13, 1997.
5. Hodge was a lead author with R. E. Gosselin, R. P. Smith, and M. E. Gleason of *Clinical Toxicology of Commercial Products*, 5th ed. (Baltimore, MD: Williams and Wilkins, 1984).

6. "Harold C. Hodge, 1904–1990, Pharmacology and Experimental Therapeutics: Oral Biology: San Francisco." In Memoriam. E. Newbrun et al., University of California, web posting.
7. Biographical details in P. Morrow et al., "Profiles in Toxicology—Harold Carpenter Hodge (1904–1990)," *Toxicological Sciences*, vol. 53 (2000), pp. 157–158.
8. A once secret document, "Detailed Duties of Dr. Harold C. Hodge," lists the problems his Pharmacology Section helped to solve. One problem, the "necessity of stated daily maximum intake of fluoride to avoid poisoning," was solved at the Conference on Fluorine Metabolism at the Hotel Pennsylvania in New York in January 1944. Hodge was one of the experts who set the maximum allowable concentration of "6 ppm as *project* allowable exposure per day" (emphasis in original). Folder 2, Box S09FO1B219, ACHRE, RG 220.

   Hodge was elsewhere also clearly conscious of the health toll the war's haste imposed upon workers. For example, in April 1945 he explained to Col. Hymer Friedell the reasons for increasing the maximum allowable concentration of uranium tetrafluoride and several other uranium compounds in bomb factories from 150 to 500 micrograms of uranium per cubic meter of factory air. It was an "emergency war measure to expedite industrial production," he explained, "a compromise between the air concentration which can be maintained during maximum production and the chance of injury to plant workers." Carl Voegtlin and Harold Hodge to Hymer Friedell, April 26, 1945. (Voegtlin was the retired head of the National Cancer Institute at the University of Rochester during the war.) This measure was implemented, directly affecting the work environment of thousands of Manhattan Project industrial workers. Col. Warren explained the new standard more bluntly: "In view of the extreme difficulty in maintaining concentrations of 150 micrograms per cubic meter in industry, it is felt that such a change will be of definite benefit in expediting the war effort." Warren to the Area Engineers, June 1945. Both documents in Mm 3, Md 700, General Essays, Lectures, Box 34, Manhattan Engineer District Accession #4nn 326-85-005, Atlanta FRC, RG 326.
9. A key text is Hodge and Voegtlin, eds., *Pharmacology and Toxicology of Uranium Compounds*. See also, J. H. Simons, *Fluorine Chemistry*, vol. IV, by Harold C. Hodge and Frank A. Smith (New York: Academic Press, 1965) supported in part by a contract with the U.S. AEC at the University of Rochester Atomic Energy Project.
10. Hodge and Voegtlin, eds., *Pharmacology and Toxicology of Uranium Compounds*, historical foreword, p. 1.
11. Hein interview with Mark Watts for Channel 4 Television in the United Kingdom. Interview recorded for "Don't Swallow Your Toothpaste," a program that aired in June 1997.
12. Lansing Lamont, *Day of Trinity* (New York: Atheneum, 1985), p. 251: "The specter of endless lawsuits haunted the military." See also Groves's memo,

cited in chapter 4, note 2, asking for toxicity data. According to the Harvard professor Phillip Drinker, a member of the AEC Stack Gas Committee and an AEC litigation consultant, "In 1947 AEC was apprehensive about damage suits from personnel allegedly injured by radiations or by exposure to various chemicals used." Phillip Drinker to Dr. Thomas Shipman, Health Division Leader, Los Alamos, November 14, 1950, Medicine Health and Safety, RG 326.

For insurance, see article for Aetna's internal magazine *The Aetna-izer*, submitted by Vice President Clifford B. Morcom to Col. K. D. Nichols, August 31, 1945, for review. "The billion-dollar atomic bomb plant at Oak Ridge, Tenn., is probably the most interesting and important of the large number of war projects on which the Aetna Casualty and Surety Company provided coverage in whole or in part, in the last few years. . . . As a result of this need for iron-clad secrecy, the representatives of the Manhattan Project could not even hint to us, or to anyone else, as to what the product of the Clinton Engineer Works was going to be, or what exposures or hazards there would be in its manufacture. It was manifestly impossible for us to provide insurance on any regular basis in view of these circumstances; but the government had asked for our help, and we were anxious to comply." The following passage is scratched out: "in essence, the plan placed the facilities of our organization at the disposal of our policyholders; and, in return for this, the Government agreed to reimburse us for any losses we might sustain." Aetna, Office of Public Information 1944–1967, Box 12, Accession #73A0898, Atlanta FRC, RG 326.

For Travelers, see memo to Col. Warren from Capt. Ferry, March 25, 1944, "Conference in Wilmington, 20th March 1944." Five DuPont officials, two majors from the Manhattan Project, and Mr. Wm. M. Worrell of the Travelers Insurance Co. "Item 3. In a number of instances, men working on construction have been exposed to fumes from processes which give off HF in concentrations sufficient to make them leave their work temporarily. In at least one case illness followed the exposure." Md 700.2, Univ. of Rochester (Medical), Box 54, Accession #4nn 326-85-005, Atlanta FRC, RG 326.

Groves, *Now It Can Be Told*, p. 57: "To facilitate the handling of claims not resulting from a major catastrophe a special fund was established. This fund was placed under the control of du Pont so that it could continue to be available for many years." And on March 28, 1944, at a conference on Extra-Hazardous Insurance attended by the military officials and industrial contractors readying the K-25 plant, Kellex management stated that they were "especially concerned" about the health risk from fluoride exposures. The K-25 employees were, accordingly, defined by a simple criterion, their exposure to fluoride, and categorized "into three (3) groups; those having regular, casual or no exposure to C616 and C216 [codes for uranium hexafluoride and fluorine gas]." At the conference Col. Warren was informed that "the decision was made by Kellex officials that the names of all employees would be submitted to the [Manhattan Project's] District Insurance Section, with estimates of the amount of their exposure." Memo to Col. Stafford Warren from Capt Ferry, April 4, 1944. "Conference on Extra-Hazardous Insurance 28 March 1944." Md 337, New York Meetings and Conferences, Box 30, Accession #4nn 326-85-005, Atlanta FRC, RG 326.

13. Several of the Manhattan Project's biggest industrial contractors had been badly exposed to worker lawsuits before the war. In the mid-1930s Union Carbide—now running the K-25 gaseous diffusion plant at Oak Ridge—had endured congressional scrutiny and legal claims following the Gauley Bridge silicosis deaths. DuPont, too, had won cruel headlines in the early 1930s from the New York press following an epidemic of death and injury at its New Jersey tetra ethyl lead plants.

    Bomb-program officials also recalled the prewar litigation and public scandal over female workers who had died following their employment at the U.S. Radium Corporation in Orange, New Jersey, their jaws eaten by cancer after licking radium from the brushes they had wetted to paint luminescent watch dials. See, for example, "Review of Document" by L. F. Spalding of the Insurance Branch to Charles A. Keller, Declassification Officer, February 5, 1948: "We have reviewed ["Biochemical Studies Relating to the Effects of Radiation and Metals" by Samuel Schwartz] from a nontechnical point of view and although it is conceivable that the contents thereof might arouse some claim consciousness on the part of former employees we are unable to predict that the Commission's interests would be unjustifiably prejudiced by its publication. However, in the event latent disabilities due to exposures reported in this document should result in publicity similar to that which arose out of the 'radium dial' industry, the public relations division would be involved." Document SO9FO1b22, File DOE 120994-A #1, ACHRE, RG 220.

    See also Hodge, *J. Dent. Res.*, vol. 26 (1947), pp. 435–439. "These women, despite all safeguards, persisted in tipping on their tongues the brushes they were using to apply radium paint to airplane dials. Those unfortunate enough to retain lethal amounts of radioactive material died of cancer from radium deposited in the bones; deaths were recorded five, ten, fifteen years later." For an excellent summary of the radium dial painters, see Eileen Welsome, *The Plutonium Files* (New York: Dial Press, 1999), p. 47.

14. See for example, "The Medical Section has been charged with the responsibility of obtaining toxicological data which will insure the District's being in a favorable position in case litigation develops from exposure to the materials," Col. Stafford Warren to Dr. John Foulgar of Du Pont's Haskell Laboratory in a letter dated August 12, 1944. Box 25, Accession #72C2386, Atlanta FRC, RG 326. Also, it appears that some studies were simply not performed, or at least that data were not published. Where are the published studies of the toxicity of oxygen fluoride, a chemical that Hodge's team referred to as "the most toxic substance known" and was listed as a high priority for bomb program investigation? Where are the chronic studies on the various fluorocarbon compounds being used in the diffusion plants? The reluctance

of Hodge's team to perform such studies, which of course better resembled the actual conditions workers faced, was a frustration of Harvard University's Harriet Hardy, a leading beryllium researcher. "The chronic disease is certainly our most pressing problem, and at present the whole weight of the Rochester work, if I understood Dr. Hodge, is on the acute manifestation.... I cannot understand the defeatist attitude about producing chronic changes in animals with beryllium compounds sufficiently approximate to the human pathology." Dr. Hardy to Dr. Warren, "Recent trips to Cleveland and Rochester," September 13, 1949, DOE Opennet #1153735.

15. Col. Stafford Warren, Memorandum to the Files, "Purpose and Limitations of the Biological and Health Physics Research Program," July 30, 1945, p. 3, Medical and Health Problems, Box 36, Accession #72C2386, Atlanta FRC, RG 326.
16. Lt. Col. Hymer Friedell, Memo, "Future Medical Research Program," February 26, 1946, is found as the third item in a file located at 0712317 in the Department of Energy's HREX electronic search engine.
17. The Rockefeller Foundation and the Carnegie Corporation had funded broad programs of dental research at Rochester, Yale, and Harvard during the Depression, seeking to improve the terrible condition of teeth in the United States. There is no indication in the files seen by this author that the prewar granting was anything other than philanthropic in nature.

    For Hodge's résumé, see his testimony before Cong. Wier. HR 2341: "A Bill to Protect the Public Health from the Dangers of Fluorination of Water." Hearings Before the Committee on Interstate and Foreign Commerce, House of Representatives, 83rd Congress, May 25, 26, and 27, 1954, p. 470. "Since 1937 I have been continuously engaged part time as a consultant toxicologist for a number of industrial companies."
18. Hodge links to Eastman from author interview with toxicologist and Rochester alumnus Robert Phalen.
19. The University of Rochester's Manhattan Project medical budget included specific funding for Rockefeller projects. Rochester Organizational Chart. Also, ESSO labs, Standard Oil, and the Rockefeller Institute were working on various projects, including the hexafluoride gas centrifuge. "PB authorizations as of March 9, 1942, 1/14/42 Standard Oil Development Co. 'Centrifuge method of separation leading to design of plant' PB #2 amount $100,000'" and "3/9/42 Standard Oil Development Co. 'Pilot Plant Building' PB #12 $250,000." Doc #310, Records of Section S-1 Executive Committee, RG 227.3.1. The Carnegie Institute of Washington had fluoride interests, as well. It investigated liquid thermal diffusion with Philip Abelson as early as 1941, in a precursor project to the Philadelphia Navy Yard project, which was itself a prototype of the S-50 complex at Oak Ridge. Amato, I., "Pushing the Horizon. Seventy-Five Years of High Stakes Science and Technology at the Naval Research Laboratory."

    See also Harold Urey, Program Chief, Columbia University to James Conant, January 19, 1942: "I wish to recommend that a contract be drawn to the Rockefeller Institute for Medical Research, New York, NY for work on the separation of the uranium isotopes by the mobility method, this work to be done under the direction of Dr. Duncan A MacInnes and Dr. Lewis G Longsworth." And November 19, 1942, to Dr. Wensel from Urey: "I have asked the Rockefeller Institute people under Dr. MacInnis to do some work on the chemical separation work . . . I wonder if it would be possible to amend their contract." Doc #336, Records of Section S-1, Executive Committee, RG 227.3.1.

20. Col. Warren to Dr. John Foulger Box 25, Accession #72C2386, Atlanta FRC, RG 326.

21. Much of this account was cowritten with Joel Griffiths and first appeared in 1997 in various alternative media outlets, including *Earth Island Journal*, eventually winning a 1999 Project Censored Award.

22. Garfield Clark was measured at 25.6 ppm blood fluoride, Ollie Danner at 31.0 ppm. Farmer Willard Kille, diagnosed by his doctor as fluoride poisoned, had 15.0 ppm. Report submitted by Philip Sadtler, December 11, 1945. In Groves papers, NARA. That these levels are high can be seen from H. Hodge and F. A. Smith, *Fluorine Chemistry*, vol. IV, p. 15. (The New York Examiner's office made available for autopsy the bodies of fatal fluoride poisonings from 1935 to 1949. Those data showed fluoride blood levels of between 3.5 and 15.5 ppm.)

23. The company's giant Chamber Works at Deepwater, New Jersey, near the mouth of the Delaware River, has long handled some of the company's most dangerous chemicals, with workers and the local community traditionally paying the price. During World War I as many as 10,000 workers had been employed there making munitions and poison gas, according to G. Colby, *Du Pont Dynasty: Behind the Nylon Curtain* (Secaucus, NJ: Lyle Stuart, 1984), p. 195.

    Referring to World War I aftermath, Colby writes, "In DuPont's Deepwater, New Jersey, plant across the river from Wilmington, workers died from poisonous fumes of the lethal benzol series, their bodies turning a steel blue. At the Penns Grove, New Jersey, plant workers were called 'canaries': picric acid had actually dyed their skins yellow. Picric acid poisons the mucous membranes of the respiratory tract, attacks the intestinal tract, and destroys the kidneys and nerve centers." In the 1920s several Deepwater workers had also been killed and hundreds injured in an horrific and months-long episode, dubbed by the New York press "the loony gas" poisoning, as DuPont began making the highly toxic gasoline additive tetra ethyl lead (TEL). Salem County, where the plant is located, had the highest rate of bladder cancer for white males in the United States from 1950 to 1969, according to the National Cancer Institute. Also, the *New York Times*' Mary Churchill learned in January 1975 that since 1919, 330 employees at the plant had contracted bladder cancer.

    See also the testimony of Willis F. Harrington, former Chair of DuPont's Kinetic Chemicals, *United States* vs. *Du Pont* (1953), p. 693. *United States of*

America vs. *E. I. Du Pont de Nemours, General Motors, United States Rubber, et al.*, Civil Action No. 49 C-1071, U.S. District Court for the Northern District of Illinois, Eastern Division, before Judge LaBuy, April 13, 1953, p. 3798. For Manhattan Project employees during World War II, William C. Bernstein, Captain Medical Corps. Memorandum To Colonel Stafford L. Warren, Chief Medical Section. November 3, 1944. Subject: Report on Medical Section in Wilmington, Delaware. November 3, 1944, Box 14, Wilmington Area, Accession #72C2386, Atlanta FRC, RG 326. (Note attached from Howland, "total engaged in work of Manhattan District 1122.")

24. William C. Bernstein, Captain Medical Corps. Memorandum to Colonel Stafford L. Warren, Chief Medical Section. November 3, 1944. Subject: Report on Medical Section in Wilmington, Delaware. November 3, 1944, Wilmington Area, Box 14, Accession #72C2386, Atlanta FRC, RG 326.
25. B. J. Mears, Captain, Medical Corps, Assistant. Medical Clearance on Terminated Madison Square Area Contracts. To: The District Engineer, Manhattan District, Oak Ridge, Tennessee. (Attention: Major J. E. Ferry). October 5, 1945, Medical Clearances, Terminated Madison Square Contracts, Box 36, Accession #4nn 326-87-6, Atlanta FRC, RG 326.
26. William C. Bernstein Captain, Medical Corps, Memorandum to Col. Stafford L. Warren, Chief, Medical Section, Subject: Occupational Disability Cases Observed. November 3, 1944, Wilmington Area, Box 14, Accession #72C2386, Atlanta FRC, RG 326.
27. To Stafford Warren, Subject: Supplementary Report of Medical Examination at X-Works [code for Chamber Works] February 2, 1945, Wilmington Area, Box 14, Accession #72C2386, Atlanta FRC, RG 326.
28. William C. Bernstein, Captain, Medical Corps. Memorandum to Col. Stafford L. Warren, Chief Medical Section. November 3, 1944. Subject: Report on Medical Section in Wilmington, Delaware. November 3, 1944. Wilmington Area, Box 14, Accession #72C2386, Atlanta FRC, RG 326.
29. "Memorandum to the files, Subject: Recapitulation of Work Accomplished During Temporary Duty at X Works." 1st. Lt. Birchard M. Brundage, February 17, 1945.
30. Memo to Capt. B. Brundage (through Col. Warren), November 23, 1945 (draft version, accompanied by handwritten notes detailing other "nuisance claims"). General Correspondence, Box 36, New York Operations Research and Medicine Division, Accession #72C2386, Atlanta FRC, RG 326.
31. Hodge to Warren, March 11, 1946. Md 700.2, Division of Rochester, Atlanta FRC, RG 326. For volume of fluoride in air pollution, see example, "In the Kinetics plant, Mr Knowles described the practice of ten years back in which SiF4 was vented to the air. SiF4 is quite poisonous." Hodge to Warren, May 1, 1946, cc Lt. Col. Rhodes, Crop Contamination (New Jersey), Box 33, Accession #72C2386, Atlanta FRC, RG 326.
32. Hodge to Warren, May 1, 1946, cc. Lt. Col. Rhodes, Crop Contamination (New Jersey), Box 33, Accession #72C2386, Atlanta FRC, RG 326.

33. Lt. Col. Cooper Rhodes memo to General Nichols, "Subject: Conference with Mr. Willard B. Kille." March 25, 1946. Groves Papers, NARA, via Griffiths and Honicker.
34. Conference on Fluorine Residues, February 12, 1946, Groves Papers, NARA, via Griffiths and Honicker.
35. Cooper B. Rhodes, Lt. Col. "Memorandum for the Files. Subject: Peach Crop Cases (*Kille et al.* vs. *Du Pont*), 2 May 1946. . . . Cc: General Groves, General Nichols." Groves Papers, NARA, via Honicker and Griffiths.
36. Groves to the Commanding General, Army Service Forces, Pentagon Building, Washington, DC, August 27, 1945, Groves Papers, NARA.
37. Gen. Groves to Sen. McMahon, February 18, 1946, Groves Papers, NARA.
38. The note to Groves's senior deputy includes a response, dated February 25, 1946. "General Groves: That firm of consulting chemists has been employed by the plaintiffs in the 'peach crop' suits against Du Pont, and Mr. Sadtler has been very active in gathering evidence to present on behalf of the plaintiffs in those suits." Groves Papers, NARA, via Honicker and Griffiths.
39. Multiple taped author interviews with Philip Sadtler, March 1993. Also, account from *The Chemist* (1965), pp. 349–350; that the Sadtler firm had testified on behalf of Coca-Cola to say that cocaine was not a chemical ingredient of the beverage.
40. Sadtler recalled that one of the agents he had met in the New Jersey orchards later gave an account of their wartime sleuthing to the media. Joseph Marshall, "How We Kept the Atomic Bomb Secret," *Saturday Evening Post*, November 10, 1945, includes the following story: "Once, in an East Coast city, Agents Harold Jensen and Harold Zindle were maintaining constant surveillance of an individual under suspicion of being involved with enemy agents." The *Post* story does not give the name of the person being tailed but reports that the government agents believed the "subject . . . apparently suspected he was under surveillance," and so they built a fence to block escape from the house via the rear. The published account concludes, "It is presumed that the subject is still wondering why his neighbor decided to put up the fence so suddenly, and his neighbor is wondering why the subject did. And Security is still wondering whether the subject is a spy." Sadtler told this writer that he had no idea he was under surveillance but that on one occasion, "I decided to take the car rather than the train and I jumped the fence so they did not see me come out." Sadtler was gutsy. He rented a plane and flew over the DuPont works, to investigate the pollution, further displeasing authorities. Author interview.
41. Interview with Joel Griffiths, first published in Griffiths and Bryson, "Fluoride, Teeth, and the Atomic Bomb," *Waste Not: The Reporter for Rational Resource Management*, September 1997.
42. File. Lt. Col. Cooper B. Rhodes, "Kille et al. (12 Separate Cases) vs. Du Pont." February 13, 1946, Groves Papers, NARA.
43. Groves to the Commanding General, Army Service Forces, Pentagon Building, Washington, DC, August 27, 1945. Groves Papers, NARA.

44. Gen. Groves to Sen. McMahon, February 18, 1946, Groves Papers, NARA.
45. Giordano interviews conducted in 1997 by Joel Griffiths and Clifford Honicker on a trip to the peach orchards. Clemente interview conducted by telephone and e-mail with the author in 2002.
46. C. A. Taney Jr., Major, Corps of Engineers, to William C. Gotshalk, September 24, 1945, cc. General Groves, in Groves file on New Jersey pollution, NARA, via Joel Griffiths.
47. William Gotshalk to Maj. C. A. Taney, U.S. Engineer Office, New York, NY, August 28, 1945, Groves Papers, NARA.
48. Maj. C. A. Taney to Gen. L. R. Groves, June 1, 1945, Groves Papers, NARA.
49. Thiessen, interviewed several times for this book, is a senior scientist with SENES Oak Ridge, Inc., Center for Risk Analysis. She is the author of *Summary Review of Health Effects Associated with Hydrogen Fluoride and Related Compounds*. U.S. Environmental Protection Agency, December 1988.
50. Lt. Col. Cooper B. Rhodes, "Memorandum for the Files. Subject: Peach Crop Cases, *Kille et al.* vs. *Du Pont*, May 2, 1946, Cc: General Groves, General Nichols." Groves Papers, NARA, via Honicker and Griffiths.

Chapter 6

1. *Time*, April 24, 1944, p. 43.
2. D. B. Ast, "A Plan to Determine the Practicability, Efficacy, and Safety of Fluorinating a Communal Water-Supply, Deficient in Fluorine, to Control Dental Caries," in W. J. Gies, ed., *Fluorine in Dental Public Health* (New York: New York Institute of Clinical Oral Pathology, 1945), p. 44.

   Ast's paper was delivered at a symposium of the New York Institute of Clinical Oral Pathology, New York City, October 30, 1944. According to the editors of *Fluorine in Dental Public Health*, "Dr. Ast's address (pp. 40–45) states the basis for, and the procedure in, the effort in the State of New York to determine, in a comprehensive and extended research, whether mass prevention (control) of dental caries (under the conditions stated in the preceding paragraph) is attainable *without inducing toxic effects elsewhere in the body*," p. 6 (emphasis in the original). See also F. McKay, *Fluorine in Dental Public Health*, p. 18, "Newburgh has become another 'biological experiment station,' in which the rationale is applied directly to humans without previous laboratory experiments on animals."
3. Memorandum "Summary of Conference with Colonel Nichols," dated New York City, July 23, 1943, notes, "5. Agreed to farming out Fl and HF toxicity experiments to Dr. Fairhall of the U.S. Public Health Service, Bethesda, Mr [left blank] through Dr. Wenzel—with experiments outlined by Drs Hodge and Ferry." Thus, the Manhattan Project is secretly directing the wartime PHS fluoride studies. Bomb-program medical planners, including Drs. Hodge, Friedell, and Warren, decided on August 31, 1943, that there was need for an "orientation conference on fluorine toxicity under auspices of the U.S. Public Health Service or OSRD." New York Operations Research

and Medicine Division, Correspondence 1945–1952, Box 36, RG 326. The conference transcripts are in a file in the same box coded "G-118."

4. James Conant, Chairman NRDC, to Mr. J. J. Townsend, Public Health Service, Bethesda, MD, September 25, 1943; and Townsend to Conant, September 29, 1943, Documents 295 and 296, Records of Section S-1 Executive Committee, RG 227.3.1.
5. Transcript of Metabolism of Fluorides Conference, main session, Hotel Pennsylvania, New York, NY, January 6, 1944, Dr. Neal, p. 24, via Pete Eisler, *USA Today*.
6. Ibid., Dr. Calvary, Chief of the Division of Pharmacology, FDA, p. 22. On animal tests, see Memo to Safety Section files, Joseph Faust, Assoc. Engineer (Safety) January 14, 1944, Oak Ridge Reading Room, ORO #1304.
7. Transcript of "Metabolism of Fluorides" Conference, main session, Hotel Pennsylvania, New York, NY, January 6, 1944, via Pete Eisler, Ast comment at p. 27. (Interviewed by me in 1997, David Ast said that he could not remember having attended the New York conference.)
8. Memorandum to The Area Engineer, Rochester Area, Rochester, NY. Subject: Funds for Incidental Expenses of Meeting on "Fluoride Metabolism," December 31, 1943. John L. Ferry, Md 123 (729.3), File labeled G-118 (c), A2, Box 36, Accession #72C2386, Atlanta FRC, RG 326.
9. "I think it would be a definite step forward if we forgot about definite limits and called them 'control limits,'" stated Helmuth Schrenk from the U.S. Bureau of Mines. Committee on Fluoride Metabolism, Round Table Discussion During Luncheon Period, continued in the Evening, January 6, 1944. All of these quotes come from the same lunchtime conference transcript. Transcript in file labeled G-118 (c), A2, Box 36, Accession #72C2386, Atlanta FRC, RG 326.
10. E. R. Schlesinger, D. E. Overton, and H. Chase, "Newburgh-Kingston Caries-Fluorine Study II. Pediatric Aspects—Preliminary Report," *Am. J. Public Health* (June 1950), p. 725.
11. "It is unknown," for example, complained Captain Peter P. Dale of Harold Hodge's Division of Pharmacology, in September 1945, "what the critical levels of T, F or 'P' storage in man are [codes for uranium, fluoride, and plutonium], or whether they may have a potentially deleterious effect. Are such factors as the age, sex and the physical and chemical properties of the reagent important?" "Dental Research Program" Memo to Stafford L. Warren, September 24, 1945, from Capt. Peter P. Dale, Capt., DC AUS.
12. D. E. Gardner, F. A. Smith, and H. C. Hodge (with D. E. Overton and R. Feltman) *UR 200 Quarterly Technical Report* (October 1, 1951–December 31, 1951), University of Rochester, "Fluoride Concentration of Placental Tissue," p. 4. "D. E. Overton of the Newburgh Fluorine Demonstration secured the samples from patients in that city." Published version in *Science* 115 (February 22, 1952), p. 208.
13. Memo to Lt. Col. Hymer Friedell from Capt., Henry L. Barnett, February 8, 1946, "Organizational Plan for Manhattan District Personnel Assigned

to Japanese Report." Barnett had also seen the Trinity explosion, and been among the first to detect the fallout cloud. Lansing Lamont, *Day of Trinity* (New York: Atheneum, 1965), p. 244. An H. "C" Barnett is listed in charge of "special studies" at the University of Rochester, "Organization Chart of the Manhattan Department, University of Rochester," in Harold Hodge and Carl Voegtlin, eds., *Pharmacology and Toxicology of Uranium Compounds, with a Section on the Pharmacology and Toxicology of Fluorine and Hydrogen Fluoride* (New York: McGraw Hill, 1949), p. 1061. Still another member of the Newburgh Technical Advisory Committee, a Columbia University biostatistician, John Fertig, may have been connected to the bomb program. Fertig did war work for the Office of Scientific Research and Development (OSRD), the same federal bureaucracy that had sired the atomic bomb. *American Men of Science*, 9th and 10th editions.

14. For Howland's fluoride work: F. A. Smith, D. E. Gardner, and H. C. Hodge, "Investigations on the Metabolism of Fluoride," *J. Dent. Research* (October 1950), p. 596: "We are indebted to Dr. J. Howland for taking the Rochester blood samples." (The study is a comparison of the fluoride levels in blood and urine in Newburgh and Rochester.) Also, Howland wrote "Studies on Human Exposure to Uranium Compounds," which investigated the Philadelphia Navy Yard blast, blaming "the fluoride ion" for injuries. Hodge and Voegtlin, eds., *Pharmacology and Toxicology of Uranium Compounds*, p. 1005: Howland had helped to assemble the atomic bombs on the Pacific island of Tinian, then surveyed the aftermath with his Rochester colleague, Capt. Barnett.
15. Eileen Welsome, *The Plutonium Files: America's Secret Medical Experiments in the Cold War* (New York: Random House, 1999). Capt. J. W. Howland was selected as Assistant Head of the Division of Biological and Health Physics Research of the Medical Section. "The Manhattan District Official History," Book 1, General, vol. 7, Medical Program, p. 5.17.
16. Similarly, Ast could not recall a documented 1946 trip to New Jersey with Harold Hodge after the war, to study the poisoned children near the DuPont uranium hexafluoride plant. Author interview, 1997.
17. "Fluoride Metabolism: Its significance in Water Fluoridation" *J. Am. Dent. Assoc.*, vol. 52 (March 1956), p. 307.
18. As medical director, Capt. Friedell had been well aware of the Manhattan Project's concern about fluorides. On January 20, 1943, for example, after visiting the War Research Laboratories at Columbia University, where a small-scale fluoride gaseous diffusion plant had already been built, he reported that "The primary potential sources of difficulty may be present in the handling of uranium compounds . . . and the coincident use of fluorides which are an integral part of the process." "Initiation of Medical Program for Project at Columbia University," January 20, 1943, Friedell to District Engineer. Friedell had also investigated fluoride poisoning in workers at the Harshaw Chemical Company in Cleveland.
19. "Information that fluorides are not hazardous" would have been especially helpful to the bomb program's Legal Division, suggested Friedell.

20. Author interview, December 3, 2002.
21. Dean's epidemiological studies in the 1930s had given key scientific support to the idea that fluoride may play a role in dental health. As the PHS's key fluoride expert, Dean attended at least one Newburgh Advisory Committee meeting.
22. G-10 file, Correspondence 1945–1952, New York Operations Research And Medicine Division, Box 38, Atlanta FRC, RG 326.
23. Also in the G-10 file is a letter from Rochester's Harold Hodge to Dr. Edward S. Rogers of the New York Department of Health, requesting more bone X-rays of Newburgh and Kingston children. "This would give us a good check on . . . whether the general development, especially the skeletal development, in the two cities is comparable," Hodge explained. Similar information was then being sought from workers in the wartime bomb factories, where bone X-rays were an early warning of fluoride poisoning. "The purpose of X-raying the Newburgh children was, to pick up any toxic effect which would manifest itself in bone changes." Conference with members of the Technical Advisory Committee on Fluorination of Water Supplies, June 1, 1944, G-10 File.
24. G-10 file, Correspondence 1945–1952, New York Operations Research And Medicine Division, Box 38, Atlanta FRC, RG 326.
25. Dean's opposition "constituted a strong minority expression" Chairman Hodge noted. New York's leading dental official, David Ast, was furious at the Public Health Service expert's behavior that day. "It upset me a great deal," commented Ast. More than half a century later, Ast stills feels double-crossed by Dean. Ast had been planning the Newburgh experiment for more than a year. "I had conferred with him about the Newburgh study and he had encouraged me," Ast told me. Taped author interview, July 31, 2002. Ast was then "almost 100." While he said that he knew Hodge had "something" to do with the Manhattan Project, he attributed Dean's flip-flop on public-health concerns to career ambition. Behind the scenes a furious scramble was taking place to be the first to add fluoride to the United States' water. Dean had been planning his own fluoridation experiment, Ast said. "He was going to do it in Michigan. He wanted to get in before I could."
26. In an enthusiastic letter to Dr. William Davis of the Michigan Bureau of Public Health Dentistry, Dean makes no further mention of the worrisome potential "toxic effects" he had feared in Newburgh. "Let me know what you think of actually getting started on this proposition," he wrote to Davis on July 14, 1944. "I still think Grand Rapids would probably be the most desirable place for the fluorination." Money would be no problem, Dean suggested. "You would probably have little difficultly in obtaining this from a foundation, for instance the Kellogg Foundation," he wrote to Dr. Davis. Frank J. McClure, *Water Fluoridation: The Search and the Victory* (Bethesda, MD: U.S. NIDR, 1970), p. 112.
27. "Let no one think that any one of us would seriously consider exposing the population of a city of 165,000 [Grand Rapids' population in 1944] to a possible hazard of an unknown risk," the Chief Dental Officer for the PHS,

John Knutson, told the Michigan State Dental Society in 1953. J. W. Knutson, "An Evaluation of the Grand Rapids Water Fluoridation Project," *J. Michigan State Medical Sc.*, vol. 53 (1954), p. 1001. Cited in McClure, *Water Fluoridation*, p. 110.

28. Multiple interviews with author. First published in Griffiths and Bryson, "Fluoride, Teeth, and the Atomic Bomb," *Waste Not: The Reporter for Rational Resource Management* (September 1997).
29. Progress Report No. 1 of Contract No. W-7401-eng-49 at the University of Rochester (report of the work for period May 1, 1943, to December 31, 1943, submitted by Andrew H. Dowdy, M.D., Director), Box S09F01B227, ACHRE, RG 220.
30. "Du Pont" File, New York Operations Research and Medicine Division, Correspondence 1945–1952, Box 28–47, Box 36, Atlanta FRC, RG 326. For Priest's fluorine work at Columbia, see multiple papers in *Industrial and Engineering Chem.* (March 1947).
31. P. Dale and H. B. McCauley, "Dental Conditions in Workers Chronically Exposed to Dilute and Anhydrous Hydrofluoric Acid," *JADA*, vol. 37, no. 2 (August 1948). First presented at the twenty-fifth general meeting of the International Association of Dental Research, Chicago, June 21–22, 1947.
32. See chapter 4, notes 72 and 73.
33. The published study acknowledges the "assistance and suggestions of Drs Harold C. Hodge, . . . in the preparation of this paper." The unpublished version is via Griffiths and Honicker.
34. Division of Safety and Hygiene, March 1, 1949, John H. Fluker, Superintendent, Division of Safety and Hygiene, Columbus, OH, In Re: Harshaw Chemical Company. Memo, via John Fedor.
35. "Tabulation of results obtained from measurements of urine samples collected from workers at the Harshaw Chemical Company from 6 to 13 December 1945." Report No. 5373. From Capt. B. J. Mears to Mr. F. A. Becker, Harshaw Chemical Company, Md 319.1, General Essays, Lectures, Medical Report, Box 34, #4nn 326-85-005, Atlanta FRC, RG 326.

Chapter 7

1. J. Marks, *The Search for the Manchurian Candidate: The CIA and Mind Control* (New York: Times Books, 1979).
2. See Subproject 46 of John Marks collection at the National Security Archive. There is a second document in these files noting an interest in fluoride. A redacted November 29, 1949, letter discusses chemicals best suited to kill people. "One of these, sodium fluoacetate, when ingested in sufficient quantities to cause death does not cause characteristic pathologic lesions nor does it increase the amount of fluorine in the body to such a degree that it can be detected by quantitative methods." See Box 4, file titled "Document Indexes Abstracts and Documents," Marks Collection, National Archive.
3. "Those present at the meeting were Drs. Dowdy, Bale, Fink, McKann, Bassett, Hodge, and others representing the Rochester Group, Capt. Bryan

representing Col. Warren's office, and W. Langham representing the Santa Fe group." Folder 4, Box SoFo1B230, ACHRE, RG 220.

4. "Distribution and Excretion of Plutonium Administered Intravenously to Man," September 20, 1950, Division of Health and Biology, Folder 5, Box SoFo1B230, ACHRE, RG 220.
5. Welsome, *The Plutonium Files*, p. 475.
6. Details of those government experiments were declassified in 1994. On June 5, 1945, for example, a University of Rochester letter marked "Secret" details plans for "increasing the human metabolism studies." Ten additional patients are scheduled for "injection with T [code for uranium]," the letter states. "The preparation and analysis will be done by Dr. Hodge," notes the author, Dr. Andrew Dowdy, to Dr. W. F. Bale, of Rochester's "Special Problems Division." The letter is cc'd to Dr. Hodge. Dr. Andrew Dowdy to Dr. W. F. Bale, Special Problems Division. University of Rochester intramural correspondence, Box SoFo1B230, ACHRE, RG 220.

   The following year, 1946, Hodge was given direct responsibility for the experiments. L. H. Hemplemann and Wright H. Langham, "Detailed Plan of Product Part of Rochester Experiment." This document has a section called "General Plan of the Rochester Experiment" at p. 5, which details Hodge's involvement in the uranium experiments. Document marked 9000528, Box So9Fo1B230, ACHRE, RG 220.
7. The document, titled "Detailed Plan of Product [code for plutonium] Part of Rochester Experiment," includes a section on other human experiments. The medical director, Stafford Warren, had determined that "fifty subjects" were needed, ten for each substance, the document explained, "in order to establish, on a statistically significant number of subjects, the metabolic behavior of the hazardous material, product [code for plutonium], radium, postum, tuballoy [code for uranium] and lead." Under the subheading "Personnel and Distribution of Responsibility," a single name is listed for the uranium experiments: "Harold Hodge." Both patient accounts are from University of Rochester Monthly Progress Reports, for April 1947 (M-1968) and February 1947 (M-1954), Box So9Fo1B230, ACHRE, RG 220.

   An internal Rochester report on the experiments, "The Tolerance of Man for Hexavalent Uranium," noted that for the final subject, the alcoholic, the experiments had been successful, and that the "rise in urinary catalase and protein" from the man's liver suggested that, for uranium exposure, "tolerance had been reached." Samuel Basset, Albert Frenkel, Nathan Cedars, Helen Can Alstine, Christine Waterhouse, and Katherine Cusson, "The Tolerance of Man for Hexavalent Uranium," Folder 4, Box SoFo1B230, ACHRE, RG 220.
8. Dr. Sweet wanted to study whether uranium could be used in "therapy of brain tumors." See Bob Bernard, interviewed by Newell Stannard in 1975, for Hodge link to uranium injections on Bill Sweet's patients at Massachusetts General Hospital. DOE Opennet #0026691. However, according to an internal report from the Union Carbide Nuclear company, the Atomic Energy

Commission at Oak Ridge was "concerned with the long-term radiological effect that enriched uranium may have upon production employees who have inhaled dusts, mists and fumes of uranium." Accordingly, uranium injections were given to Massachusetts General Hospital patients following their tumor operations, in order to obtain "data on the distribution and excretion of uranium in these patients" and to "determine the permissible intravenous dose." S. R. Bernard and E. G. Struxness, *A Study of the Distribution and Excretion of Uranium in Man, An Interim Report*, ORNL-2304, Box S09F01B294, ACHRE, RG 220.

9. Morgan interview, January 6, 1995, by Gil Wittemore and Miriam Bowling, p. 109, ACHRE, RG 220.
10. And as an early member of a group of scientists known as the American Conference of Government and Industrial Hygienists (ACGIH), Hodge helped set standards for the "threshold" levels of chemicals and contaminants that millions of citizens breathe in factories and mills.
11. Advisory Committee on Human Radiation Experiments, *Final Report* (Washington, DC: Government Printing Office, 1995).
12. The biomedical work had continued during the cold war at the Rochester Atomic Energy Project (AEP), funded with millions of dollars from the federal Atomic Energy Commission.
13. In the 1930s generally workers had contracted jaw cancer from licking their brushes as they painted radium onto watch dials. The poisoning was widely reported in the press and guided the Manhattan Project in its insistence on secrecy to prevent similar lawsuits or bad publicity. For example, in a letter to Charles Keller of the AEC Declassification Branch, L. F. Spalding of the Insurance Branch contemplates declassifying a medical document "Biochemical Studies Relating to the Effects of Radiation and Metals" by Samuel Schwartz. Spalding warns that "the contents thereof might arouse some claim consciousness on the part of former employees" and writes that "in the event latent disabilities due to the exposures reported in this document should result in publicity similar to that which arose out of the "radium dial" industry, the public relations section would be involved." When Guttman's team asked in 1995 for the files of the AEC Insurance Branch, he recalled, nobody at today's DOE had even heard of the Insurance Branch. Finding the old documents was like "asking my nephew for his grandfather's stamp collection," Guttman said.
14. AEC Memorandum dated October 8, 1947, to Advisory Board on Medicine and Biology, "Subject: MEDICAL POLICY," Document DOE #1019707, also marked RHTG Classified Docs, Box RHA 248-7 2 of 3, Building 2714.H, Vault. Via Peter Eisler, *USA Today*.
15. "Questions of General Policy," November 16, 1943, Md 319.1 Report Medical Ferry Box 25, Manhattan Engineer District Accession #4nn 326-85-005, Atlanta FRC, RG 326. See also Jack Hein interview with UK journalist Bob Woffinden, at timecode 04:18:55 1997; "They also did extensive studies on

the people working in the atomic energy plants that might be exposed to fluoride."

16. "Memorandum to Major J. L. Ferry, Manhattan District Oak Ridge, from Capt. B. J. Mears, July 5, 1945, subject, Visit to E. I. Du Pont de Nemours & Co." "Preparation of the IBM cards will be done by Dr. Evans [Du Pont] after he has received his new equipment and the operators have been instructed by Mr. M. Wantman [Rochester]. . . . The results of the statistical survey will be available only to the Medical Section of the Manhattan District," Md 701, Medical Attendance, Box 54, Accession #4nn 326-85-005, Atlanta FRC, RG 326.
17. Whether fluoride damaged kidneys, and whether fluoride in urine would therefore be a good measurement of occupational fluoride exposure, was key information sought by the bomb program. (Extra fluoride was stored in the bones of those injured patients, the government scientists found.) AEC No. UR-38, 1948, Quarterly Technical Report. Also cited in Kettering Lab unpublished report, "Annual Report of Observations on Fluorides. October 25, 1954." Kettering did similar experiments on patients with damaged kidneys, according to this report.
18. Special Report 454, "Report on the Work of the Pharmacology Division," included in summarized subsection "The Toxicology of Special Materials," via Joel Griffiths.
19. Roholm to Col. J. P. Hubbard, Public Health Section, Dagmarhus, July 20, 1945. Hubbard is probably an Allied occupation official. The letter is in the files of the Rockefeller Archive, Folder 2102, Box 310, RF RG2 713. Roholm explained that he wanted to recontact H. T. Dean at the National Institute of Health and Margaret C. Smith at the University of Arizona, who had discovered that fluoride causes dental mottling.
20. Roholm to Frank J. McClure (U.S. National Institutes of Dental Research), June 13, 1946. On Roholm's attitudes toward American health care: Danish newspaper clipping in Roholm family scrapbook, translated by daughter-in-law Karin Roholm. Personal meeting in New York, May 2001.
21. "Fluorine interferes with the normal calcification of the teeth during the process of their formation," the U.S. Department of Agriculture claimed in 1939, "so that affected teeth, in addition to being unusually discolored and ugly in appearance, are structurally weak and deteriorate early in life. For this reason, it is especially important that fluorine be avoided during the period of tooth formation, that is from birth to the age of 12 years . . . this dental disease is found when water containing even as little as 1 part per million is used." *Yearbook of Agriculture* (1939), p. 212.

    "Fluorides are general protoplasmic poisons," the American Medical Association warned in 1943, "probably because of their capacity to modify the metabolism of cells by changing the permeability of the cell membrane and by inhibiting certain enzyme systems. . . . The sources of fluorine intoxication are drinking water containing 1 part per million or more of fluorine. . . . Another source of fluorine intoxication is from the fluorides used in the smelting of many metals, such as steel and aluminum, and in production of glass, enamel and brick." *JAMA*, vol. 123 (September 18, 1943), p. 150. Even the American Dental Association had editorialized in October 1944, "our knowledge of the subject certainly does not warrant the introduction of fluorine in community water supplies," the association's magazine stated, "we do know that the use of the drinking water containing as little as 1.2 to 3.0 parts per million of fluorine will cause such developmental disturbances in bones as osteosclerosis, spondylosis and osteopetrosis, as well as goiter." (Today, the EPA permits 4 parts per million of fluoride in water, a standard vigorously resisted by some EPA scientists, including the former senior toxicologist of the Office of Drinking Water, Dr. William Marcus.) Marcus interview with author.

22. K. Roholm, *Rejsebreve Indtryk Fra USA* (Efteraar 1945); *Ugeskrift For Laeger*, vol. 108 (1946), pp. 234–243.
23. Ibid. Before the war, Roholm recalled, "it was discovered that the concentration of fluoride; 1 milligram of fluoride per 1 liter drinking water; causes mottled teeth amongst those who drink the water, while the permanent teeth calcify, i.e., during infancy. The enamel become indistinct, chalklike and sometimes dark colored and fragile. The disease has since been discovered throughout the entire world and continues to be a serious problem of sanitary reasons, which makes it necessary to change the water supply."
24. Ibid., pp. 234–243.
25. In early 2001 Roholm's daughter-in-law, Karin, showed me a scrapbook of news stories collected by a family friend during his lifetime. She translated them for me over coffee at the New York YMCA at West Sixty-third Street. In his address Roholm made a single reference to fluoride. "In recent years, we have learned that a small quantity of the element fluoride in the drinking water significantly seems to protect against caries," he said. *Ugeskrift For Laeger*, vol. 110 (1948), pp. 221–226.

Chapter 8

1. Jamie Lincoln Kitman, "The Secret History of Lead," Part 1, *The Nation* (March 20, 2000), in which a 1985 EPA study is cited for heart-disease deaths. Kitman wrote, "According to a 1988 report to Congress on childhood lead poisoning in America by the government's Agency for Toxic Substances and Disease Registry, one can estimate that the blood-lead levels of up to 2 million children were reduced every year to below toxic levels between 1970 and 1987 as leaded gasoline use was reduced. From that report and elsewhere, one can conservatively estimate that a total of about 68 million young children had toxic exposures to lead from gasoline from 1927 to 1987."
2. Humor and ancestry; Interview with Edward Largent Jr. Arrogance: interview with Dr. Albert Burgstahler.
3. Kehoe testimony at Martin trial, p. 965.
4. For example, he was an associate editor of the American Medical Association's *Archives of Industrial Hygiene and Occupational Medicine.*

5. From 1925 to 1958 Kehoe was the medical director of the Ethyl Corporation, the partnership between Standard and General Motors that distributed the DuPont-manufactured antiknock gasoline additive known as tetra ethyl lead (TEL). In 1966 he told Congress that he "had been looking for 30 years for evidence of bad effects from leaded gasoline in the general population and had found none." Kitman, "The Secret History of Lead." Kehoe's work would take him to Germany immediately after World War II, from which he sent home photographs of the Nazi death camps. See also diary, RAK Collection.

   The German industrial conglomerate I. G. Farben had operated the Auschwitz camp with Hitler's SS. Before the war Farben had partnered in Germany and the United States with Standard Oil. Shortly before European hostilities broke out, Ethyl Corporation transferred the technology for making TEL to its German partner, greatly aiding the Nazi war effort. According to Farben official August von Knieriem at the Nuremberg war crimes trial, "Without tetraethyl lead the present method of warfare would have been impossible. The fact that since the beginning of the war we could produce tetraethyl lead is entirely due to the circumstance that shortly before, the Americans presented us with the production plans, complete with their know how." J. Borkin, *The Crime and Punishment of I. G. Farben* (New York: The Free Press, 1978), p. 78.
6. On April 17, 1952, Kehoe wrote to Seward Miller—medical director of the Division of Industrial Hygiene, Public Health Service—on behalf of nine corporations then sponsoring his fluoride research, to request that the PHS perform some fluoride safety studies on animals. The industry groups, Kehoe noted, "are concerned mainly with the results of exposure to fluorides in various occupations." These industries included "The Pennsylvania Salt Manufacturing Company, Aluminum Company of America, Reynolds Metals Company, Universal Oil Products Company, American Petroleum Institute, Kaiser Aluminum and Chemical Corporation, Tennessee Valley Authority, The Harshaw Chemical Company, [and] Minnesota Mining and Manufacturing Corporation." RAK Collection.
7. A great number of claims were settled out of court. The following is a partial listing of legal actions against U.S. corporations following the war, and during the early cold war, in which fluoride was suspected as a poison. These data are culled from press accounts and this author's research. See also E. J. Largent, "Fluorosis—The Health Aspects of Fluorine Compounds," for the difficulty of comprehensively tracking the frequency and number of fluoride lawsuits. Also, M. J. Prival and F. Fisher, "Fluorides in the Air," *Environment*, vol. 15, no. 3 (April 1973), pp. 25–32. "The number of out of court settlements of claims of fluoride damage to vegetation is impossible to determine, although it certainly exceeds the number of court-ordered payments."
   - 1946. The "Peach Crop Cases" by New Jersey farmers in Gloucester and Salem County, claiming $430,000 against DuPont and the U.S. government.

- *1946. Suit "exceeding half a million dollars" mounted against the Pennsylvania Salt Company, Sun Oil, and the General Chemical Company by some 41 farmers near the town of Delran, New Jersey, on the Delaware River.* Pennsylvania Salt was being sued along with Sun Oil and General Chemical for more than a half-million dollars by as many as forty-one different farmers in New Jersey and Pennsylvania. The farmers claimed that they had been poisoned by fluoride—their crops and farm animals killed. Downwind of the Pennsylvania Salt Company's plant in Cornwall Heights, built by the government during the war, lay a half-mile-square zone just across the Delaware River, "where all trees have been killed." Another of the company's fluoride plants in Easton, Pennsylvania, "revealed an almost identical picture of damage." John H. Claypool to Edward Largent, 10/19/45; "Recently the first actions in bringing suit have been taken in behalf of 26, out of an original 41, peach growers." Also Largent to R. W. Champion, Harshaw Chemical, 4/25/1946, File 13, Box 32, RAK Collection.
- *Immediately postwar. A Philadelphia gun club filed suit against the nearby Pennsylvania Salt Company.* According to Philip Sadtler: "The Plant had damaged the Philadelphia gun club which was next door—that was a relatively simple case. The gun club won because of my testimony, and all I had done was gather some of the vegetation and measured the fluorine." Taped author interview, March 23, 1993.
- *1948. Claims filed by a group of horticulturist farmers against phosphate fertilizer manufacturers in Bradenton, Florida, on the Gulf Coast, alleging agricultural damage.* "They won a large settlement," according to lead investigator Philip Sadtler. "The vegetation showed [damage] around the edges. One farmer named the (claprood?) family grew a large number of gladioli which were shipped all over the United States. For at least two years they were ruined by the phosphate roasting. Therefore, I was asked to go down to Bradenton to investigate the problem. I took samples and came home and analyzed them. They were no different from [what Sadtler had found in the fluorine poisoning from industry in] New Jersey. They won a large settlement. It took several years but they got repaid for what they had lost."
- *October 1948. Donora, Pennsylvania. Four and half million dollars in legal claims against U.S. Steel following some two dozen fatalities and thousand of injuries, blamed by one investigator on fluoride.* The legal action did not focus on fluoride.
- *1949. Lawsuits filed against the Tennessee Valley Authority (TVA) alleging fluoride pollution.* For example, "In connection with the plaintiffs living in the Columbia area who should be examined for possible fluorosis," Edward Largent to Joseph C. Swidler, General Counsel, TVA, Knoxville, Tennessee. Also, Kettering's William Ashe performed a pilot study in 1950 of conditions at TVA's phosphate fertilizer plant at the Wilson Dam. While most of the men had worked in the plant "a relatively short time

(a few months to 7 years; av. 2.6 years)," X-ray and urine analysis of the men found widespread bone fluorosis, urine values as high as 27.28 mgs of fluoride per liter, and concluded "1) There was a fluoride problem in the fertilizer plants at Wilson Dam 2) Some workmen are absorbing abnormal amounts of fluoride in quantities sufficient to produce fluorosis of the bone." Ashe to Dr E. L. Bishop, Director of Health TVA, File 14, Box 15, RAK Collection.

- *1950 Alcoa was fined for dumping fluorides into the Columbia River. Airborne fluorides heavily contaminated the grass and animal forage "which resulted in injury and death to cattle" and a claim for $200,000 compensation, according to newspaper accounts.* "Oregon Rancher asks $200,000 of Aluminum Company," *Seattle Times*, December 16, 1952. Cited in G. L. Waldbott et al., *Fluoridation: The Great Dilemma*, p. 296. Alcoa had dumped between 1,000 and 7,000 pounds of fluorides per month into the Columbia before 1950, according to *National Fluoridation News* (March–April, 1967), p. 3.
- *1950. Mr. and Mrs. Julius Lampert won suit against Reynolds' Troutdale, Oregon, plant for fluoride damage to gladiolus crops.* Cited in Waldbott et al., *Fluoridation*, p. 298, "Damages for Crop Burns," *Lewiston (ID) Morning Tribune*, February 6, 1962.
- Alcoa had compensated 141 farmers and cattle raisers in Blount County, Tennessee, prior to January 1, 1953, when another suit charged that fluoride fumes had damaged farmlands and injured cattle. Cited in Waldbott, *Fluoride*, p. 298, "Jury Decides Alcoa Liability Ended in 1955," *Knoxville (Tenn.) Journal*, July 30, 1955. Cited in Waldbott et al., *Fluoridation*, p. 298, "Alcoa Sued for Nearly $3 Million," *Knoxville (TN) Journal*, October 29, 1970.
- Also in Tennessee, by 1953 Monsanto was "faced with a number of claims for personal and property damage which total a considerable amount" including "claims for personal injury due to fluoride-containing effluents released from the stacks of the plant at Columbia owned by Monsanto." ("Last week when Mr. Wheeler was in Cincinnati he talked briefly with Dr. Heyroth about Monsanto's fluoride problems. As you know, Monsanto is faced with a number of claims for personal and property damage which total a considerable amount. These cases have accumulated over quite a period and have been pending for three or four years. It now appears that they may come to trial this fall." R. Emmet Kelly, M.D., Monsanto's medical director, to Robert Kehoe, July 7, 1953, File 26, Box 38, RAK Collection. Also: "Two couples, a man and a wife in each case, have filed claims for personal injury due to fluoride-containing effluents released from the stacks of the plant at Columbia owned by Monsanto . . . Symptoms described by the plaintiffs in part fit the description of acute fluoride poisoning, in part fit the description of chronic fluoride poisoning, and in part they appear so bizarre as to fit neither." Memorandum of meeting held August 19, 1953 between Edward Largent, Dr.

Francis Heyroth, Mr. John Jewell. Monsanto Chemical Company, and their attorney, Mr. Lon McFarland, August 20, 1953, File 26, Box 38, RAK Collection.

- *In Utah, by 1957, U.S. Steel had settled 880 damage claims totaling $4,450,234 with farmers in Utah County. An additional 305 claims for a further $25,000,000 were filed against the company.* D. A. Greenwood, "Background for Studies in Utah County." Unpublished paper given at the 1957 Kettering Fluoride Symposium, File 17, Box 42, RAK Collection. Another figure states that the legal claims against U.S. Steel in Utah were for $30 million. C. Butler, Discussion in: *Proceedings: Nat'l. Conf. on Air Pollution, Nov. 18–20, 1958* (Washington, DC: Government Printing Office), p. 268. Also, Prival and Fisher: "U.S. Steel paid $4 million to cattle ranchers around its steel mill near Provo, Utah, before spending $9 million on pollution control devices," citing *Chemical and Engineering*, vol. 65, no. 4, p. 66, February 24, 1958, and W. T. Purvance, *Chem. Eng. Prog.*, vol. 55, no. 7 (July 1959), p. 49, . This writer did not delve into the legal papers surrounding these cases. However, a clue as to their ultimate fate may be found in an essay by Keith E. Taylor Esq., senior partner, Parsons, Behle, and Latimer, Salt Lake City. He writes in 1982 of a proceeding "of nearly 25 years ago [in which] farmers and ranchers, approximately 300 strong, sought damages in a Federal Court for claimed injury to thousands of cattle and sheep and to numerous types of vegetation from fluorides emitted from an industrial facility." According to Taylor, Utah State scientists examined a dairy cow, Ms. Penelope, "ear tag No. G-571023," that plaintiffs claimed had been poisoned by fluoride; these scientists then "testified on behalf of the defendant, [and] came up with opposite conclusions. They found no evidence of fluorosis. The cause of her poor health was a wire that she had ingested, which had punctured her heart. . . . Except for that research . . . the result would probably have been different. Cows like Penelope would have continued to be diagnosed as dying of fluorosis. The farmers would not have had a compelling reason to clean the nails and wire from cattle feed, and to correct the various other problems that were contributing culprits. In the long run even the farmers would have been the losers." K. E. Taylor "Research Needs—A Lawyer's View" in J. L. Shupe, H. B. Peterson, and N. C. Leone, eds., *Fluorides: Effects on Vegetation, Animals, and Humans* (Salt Lake City, UT: Paragon Press, 1983), p. 359.

8. At a gathering of industry scientists and profluoride dental researchers in 1983, Seamans explained how wartime production had propelled a wave of fluoride pollution lawsuits against industry. "After the German bombing of Coventry had knocked out the English aluminum production," Seamans began, "President Roosevelt announced that America would build 50,000 planes. This was an unbelievable number and required a tremendous amount of aluminum, far more than existing capacity could produce. Accordingly, through a government agency known as the Defense Plant

Corporation, aluminum smelters were built wherever the needed electricity could be obtained . . . one DPC plant was built in the San Joaquin Valley of California. . . . There were, of course, no controls of any kind on this plant. As you can expect, there was a great consternation in the San Joaquin Valley. Vigilante committees were formed, and an injunction suit was filed. In August 1943, as a young lawyer for Alcoa I was sent out there to find out what the problem was all about. . . . Fortunately, Dr. Francis C. Frary, who was then director of research at Alcoa, had seen Roholm's book describing some of the consequences of cryolite mining in Greenland and this led him to wonder whether fluorides were the culprit . . . we all finally became convinced that there had been undue exposure to fluorides. Because we had the injunction suit and other claims to handle, as soon as possible we persuaded the Defense Plant Corporation to close the San Joaquin Plant. Thereafter, over a period of years we were able to settle all the cases, and thus the 'Riverbank, California' nightmare came to an end. After this experience however, knowledge quickly spread and soon we had claims and lawsuits around aluminum smelters from coast to coast. These required prodigious effort and great expenditures of time and money to settle. During the course of events, many significant and extended lawsuits were tried. Some of the more crucial were the Fraser case involving the Vancouver, Washington, plant and the Hitch case involving the Alcoa, Tennessee, Plant." Seamans continued, "There was very little solid information on the subject about what harm fluorides could do, what harm they did not do and what the tolerance levels were for people." Accordingly, "research was encouraged and supported at the University of Wisconsin, Utah State, Stanford Research Institute, University of Tennessee, Kettering Institute, the Boyce Thompson Institute for Plant research and other noted scientific centers." F. L. Seamans, "Historical, Economic and Legal aspects of Fluoride," in Shupe et al., eds., *Fluorides*, p. 5.

9. Frank Seamans to attorney Theodore C. Waters, August 30, 1956. "You will recall the occasion of our meeting together in Washington with a group of lawyers who have clients interested in the fluoride problem, at which time we were discussing the U.S. Public Health Service. The group, which in the past has consisted of representatives of Aluminum Company of Canada, Food Machinery and Chemical Corporation, U.S. Steel, Kaiser Aluminum and Steel, Tennessee Corporation and subsidiaries, Monsanto Chemical, Victor Chemical, Reynolds Metals Company, T.V.A. and Alcoa has had some discussions with Dr. Kehoe relative to some research and regarding the effect of fluorides on human beings." File 5, Box 76, RAK Collection.
10. On the relationship of Medical Advisory Committee to the Fluorine Lawyers, Seamans to Medical Advisory Committee, April 16, 1957: "The legal representatives of the several companies interested in the Kettering Research project have agreed that it would be advantageous if the principal liaison with Kettering were undertaken by persons of competent technical background . . . [to] conduct the necessary liaison between the Kettering Insti-

tute and the lawyers' group by a system of regularly scheduled visitations to Kettering and regular reports to the lawyer's group." File 17, Box 42, RAK Collection.

11. Memorandum on the Meeting of the Pittsburgh Section of the Industrial Hygiene Association, April 30, 1946, marked "Confidential."
12. During the key wartime Manhattan Project–sponsored "Conference on Fluoride Metabolism" at New York's Hotel Pennsylvania on January 6, 1944, Largent was a member of an inner-sanctum group of experts—along with Harold Hodge from the University of Rochester—that had decided how much fluoride U.S. workers could be "safely" exposed to inside the giant wartime atomic-bomb factories.
13. The phrase is from Francis McClure of the National Institutes of Dental Research. Largent's human experiments, McClure said, "provided much basic information not only for appraisal of industrial fluoride hazards but for resolution of a public health hazard which might be associated with use of fluoride drinking waters." F. J. McClure, *Fluoridation* (NIH publication, 1970), p. 200.
14. From the 1933 level of 1.43 mg F/Kg, raised in 1944 to 7 mg F/Kg. K. Roholm, *Rejsebreve Indtryk Fra USA* (Efteraar, 1945); *Ugeskrift For Laeger*, vol. 108 (1946), pp. 234–243.
15. Mellon guests were told that fluoride air concentrations of up to 4 parts per million had been found inside Alcoa plants, according to Dr. Lester Crawley of Alcoa. Memorandum on the Meeting of the Pittsburgh Section of the Industrial Hygiene Association, April 30, 1946. Stamped "Confidential." File 13, Box 38, RAK Collection.
16. See note 7 above. Also, E. J. Largent, *Fluorosis: The Health Aspects of Fluorine Compounds* (Columbus, OH: Ohio State University Press, 1961), p. 124.
17. "We have on file in our laboratory evidence of bone changes in employees in manufacturing operations where there are known atmospheric contaminations from fluorides," Largent noted. Attending the Mellon conference might help industry confront such threats, Largent added. The aluminum industry, in particular, had long ago seen the danger of workers' lawsuits for fluoride exposure and had taken preemptive action. "It was in anticipation of such an eventuality that Aluminum Company of America set out several years ago to obtain all possible data with which to meet such a situation," Largent told the Harshaw Chemical Company. Edward Largent to R. W. Champion, assistant sales manager, Harshaw Chemical Company, April 25, 1946, File 13, Box 38, RAK Collection.
18. Bovard announced at the Pittsburgh conference that there was "no evidence to prove that there was any relation between ankylosing spondylitis [the fusing of spinal vertebra] and the deposition of fluorides in the osseous tissue," Largent reported. Memorandum on the Meeting of the Pittsburgh Section of the Industrial Hygiene Association, April 30, 1946. Stamped "Confidential." RAK Collection Box 38 File 13. Bovard would regularly consult for the Kettering Laboratory and industry during the cold war, helping the TVA, for

example, in preparing its 1953 report, "Study of Fluoride Hazards—Final Report—Project Authorization 408."

19. Edward Largent to Dr. S. C. Ogburn Jr., manager, Research and Development Department, Pennsylvania Salt Company, May 8, 1946, File 13, Box 38, RAK Collection.
20. Memorandum on the meeting of the Pittsburgh Section of the Industrial Hygiene Association, April 30, 1946. Stamped "Confidential," File 13, Box 38, RAK Collection.
21. "Suggestions have been made both by Dr. Frary and by some of the du Pont group, including their medical director . . . that it might be advisable for representatives of du Pont, Aluminum Company, and Pennsylvania Salt to get together and to discuss carefully the whole problem." Robert Kehoe to S. C. Ogburn Jr., Pennsylvania Salt Company, May 25, 1946, File 13, Box 38, RAK Collection.
22. Joel Griffiths interview.
23. George Blakstone said that Maurice and Elmo had both participated in the lead experiments. He recalled that Maurice "would go in a chamber and inhale." Gentry Blackstone, who inhaled hydrogen fluoride gas, was also "drinking something, I think," according to George.
24. "Summary of Investigations of the Metabolism of Fluorides by Man and Dogs," Nov. 1, 1950, Unpublished Reports, vol. 24 b, RAK Collection.
25. Gerald Markowitz and David Rosner, *Deceit and Denial: The Deadly Politics of Industrial Pollution* (Berkeley: University of California Press, 2002), p. 110.
26. E. K. Largent, P. G. Bovard, and F. F. Heyroth, "Roetgenographic Changes and Urinary Fluoride Excretion among Workmen Engaged in the Manufacture of Inorganic Fluorides," *Amer. J. Roentgenol*, vol. 65 (1951), p. 42.
27. The dueling European and American medical theories had an odd transatlantic symmetry. Both scientists had studied workers handling cryolite, mined in the Danish colony of Greenland. Most of Europe's cryolite arrived via Roholm's hometown port of Copenhagen, while an old Philadelphia Quaker firm, the Pennsylvania Salt Manufacturing Company, whose workers Largent studied, had been granted sole rights to sell Danish cryolite in the U.S. During World War II, when the Nazis occupied Denmark, Greenland was governed by the Danish minister in Washington and a committee of five advisers, one of whom was Leonard T. Beale, the President of Pennsylvania Salt. R. K. Leavitt, *Prologue to Tomorrow: A History of the First Hundred Years in the Life of the Pennsylvania Salt Manufacturing Company* (The Pennsylvania Salt Company, 1950), chapter on fluorine, "Bad Actor turns Patriot," p. 78.
28. From 1939 to 1944, for example, industrial consumption of the most voluminous fluoride mineral, fluorspar, had more than doubled. It rose from 176,000 tons of fluorspar in 1939 to 410,000 tons in 1944. See Largent, Table 4, "The Occurrence and Use of Inorganic Fluorides." Paper given at 1953 Fluoride Symposium, in Unpublished Reports 32b, RAK Collection.

29. National Institute for Occupational Safety and Health, 1975 DHEW/NIOSH-76-103. Cited in "Summary Review of Health Effects" EPA/600/8-89/002F, December 1988, pp. 3–5.
30. Largent spent a career doubting Roholm. Roholm's findings were not "authenticated" and "cannot be generally accepted," Largent insisted. The Dane had failed to show "a causal relationship" between fluoride and injury, he told a Kettering roundtable of industry doctors. 1957 Kettering Fluoride Symposium, Box 63, RAK Collection.
31. H. C. Hodge and F. A. Smith, *Fluorine Chemistry*, vol. IV, p. 385. Also, "Largent's research is often quoted as evidence that bone changes, of the kind encountered in high fluoride areas and in industry, are never associated with harm elsewhere in the human organism and therefore have no significance." G. L. Waldbott, *A Struggle with Titans* (New York: Carlton Press, 1965), p. 289.
32. "When one finds, in cases of severe fluorosis of the bone, limitation of motion of the elbow and the X-ray reveals exostoses of unusual density about the elbow, one is probably entirely justified in concluding that the deformity and dysfunction are due to fluorosis, and that disability exists in association with and because of this disease, whether or not the man is aware of it, and whether or not he continues to do his job at the plant." Aluminum Company of America, Niagara Falls Works Health Survey, File 4, Box 82, RAK Collection.
33. "An exostoses (a bony outgrowth from the surface of the bone) on one of the bones of the right forearm and some calcification of the ligaments of the lower vertebrae were noted" in Ira Templeton's X-rays, according to Dr Smyth. He also found "In several instances bony outgrowths which seemed very much like the bone changes seen by Roholm in the monograph, 'Fluorine Intoxication,'" Largent told the Pennsylvania Salt Company. "On the basis of the data and the conclusions of that book alone, one would accept the presence of these outgrowths as evidence of the existence of fluorine intoxication. The conclusions of Dr. Smyth, who used the expression 'fluorine intoxication,' in the interpretation of his findings, would seem to follow this thesis," Largent added. Edward Largent, "Report to the Pennsylvania Salt Company," May 8, 1948, File 13, Box 38, RAK Collection.
34. Bovard to Kehoe, February 28, 1946. Also, Bovard X-ray interpretation, February 19, 1946, File 13, Box 38, RAK Collection.
35. J. Russell Davey, M.D., to Pennsylvania Salt Co., In Re: Ira Templeton. January 31, 1947, File 13, Box 38, RAK Collection.
36. S. C. Ogburn Jr., manager, Research and Development Department, Pennsylvania Salt Company, to Kehoe, February 10, 1947, File 13, Box 38, RAK Collection.
37. Kehoe to S. C. Ogburn Jr., February 12, 1947, File 13, Box 38, RAK Collection.
38. "Final Report of the Results of Investigations Relating to Fluoride Metabolism Conducted Under the Sponsorship of the Pennsylvania Salt Company." Unpublished Reports vol. 24-a, Kettering Laboratory, p. 13, RAK Collection.

39. Largent was familiar with Roholm's research, of course, and knew about the subtle effects of fluoride poisoning. During the war, for example, Largent told a 1943 industry conference at the Mellon Institute that "it seems probable that exposure to fluoride dusts may be capable of lowering the efficiency and well-being of workmen without inducing any very specific and dramatic symptoms." Proceedings of the Eighth Annual Meeting of Industrial Hygiene Foundation of America, Inc., November 10–11, 1943, p. 32.
40. E. J. Largent, P. G. Bovard, and F. F. Heyroth, "Roetgenographic Changes and Urinary Fluoride Excretion among Workmen Engaged in the Manufacture of Inorganic Fluorides," *Am. J. Roentgenol.*, vol. 65 (1951), p. 42.
41. See chapter 6 for the January 1944 fluoride conference held at the Hotel Pennsylvania in New York.
42. The Kettering Laboratory's "investigation of the metabolism of fluorides in the human body" was funded in 1953 by Alcoa, Reynolds, Kaiser, Harshaw Chemical, Minnesota Mining and Manufacturing, Universal Oil Products, the Tennessee Valley Authority, and the American Petroleum Institute. Kehoe to Seward E. Miller Medical Director, Division of Industrial Hygiene, Public Health Service, April 17, 1952, RAK Collection.
43. H. E. Stokinger, "Toxicity Following Inhalation of Fluorine and Hydrogen Fluoride," *The Pharmacology and Toxicology of Uranium Compounds* (New York: McGraw Hill, 1949), p. 1021. (Stokinger was a former Kettering scientist who went to Rochester during the war. Largent called him Herb). Also, in 1909 Ronzani had done HF inhalation studies on animals. He found no harm at 3 ppm, during a month's exposure, but was unable to report the same at 5 and 7.5 ppm. The Kettering Laboratory had abstracted Ronzani in their Kettering Abstracts series.
44. Among the attendees were the medical directors of DuPont, Alcoa, and TVA. Alcoa's attorney, Frank Seamans, from the Pittsburgh firm of Smith, Buchanan, Ingersoll, Rodewald, and Eckert, was also in attendance. File 13, Box 38, RAK Collection.
45. Largent had suggested in 1943 that 1.5–2.00 mg/liter of fluoride in urine might be associated with deposition in worker's bones. "Proceedings of the Eighth Annual Meeting of Industrial Hygiene Foundation of America, Inc," November 10–11, 1943, p. 32.
46. "If there were any changes in the bone as a consequence of 3 ppm it was beneficial deposition of fluoride, not harmful," he told writer Joel Griffiths. Griffiths interview.
47. Largent, "Absorption and Elimination of Fluorides by Man," Kettering Fluoride Symposium 1953, p. 92. Also, Largent reported in the unpublished "Industrial Health Surveys in Plants Processing Inorganic Fluorides," that in "a plant dealing with hydrogen fluoride . . . One man, who had an average urinary fluoride concentration of 9 mg. per liter, gave evidence of a moderate increase in radiopacity." He continued, "If all threats of medico-legal problems are to be avoided it seems probable that average urinary fluoride levels must be kept below 10 mg. per liter." Fluoride Symposium, 11.1, RAK Collection.

48. See 2001 ACGIH TLV, data summary for HF, "based on results of controlled inhalation studies in human volunteers" (Largent cited). ACGIH also cite E. Ronzani, "Influence of the Inhalation of Irritant Industrial Gases on the Resistance of the Organism to Infectious Disease. Experimental Investigations. II. Hydrofluoric Acid Gas, Ammonia, Hydrochloric Acid Gas," *Arch. Hyg.*, vol. 70 (1909), pp. 217–269. Ronzani was prompted to his studies because "disputes about the duties of factory and workshop owners towards their neighbors are brought to the court in rising frequency." He therefore sought a "no effect" level to help resolution of such disputes. He studied animals at various concentrations of HF, including 7.5 ppm and 5.0 ppm, but was forced to go to 3 ppm to find a no-effect level over 31 days—little comfort surely almost a century later for workers breathing HF today at 3 ppm *for all of a working life*.

In the NIOSH "Criteria Document for a Recommended Standard: Occupational Exposure to Hydrogen Fluoride," Publication 76-143, it is noted that Elkins had found "workers in the etching process had nosebleeds as did welders exposed to 0.4–0.7 mg F/cu m who were excreting 2–6 mg F/liter of urine . . . other workers exposed to 0.1–0.35 mg F/cu m and excreting, on the average, 4.5 mg F/liter reportedly experienced sinus trouble. The ACGIH suggested that the urinary excretion values reported by Elkins seemed 'inconsistently high' relative to airborne HF levels, and that dietary F was suggested as a possible factor." Citation, H. B. Elkins, *The Chemistry of Industrial Toxicology*, 2nd ed. (New York: John Wiley and Sons, Inc., 1959), pp. 71–73.

The ramifications of the ACGIH reliance on Largent and Ronzani can perhaps be seen in the U.S. standard for HF occupational exposure of 2.5 mg HF/cu m, compared to other countries (cited in NIOSH, above document, 1976): The (former) Soviet Union, 0.5 mg HF/cu m, Hungary and Poland 0.5 mg HF/cu m, (the former) East Germany and Czechoslovakia, 1 mg HF/ cu m, and Bulgaria 1 mg HF/cu m.

49. In that second interview, Largent became aware that the interviewer Joel Griffiths might not view his experimental work favorably. The verbatim exchange continued as follows:

EL: I never did develop osteofluorosis.

JG: Excuse me?

EL: I never developed personally any aspect of osteofluorosis—you just got through saying I developed osteofluorosis.

JG: Because I think that is what you told me the last time we talked.

EL: No—I would have talked about skeletal deposition, and that is not osteofluorosis.

JG: Well, skeletal deposition, right—that led to some difficulties with your knees.

EL: Not in the slightest.

JG: Well, this doesn't seem to jibe with what you told me the first time.

EL: That's not true—I was developing more like osteoporosis—I have arthritic difficulties in my extremities serious enough that the right knee was replaced with a prosthesis but that was more on the side toward osteoporosis than fluorosis—I didn't get enough to do me any good, I can tell you that. [Osteosclerosis, thickening of the bone, is a sign of a small amount of additional fluoride exposure; osteoporosis is an indication of massive fluoride exposure, Roholm and others reported.]

JG: Because you said to me quite distinctly the first time that it was osteofluorosis.

EL: No.

JG: And that fluoride can cause this condition.

EL: No.

JG: And that as far as you were concerned that was what it was.

EL: No.

JG: And that you believed it could have possibly come from the drinking water in the high school you attended in Fort Ames, Iowa, back in nineteen-whatever-it-was.

EL: Yeah.

JG: And also that the fluoride that you absorbed in your experiments might possibly have been a contributing factor.

EL: Factor—what factor?

JG: To the osteofluorosis.

EL: I didn't have osteofluorosis—at any time.

JG: I see, because the first time I'm certain that you said you did.

EL: No—I don't think that I did.

JG: In other words, you're not saying it now.

EL: I don't know what I said then, but if I said it then I was wrong. . . . If you say I developed osteofluorosis I will challenge that . . . I didn't get enough fluoride to do me any good.

JG: Well, let me see if I can find the tape and see I'll see if I misheard you.

EL: You may not have misheard me, but you may be able to correct me if I misspoke.

50. Fluoride appears to carry aluminum over the blood-brain barrier; the aluminofluoride complexes then damage the brain structure. See esp. J. A. Varner, K. F. Jensen, W. Horvath, and R. L. Isaacson, "Chronic Administration of Aluminum-Fluoride or Sodium-Fluoride to Rats in Drinking Water: Alterations in Neuronal and Cerebrovascular Integrity," *Brain Research*, vol. 784 (1998), pp. 284–298. "There are striking parallels between al-induced alterations in cerebrovasculature [and] those associated with Alzheimer's disease and other forms of dementia."

Chapter 9

1. Collected by Dan Hoffman, "Three Ballads of the Donora Smog," *New York Folklore Quarterly*, no. 5 (spring 1949), pp. 58–59. Quoted in Lynne Page Snyder, "'The Death Dealing Smog over Donora, Pennsylvania': Industrial Air Pollution, Public Policy, and Federal Policy, 1915–1963" (1994). Available from UMI Dissertation Services.

2. *Bulletin No. 306, Air Pollution in Donora, PA, Epidemiology of the Unusual Smog Episode of October 1948* (Public Health Service).
3. Donora is often referred to as the worst recorded air-pollution disaster in U.S. history. This may or may not be entirely true. During a similar seasonal temperature inversion from November 12 to 22, 1953, between 175 and 260 people were killed in New York City from air pollution, according to Howard R. Lewis, *With Every Breath You Take* (New York: Crown Publishers, 1965), p. 19. Although there were numerous complaints of eye irritation and coughing, the total number of New York deaths from the smog incident was only revealed later by statistical analysis, Lewis writes.
4. A key source for this chapter is Lynne Page Snyder's excellent "'The Death Dealing Smog over Donora.'"
5. The 1949 official Public Health Service report, *Bulletin No. 306, Air Pollution in Donora, PA*, lists twenty deaths. However, Snyder refers to "dozens" of deaths, p. viii. Residents report many additional deaths in the weeks after the disaster. For example, "The death of an estimated 100 people in the following year was attributed to the smog. Also, there were a lot of people who were affected in other ways. They were sick with respiratory problems. Internal illness and a couple of cases of blindness occurred." Account of former resident Joe Battilana, submitted as a 1970 report to Professor Gerard Judd of Phoenix Community College.
6. Berton Roueché, article in *The New Yorker*, September 30, 1950.
7. Roueché, case 11, p. 51; and from PHS Bulletin No. 306. Ceh's name is from *The New Yorker* article.
8. Roueché, case 9, p. 50 ; PHS Bulletin No. 306.
9. Author's taped interview, March 24, 1993.
10. Snyder, p. 25.
11. Roueché, p. 41.
12. Author's taped interview, March 24, 1993.
13. Recollections of Mayor John Lignelli, who attended the game, in "Donora's Killer Smog Noted at 50," *Pittsburgh Tribune*, October 25, 1998. See also PHS report and Snyder, p. 27, for death tally.
14. Snyder, p. 28.
15. *New York Times*, November 1, 1948; cited in Snyder, p. 29.
16. Snyder, p. 33.
17. For employment data, see Snyder, p. 35. For profit data, see Ross Bassett, "Air Pollution in Donora, PA" (December 6, 1990), unpublished paper, pp. 11, 21–41. Paper from Allen Kline. See also Paul A. Tiffany, *The Decline of American Steel: How Management, Labor, and Government Went Wrong* (New York: Oxford University Press, 1988).
18. The main thoroughfare, McKean Avenue, was named for Andrew Mellon's banker James S. McKean, who had brought Mellon and Donner together with coke baron Henry Clay Frick and whose combined investment of $20 million raised the first steel works on the virgin site in 1901. *Pittsburgh Press*,

March 18, 1934, Society Section, p. 11. Also, H. O'Conner, *Mellon's Millions* (New York: Blue Ribbon Books, 1933).

19. Bassett, "Air Pollution in Donora, PA."
20. Snyder, p. 71. Also, author interview with Bill and Gladys Shempp.
21. E. K. Roholm, "The Fog Disaster in the Meuse Valley, 1930: A Fluorine Intoxication," *J. Hygiene and Toxicology* (March 1937), p. 126. Also, W. S. Leeuwen, "Fog Catastrophe in Industrial Section South of Liege," abstracted in *J. Ind. Hygiene*, vol. 13, no. 7 (September 1931), pp. 159–160 (abstract section).
22. Sadtler gathered vegetation from across the region, tested it, and found that fluoride pollution was endemic and serious. "Buttonwood leaves had anywhere up to twelve hundred parts per million of fluorine," Sadtler noted. Further afield there was much less fluoride in the environment. "To get clean air with no fluorine damage, I had a friend who was a professor at Penn State University and he picked up leaves for me and they had ten parts per million," Sadtler said. Author interview.
23. Although coal was a source of fluoride, this knowledge was poorly disseminated. (Francis Frary announced the discovery to the Air Hygiene Foundation in 1946, as we saw in chapter 8.) Roholm makes no mention of coal in his discussion of the Meuse Valley disaster, for example. And the role that fluoride from coal may have played in the London smog disasters is almost entirely ignored.
24. E. K. Roholm, *Fluorine Intoxication: A Clinical-Hygienic Study, with a Review of the Literature and Some Experimental Investigations* (London: H. K. Lewis and Co. Ltd, 1937), chapter 13.
25. When the senior U.S. Steel metallurgist Glen Howis, who was born in Donora, had a routine medical exam before attending Penn State, a college doctor told him, "I can always tell you boys from the valley from the looks of your X-rays. Your lungs are always clouded," Howis recalled. Author interview in Donora.
26. The U.S. steel industry emitted 64,600 tons of fluoride in "1968 or 1972," according to EPA figures, cited by the Canadian National Research Council, NRCC #16081, ISSN 0316-0114. "Coal for power" is next at 26,000 tons, phosphate rock processing at 21,200 tons, and then aluminum smelting, at 16,230 tons. See similar data in "Summary Review of Health Effects Associated with Hydrogen Fluoride and Related Compounds" (U.S. Environmental Protection Agency, December 1988). For characterization of fluoride as "worst," see citations for fluoride toxicity and damage in chapter 15. For example: "From 1957 to 1968, fluoride was responsible for more damage claims than all twenty other air pollutants combined." N. Groth, "Air Is Fluoridated," *Peninsula Observer*, January 27–February 3, 1969. See also chapter 8, citations on lawsuits against the steel industry, and chapter 15's reference to fluoride's synergistic potential to worsen the toxicity of such pollutants as sulfur dioxide.
27. For fluoride's chronic health effects in Donora, see account of resident Devra Lee Davis, *When Smoke Ran Like Water* (New York: Perseus Press, 2002).

28. B. Davidson, *Collier's* (October 23, 1948). But other air pollution experts, such as Harvard Professor Philip Drinker, had scorned the idea that a Meuse Valley–type disaster could occur in the United States. "We have no districts in which there is even a reasonable chance of such a catastrophe taking place," he asserted. P. Drinker, *Industrial and Engineering Chem.* (November 1939).
29. Medical exams of plaintiffs by Kettering physicians, July 1950, William Ashe physician in charge. Box 5, RAK Collection.
30. Snyder, p. 28.
31. Dudley A. Irwin, Aluminum Company of America, minutes of meeting, Air Pollution Abatement Committee, the Chemists Club, New York City, January 11, 1950. Minutes of Manufacturing Chemists Association, from searchable database of the Environmental Working Group.
32. Oscar Ewing also led a semisecret group of administration insiders known informally as the Monday Night Steak Group. These men met most weeks at Ewing's Wardman Park apartment in Washington, DC, to plot strategy and discuss government policy over dinner and cigars. Clark Clifford, a military confidante and Truman favorite, was a regular at the Monday night meetings. See Ewing interview and Clark Clifford's in the Truman Library (available online).
33. P. Healy, "The Man the Doctors Hate," *Saturday Evening Post*, July 8, 1950.
34. Ewing's war years were spent in a Washington hotel suite with Alcoa senior management, defending the company's strategic interests from upstart companies such as Reynolds and Kaiser, who were fighting Alcoa's nearly fifty-year monopoly on aluminum production. After the war Ewing was invited to an intimate Washington dinner with Alcoa's president, Arthur Vining Davis, and senior officials from the Alcoa "family." Arthur Hall to Ewing, September 4, 1945. Personal Correspondence: August 1, 1944–September 20, 1945.
35. In early 1947 Ewing was a special assistant to the attorney general. He became FSA administrator in August 1947. See oral history interview, Truman Library, available online.
36. Ewing to Ingersoll, June 30, 1947, Political File, Correspondence, Ewing Collection. Ewing helped family members gain from trading fluoride. On July 8, 1946, he arranged a meeting for a relative, Thomas Batchelor, and Paul Collom, president of the Farmers Bank of Frankfort, Indiana, with President Allen B. Williams of the Aluminum Ore Company, in regard to "some fluorspar property in Kentucky" that Collum had acquired. Personal Correspondence, Ewing Collection.
37. *Pittsburg Press*, November 3, 1948. John Bloomfield was no stranger to Donora. Twenty years earlier, as a public official, he had helped American Steel and Wire attorneys to prepare a legal defense against pollution-damage claims by area residents. His job had been to test air quality. Bloomfield now told the newspaper that he recalled that his old measurements in Donora had shown that industrial emissions were safely diluted. Snyder, p. 40.

38. "Therefore the Company comes to us." Kehoe handwritten note, with the word "Mr. Jordan" (president of American Steel and Wire) at the top. Box 5, RAK Collection.
39. Snyder, p. 148.
40. On Schrenk's participation, see "Committee on Fluoride Metabolism, Round Table Discussion During Luncheon Period, Continued in the Evening, January 6, 1944." Conference on Fluoride Metabolism, Hotel Pennsylvania, New York. File Labeled G-118 (c), A2, Box 36, Accession #72C2386, Atlanta FRC, RG 326. Also, James Conant wrote to the Bureau of Mines, at Col. Warren's request, to have Schrenk go to the Rochester bomb program during the war. Conant to R. R. Sayers, February 3, 1944, Document #0291, Records of Section S-1 Executive Committee, RG 227.3.1.
41. Snyder, p. 152.
42. Snyder, p. 152.
43. *Chemical and Engineering News* (December 18, 1948) and author interview. The PHS report on Donora did not find excessive dental mottling. Author visit to Donora in 1993 noted severe mottling. For preexisting community health problems, see Donora resident Devra Lee Davis's *When Smoke Ran Like Water: Tales of Environmental Deception and the Battle Against Pollution* (New York: Basic Books, 2003).
44. U.S. Steel officials knew of the Meuse Valley disaster and of Roholm's report that blamed fluoride. Court Brief, "Evidence of Foreseeability," Box 5, RAK Collection.
45. Snyder, p. 29.
46. Pete Eisler, "Poisoned Workers and Poisoned Places," multipart series, *USA Today*, September 6–8, 2000.
47. All correspondence, Box 5, RAK Collection.
48. W.F.A to Dr. Kehoe, undated, RAK Collection.
49. Box 5, RAK Collection.
50. Ashe, of course, was well acquainted with Alcoa officials and their concerns with fluoride. That summer he had performed an investigation of health conditions for Kettering in Alcoa's Massena aluminum plant and found widespread injury and disability in workers that he attributed to fluoride. Aluminum Company of America, Niagara Falls Works Health Survey, File 4, Box 82, RAK Collection.
51. Not a conclusion shared by Phyllis Mullenix, who said that if the fluorine had been in soluble gaseous form, then it might readily have passed into the blood, leaving no trace in the lung tissue.
52. The meeting had been arranged in advance through a family friend, Sadtler explained. Author interview.
53. After meeting directly with the FSA in Washington, the CIO allocated $10,000 for the investigation. Oscar Ewing was close to labor leaders and had been an associate of Sidney Hillman, boss of the CIO. Hillman died in 1946.

54. PHS memorandum, November 16, 1948: "Report of Investigation at Donora, Pennsylvania," to Chief of Industrial Hygiene Division from Chief of the Field Unit, Duncan A. Holaday. PHS, Air Pollution Medical Branch, Special Projects, Folder 542.1 (1956). National Archives.

Chapter 10

1. Snyder, p. 70.
2. Author interview with Allen Kline, March 23, 1993.
3. "The Donora Smog Disaster," *Hygia, The Health Magazine* (AMA), October 1949.
4. Thomas Bell, *Out of This Furnace* (Boston: Little Brown and Co., 1941; reprinted Pittsburgh: University of Pittsburgh Press, 1976), pp. 356–357, cited in Ross Bassett, "Air Pollution in Donora, PA," unpublished paper, December 6, 1990.
5. Snyder, p. 217.
6. Following the smog, "Not a single adjustment was made in the production system—no pollution control devices, nothing, and there was nothing ten years later," Allen Kline told me.
7. Snyder, p. 219.
8. Kehoe to J. G. Townsend, Townsend to Kehoe, and data for Ashe, Box 5, RAK Collection. Also, Snyder, p. 258.
9. See chapter 9.
10. *Air Pollution in Donora, PA*, Bulletin 306, USPHS, p. 161.
11. Box 5, RAK Collection.
12. W. F. Ashe to E. Soles, July 11, 1949, and Largent's report from August 8, 1949, which found 110 mgs f/kilo (dry basis) in elm leaves three quarters of a mile opposite the open-hearth furnace. Box 5, RAK Collection.
13. F. A. Exner, "Economic Motives Behind Fluoridation," address to the Western Conference of Natural Food Associates, Salt Lake City, Utah, October 27, 1961.
14. *Monessen Daily Independent*, November 18, 1949, cited in Snyder, p. 170.
15. She has evaluated the health threat from several government Department of Energy nuclear sites, including Oak Ridge. She wrote the government monograph "Summary Review of Health Effects Associated with Hydrogen Fluoride and Related Compounds." (U.S. Environmental Protection Agency, December 1988).
16. So did the Society for Better Living, which charged the mill with reducing effluents during the test smog by staggering the run of mill processes rather than performing them simultaneously, as was standard practice. Snyder, p. 170.
17. Drinker, too, was well aware of the fluoride problem facing the AEC, since he serviced the agency as a litigation consultant on stack "waste gases." Phillip Drinker to Dr. Thomas Shipman, Health Division Leader, Los Alamos, November 14, 1950. RG 326. Medicine Health and Safety. NARA.

18. A. Ciocco and D. J. Thompson, "A follow-up of Donora ten years after: methodology and findings," *Am. J. Public Health*, vol. 51 (1961), pp. 155–64. H. Lewis, *With Every Breath You Take*, p. 201
19. Memorandum: "Discussion with Mr. Rumford on his 'Study of the Correlation of Meteorological Conditions and Morbidity in Donora' during his recent one day visit to Washington," from Nicholas Manos, Chief Statistician, Air Pollution Medical Program, Division of Special Health Services, to Records File, November 18, 1957, Box 13, File 542.1, RG 90 Records of the Public Health Service, Air Pollution Medical Branch Project Records, 1953–1960.
20. Snyder describes Rumford as a "consultant." He is described as being "assigned" to Dr. Ciocco in the memorandum "Mr. Rumford's Report on Donora" from Nicholas Manos, Chief Statistician, Air Pollution Medical Program, to Chief, Air Pollution Medical Program, January 29, 1958, Box 13 File 542.1, RG 90 Records of the Public Health Service, Air Pollution Medical Branch Project Records, 1953–1960.

Chapter 11

1. In 1946 Congress passed the Strategic and Critical Materials Stockpiling Act. In 2001 the Defense Nation Stockpile, maintained by the Pentagon, held 112,000 tons of fluoride in sites around the country. U.S. Geological Survey, *Minerals Yearbook*, 2001.
2. The task force was known formally as the President's Materials Policy Commission; its report was published in five volumes as *Resources for Freedom* in June 1952. (An annex to this report, referred to in several commission documents found in the Truman Library, is not in the National Archives holdings, and researchers were not able to find a reference for it. "The final report of the Paley Commission consists of only 5 volumes, all of which are open," stated an e-mail from NARA archivist Tab Lewis, June 26, 2001.)
3. D. M. Lion, "Fluorspar, Draft Commodity Study," marked "RESTRICTED," Box 12, Folder "Fluorspar," PMPC, Truman Library. Fluorspar's use increased a hundredfold from 1887 to 1950—annual consumption from 5,000 short tons to 426,000 thousand short tons. D. M. Lion, "Commodity Studies, Fluorspar," NSRB 6109, Paley Commission, Truman Archive.
4. Haman and Anderson to PMPC. Haman noted especially "the comparatively new use of fluorspar in the production of uranium hexafluoride for the manufacture of the atomic bomb." PMPC, Truman Library, Box 113, Fluorspar.
5. D. M. Lion, "Fluorspar, Draft Commodity Study." Also, H. Mendershausen, "Review of Strategic Stockpiling." Only 28,671 tons of bomb-quality "acid grade" fluoride was stockpiled in October 1951, just 11 percent of desired levels, Menderhausen reported. PMPC, Truman Library.
6. Analysts were enthusiastic about the phosphate beds as a source of fluorine. "If economic methods can be developed and applied for recovering most of this fluorine as a byproduct of phosphate processing, the yield would amount to the equivalent of about 600,000 to 700,000 tones of 100 percent

calcium fluoride . . . Such an annual increment would more than make up our potential deficit 10 years hence," the Paley Commission stated. "All the resources of technology must be enlisted to solve the problems of assuring ample supplies of fluorspar, or fluorine containing materials," the report added. *Resources for Freedom*, June 1952.

7. A. F. Blakey, *The Florida Phosphate Industry* (Cambridge, MA: Harvard University Press, 1973), p. 112 citing the Florida Air Pollution Control Commission.
8. Paul Manning to Donor M. Lion, August 13, 1951, PMPC, Truman Library, Box 113, Fluorspar.
9. For an account of this longstanding battle, see A. F. Blakey, *The Florida Phosphate Industry*, citing the Florida Air Pollution Control Commission. Also interview with Philip Sadtler and Congressional hearings chaired by Senator Ed Muskie. Subcommittee on Air and Water Pollution of the Committee on Public Works of the United States Senate, 59th Congress, June 7–15, 1966 (Washington, DC: U.S. Government Printing Office), pp. 113–343.

   Paul Manning was well aware of the scale of the fluoride-pollution problem around the country. He was an associate of both Robert Kehoe and the head of the Fluorine Lawyers Committee, Frank Seamans, who invited him to participate in the sponsored research at the Kettering laboratory. "On behalf of our client, Alcoa, I have for some time been participating in an informal group of lawyers, all of whom have clients involved in fluoride claims of one kind or another. . . . I know that your company is interested in this problem to some extent and that conversations have occurred between myself and other Alcoa representatives and personnel of your company." Seamans to Paul D. V. Manning, International Minerals and Chemical Corporaton, August 30, 1956. File 76, Box 5, RAK Collection.
10. The Tennessee Valley Authority was also interested. "A recovery system that would pay its own way should be attractive . . . the present price of sodium fluosilicate and increased demand for it will very likely encourage more manufactures to recover it" said TVA's T. P. Hignett. PMPC, Truman Library, Box 113, Fluorspar.
11. Elias, Technology Reports, p. 9, PMCC, Truman Library, Box 130.
12. It was beyond my resources to probe deeply the Florida cold-war uranium story. How much additional fluoride was produced by such production, and whether money was saved by using fluorsilicic acid as a water fluoridation agent remains to be reported. For Florida as source of cold-war uranium, see P. Eisler, "Poisoned Workers and Poisoned Places," *USA Today*, multipart series, September 6–8, 2000. Of interest, two companies producing uranium from phosphate included International Minerals and Chemical Corporation and the Olin Mathieson Corporation. The former is cited in the text and notes above, while Olin was one of the companies that joined Reynolds in the amicus curie brief for the Martin trial appeal (see chapter 13).
13. Rebecca Hanmer, Deputy Assistant Administrator for Water, to Leslie A Russell, DMD, March 3, 1983.

14. D. McNeil, *Fight for Fluoridation*,(New York: Oxford University Press, 1957), p. 209, n. 22.
15. "By the end of the fifth year a reexamination of the school pupils in Newburgh and Kingston showed that the Newburgh children had approximately 65 percent fewer cavities than the children of Kingston. The report of these findings was made public over my name . . . ," Oscar Ewing, Oral History Interview with J. R. Fuchs of the Truman Library in Chapel Hill, North Carolina, April and May 1969. Truman Library.
16. Philip R. N. Sutton, *Fluoridation; Errors and Omissions in Experimental Trials* (Carlton: Melbourne University Press, 1959), p. 49, for calcium in water, citing E.W. Lohr and S. K. Love., 1954, "The Industrial Utility of Public Water Supplies in the United States, 1952" (U.S. Geological Survey), and United Kingdom Mission Report, *The Fluoridation of Domestic Water Supplies in North America as a Means of Controlling Dental Caries* (London: HM Stationary Office, 1953). John A. Forst, MD, The University of the State of New York, the State Education Department, Albany, NY, Division of Pupil Services, to Dr. James G. Kerwin, the Department of Health, Passaic, New Jersey, October 26, 1954. Via Martha Bevis.
17. See chapters 9 and 10 for Ewing's profile. Clifford had helped to write the National Security Act of 1947, which had authorized the CIA. See his interview at the Truman Library.
18. Letter from H. V. Smith to George Waldbott, January 6, 1964, cited in *A Struggle with Titans*, p. 65 . See also, "Beyond certain limits, fluorides are toxic and that the first evidence of toxicity manifests itself in the form of mottled enamel," B. Bibby, "Effects of Topical Application of Fluorides on on Dental Caries." In *Fluorine in Dental Public Health* (New York Institute of Clinical Oral Pathology Inc, A Symposium, 1944).
19. D. McNeil, *Fight for Fluoridation*, p. 74.
20. M. C. Smith and H. V. Smith, "Observations on the Durability of Mottled Teeth," *Am. J. Public Health*, 30 (1940), p. 1050, cited in Waldbott, *A Struggle with Titans*, p. 12.
21. B. Lee, "Boon or Blunder?" *Toronto Globe and Mail*, January 1954, cited in G. L. Waldbott, *A Struggle with Titans*, p. 11.
22. Waldbott had emigrated to the United States in 1923. His father, Leo Waldbott, barely escaped the terror of Hitler's regime, in December 1938 joining George and an elder brother Emil in Detroit, Michigan. For several generations the Waldbotts had been important members of the community of Speyer, on the Rhine. Leo Waldbott was chairman of the Speyer teachers' and cantors' club, and treasurer of the local Jewish home for the elderly, which was burned to the ground by the Nazis on November 10, 1938. "My Life Before and After Jan. 30, 1933," by Leo Waldbott, via Elizabeth Ramsey, George Waldbott's daughter.
23. For pollen, "In Memoriam—G. L. Waldbott (January 14, 1898–July 17, 1982)," *Fluoride* vol. 15, no. 4 (1982); *Contact Dermatitis* (Springfield, IL: Charles C. Thomas, 1935); "Anaphylactic Death from Penicillin," *J. Am. Med. Assoc.*,

vol. 139 (1949), pp. 526–527; *Time*, March 7, 1949; "Smoker's Respiratory Syndrome," *J. Am. Med. Assoc.*, vol. 151 (1953), pp. 1398–1400.

24. Edith Waldbott referred her husband to the hearings chaired by New York Congressman James Delaney (Dem.) in February 1952, before the House Select Committee to Investigate the Use of Chemicals in Food and Cosmetics. (Congressman A. L. Miller here exposed Oscar Ewing's vested interest as a former Alcoa attorney.) She pointed to a January 1954 eight-part series called "Boon or Blunder" in the *Toronto Globe and Mail*. She had also seen a *Seattle Times* story of December 16, 1952, which detailed Alcoa's efforts to fund fluoride research, according to Waldbott, *A Struggle with Titans*. The first article Edith Waldbott gave her husband was James Rorty's "The Truth About Fluoridation" in *The Freeman* (Irvington-on-Hudson, NY: June 1953).
25. G. L. Waldbott, "Chronic Fluorine Intoxication from Drinking Water," *Int. Arch. Allergy Appl. Immunol.*, 7 (1955), pp. 70–74. "Incipient Chronic Fluorine Intoxication from Drinking Water," *Acta Med. Scand.*, 156 (1956), pp. 157–168. "Tetaniform Convulsions Precipitated By Fluoridated Drinking Water," *Confin. Neurol.*, 17 (1957), pp. 339–347.
26. G. L. Waldbott, "Allergic Reactions to Fluorides," *Intern. Arch. Allergy*, 12 (1958), p. 347, and "Urticaria Due to Fluoride," *Acta Allergologica*, 13 (1959), p. 456.
27. R. Feltman and G. Kostel, "Prenatal and Postnatal Ingestion of Fluorides—Fourteen years of Investigation—Final Report," *J. Dent Med.*, 16 (1961), pp. 190–199. See also double blind tests in Haarlem, Holland, by Moolenburgh and others. G. W. Grimbergen, "A Double Blind Test for Determination of Intolerance to Fluoridated Water (Preliminary Report)," *Fluoride*, 7 (1974), pp. 147–152.
28. The director of the Forsyth Dental Infirmary for Children in Boston, V. O. Hurme, warned in 1952, "Medical researchers have paid relatively little attention to the problem of chronic fluoride toxicosis." He worried about fluoride's potential effect on teeth and gums. "Among the very inadequately studied physical signs of fluoride toxicosis are inflammation and destruction of gingival and periodontal (gum) tissue. Published and unpublished observations by many men suggest rather strongly that periodontoclasia (gum disease) may be induced by certain chemicals, including fluoride," noted Hurme. V. O. Hurme, "An Examination of the Scientific Basis for Fluoridating Populations," *Dent. Items of Interest*, 74 (1952): pp. 518–534.
29. G. W. Rapp, "The Pharmacology of Fluoride," *The Bur* (April 1950). Cited in Walbott, *A Struggle with Titans*, p. 19.
30. H. T. Dean,"Chronic Endemic Dental Fluorosis," *JAMA*, 107 (October 17, 1936), pp. 1269–1273. Also, H. T. Dean, F. S. McKay, and E. Elvove, "Mottled Enamel Survey of Bauxite, Arkansas Ten Years After Change in the Common Water Supply," *Pub. Health Rep*, 53 (September 30, 1938), pp. 1736–1748.
31. H. C. Hodge and F. A. Smith, "Some Public Health Aspects of Water Fluoridation," in James H. Shaw, ed., *Fluoridation as a Public Health Measure* (AAAS, 1954), and *The Problem of Providing Optimum Intake for Prevention*

of Dental Caries: A Report of the Committee on Dental Health of the Food and Nutrition Board, Prepared by the Subcommittee on Optimal Fluoride Levels (NRC Publication 294, 1953).

32. Hodge relied on the unpublished personal assertions of Alcoa's top fluoride expert, Dr. Dudley Irwin, for his frequently reiterated assurances that industrial workers, and by extension the general population, were not being injured by fluoride. "In industrial populations, Irwin (REF: Irwin, Dudley. Personal communication) has come to the conclusion that if the urine contains less than 5 mg. per liter (presumably indicating a fluoride intake of less than 5 to 10 mg per day), osteosclerosis never develops. *On this basis*, it can be predicted that persons drinking fluoridated water and excreting approximately 1 mg of fluoride per day will never develop demonstrable osteosclerosis." (Emphasis added)–H. C. Hodge, "Fluoride metabolism: its significance in water fluoridation," *JADA*, vol. 52 (1956) pp. 307–314. Such reassurances flew in the face of the work of Siddiqui (1955), for example, who measured the urine F in skeletal fluorosis: "The urinary fluoride excretion varied between 1.2 and 5.8 ppm . . . The mean values for blood and urinary fluoride were 0.34 and 2.75 ppm respectively." A. H. Siddiqui, "Fluorosis in Nalgonda district, Hyderabad-Deccan," *British Medical Journal* (December 10, 1955), pp. 1408–1413. There are a number of additional and obvious problems with relying on Dr. Dudley Irwin for medical reassurances. The aluminum industry was one of leading fluoride polluters in the country, with an enormous interest in 'proving' fluoride safe. And Alcoa in particular had failed to disclose a great deal of health information about fluoride. For example, their discovery of high fluoride levels in the blood of one of the Donora dead was never made public (see chapter 9). Their 1948 study of aluminum workers in Massena, NY, in which high incidence of disability was reported was also never disclosed (see chapter 3). Additionally, Dr. Irwin was the head of the Medical Advisory Committee, which had been constituted by the Fluorine Lawyers Committee in order to help industry fight and defend against legal claims of fluoride injury from workers (see chapter 8). Nor would the threshold injury level for skeletal fluorosis be the only serious misstatement by Hodge relating to fluoride analysis. Astonishingly, and with surely devastating consequences for public health, he claimed, "Serious kidney injury or disease does not interfere with fluoride excretion, e.g., in rabbits given near-fatal doses of uranium (a kidney poison), in rats poisoned with fluoride, in elderly patients and in children suffering from kidney diseases (Hodge and Smith 1954)." H. C. Hodge. "Safety factors in water fluoridation based on the toxicology of fluorides." *Proceedings of the Nutrition Society* 22 (1963), pp. 111–117.
33. D. McNeil, *The Fight for Fluoridation*, p. 184.
34. HR 2341 "A Bill to Protect the Public Health From the Dangers of Fluorination of Water." Hearings Before the Committee on Interstate and Foreign Commerce. House of Representatives, 83rd Congress. May 25–27, 1954.

Frederick Exner of Portland and Dr. Veikko Hurme of the Forsyth Dental Infirmary also testified against fluoridation.

35. Hearings, HR 2341, p. 472.
36. *Dietary Reference Intakes for Calcium, Phosphorus, Magnesium, Vitamin D, and Fluoride*. Institute of Medicine (Washington, DC: National Academy Press, 1997), p. 311, citing H. C. Hodge and F. A. Smith, "Occupational Fluoride Exposure," *J. Occup. Med.*, 19 (1977), pp. 12–39.

Chapter 12

1. "Outlook," November 22, 1991, BBC Word Service.
2. Although a favorite Bernays strategy was to harness liberal ideals such as women's suffrage for clients, it was often done with cynical or mercenary intent. Privately he was contemptuous of those with average intelligence and won corporate clients by warning them of the dangers of democracy and socialism. See the account of his speech to oil executives: "Eddie led the oil boys up to the brink of the public ownership precipice, and let them look into the yawning abyss. Oh my, hold on tight!" E. L. Bernays, *Biography of an Idea: Memoirs of Public Relations Counsel Edward L. Bernays* (New York: Simon and Schuster, 1965), p. 780.

   Bernays's "hallucination of democracy" was described by writer Stuart Ewen as a hierarchical world in which "an intelligent few" had the responsibility of "adjusting the mental scenery from which the public mind, with its limited intellect, derives its opinions." S. Ewen, *PR: A Social History of Spin* (New York: Basic Books, 1996), p. 10.
3. Bernays does not mention the fluoride campaign in his autobiography, *Biography of an Idea*. He was reluctant to discuss it with this author, at first denying his involvement. When confronted with his own prior admissions to the medical writer Joel Griffiths, Bernays agreed to discuss several aspects of his involvement. Second taped author interview at Bernays home, December 11, 1993.
4. L. Tye, *The Father of Spin: Edward L. Bernays and the Birth of Public Relations* (New York: Crown, 1998); for Crisco: Procter and Gamble file in ELB Archive, Library of Congress.
5. E. L. Bernays, *Propaganda* (New York: Horace Liveright, 1928).
6. Chapter 11.
7. Bernays to Dr. Leona Baumgartner, commissioner of health, the City of New York, December 8, 1960. City officials were also advised by Dr. Edward A. Suchman, a social scientist employed by the Health Department, on how "to obtain fluoridation by edict rather than referendum" because, he explained, "the opposition seems to get a better reception from the public in any battle of propaganda or public debate." E. Suchman to Dr. Paul Densen, December 13, 1960.

   Fluoridation could be achieved, Suchman added, by "a systematic targeted campaign directed at those specific officials who had voted against fluoridation" and by lobbying "ethnic groups of political importance." E. Suchman to Dr. Arthur Bushel, Director of the Bureau of Dentistry, New York Health Department, February 9, 1961. Suchman suggests several "courses of action." The first: "Remove the fluoridation issue from the arena of public opinion. Make the decision a health ruling form the Board of Health and/or secure enough votes from the Board of Estimate to back up this ruling." The second item: "Change the balance of public opinion so that the political leader can be convinced that a large majority of his supporters favors this action. This is difficult to do on a mass basis, but should be possible in terms of specific group pressures, especially from those groups carrying political weight. This is behind our current attempt to determine the major ethnic groups of political importance in particular communities." This memo is cc'd to Leona Baumgartner, Paul Densen, and Edward Bernays. Leona Baumgartner file, ELB Collection, Library of Congress.

8. Baumgartner wrote Bernays: "The problem of equal time has been a continual headache with the networks. I don't know what to do about this." Baumgartner to Bernays, February 14, 1961, Baumgartner file, ELB papers, Library of Congress. He responded on February 16, 1961.
9. Robert Kehoe wrote, "The question of the public safety of fluoridation is non-existent from the viewpoint of medical science." *Our Children's Teeth* (New York: Committee to Protect our Children's Teeth, 1957), p. 31.
10. The committee received a $25,350 grant from Kellogg Foundation and "a second" $2,500 grant from the Rockefeller Brothers Fund. Bethuel Webster to Detlev Bronk, January 13, 1958. Folder 42, Correspondence, Box 21, Detlev Bronk Collection, Rockefeller Archive, RG 303-U.
11. Beginning in May 1940, Webster had been present at luncheon meetings of the prestigious Century Association, an elite club of powerbrokers and wealthy families, whose members have included eight presidents of the United States. Those luncheons, "resulted in the organization of the wartime Office of Scientific Research and Development, led by Centurions Vannevar Bush and James Conant," according to a chronicler of the group, William J. Vanden Heuvel. See "Franklin Delano Roosevelt: A Man of the Century," an address by William J. Vanden Heuvel to the Monthly Meeting of The Century Association, April 4, 2002, press release of the Franklin and Eleanor Roosevelt Institute.

Also, Webster worked with Vannevar Bush and Carroll L. Wilson (the AEC's first general manager) to help shape the direction of scientific research in the immediate postwar period. See J. D. Kevles,"The National Science Foundation and the debate over Postwar Research Policy—A Political Interpretation of Science—the Endless Frontier," in R. L. Numbers and C. Rosenberg, eds., *The Scientific Enterprise in America:Readings from Isis* (Chicago: University of Chicago Press, 1996), p. 313.

Webster wanted to win the public battle for fluoridation in New York and thus persuade the entire nation. See Webster's letter to Robert Kehoe, asking him to contribute to the Committee's booklet: "Local authorities must be more than convinced. . . . For the good of the cause I hope you will not be unduly modest . . . they must be furnished with irrefutable evidence in a popular form to protect them from criticism of the favorable action which we hope they will take. . . . By conclusively demonstrating to local authorities our ability to meet and demolish the opposition, fluoridation and future public health measures may be saved from the impossible requirements of mass scientific education, popular referenda, etc." Webster to Kehoe, December 11, 1956, Box 42. File: "Committee to Protect Our Children's Teeth," RAK Collection.

12. Listening as Bronk dictated his contribution to *Our Children's Teeth* was Shields Warren, the former director of the AEC's Division of Biology and Medicine. Dr. Warren "was in agreement with all that I have said," wrote Bronk in his contribution. Bronk was president of the Rockefeller Institute for Medical Research.
13. Bernays, *Biography of an Idea*, p. 380.
14. New York's Board of Estimate voted for fluoridation after a marathon public hearing, which lasted for twenty hours.
15. Dr. J. Knutson, Assistant Surgeon General, Chief Dental Officer PHS to E. L. Bernays, February 14, 1961, and Bernays to Knutson, February 16, 1961, in Baumgartner file, E. L. Bernays papers, Library of Congress.
16. In the summer of 1951, just one year after the PHS endorsement, Dr. John Knutson—then the top official at the National Institutes of Dental Health—summoned state dental directors to Washington. It was time for a sales pitch. The officials gathered at 9:40 AM in the Washington, DC, offices of the Federal Security Administration on Wednesday, June 6, 1951. Sitting in the meeting were the surgeon general, Leonard Scheele, Katherine Bain of the Children's Bureau, Phil Phair from the American Dental Association, and a top official from the Kellogg Foundation, Phil Blackerby.

    The keynote speaker was Wisconsin's state dental director, Dr. Frank Bull. He had recently fought a furious but losing battle for fluoridation in the town of Seven Points. He now outlined a game plan for state authorities. "Keep fluoridation from going to a referendum," advised Bull. "Are we trying to promote this thing, or do we want to argue about it? When we are inviting the public in and the press in, don't have anybody on the program who is going to go ahead and oppose us because he wants to study it some more. . . . You are like any salesman," Bull told his fellow dental directors, "You have got to be positive." He added, "Don't put any ifs, ands or buts, or maybes in the thing . . . you have got to get a policy that says 'Do it.' That is what the public wants, you know."

    Health officials had to choose their words carefully, Bull advised. If asked, "Isn't fluoride the thing that causes mottled enamel or fluorosis?" Bull suggested, "Tell them this, that at one part per million dental fluorosis brings about the most beautiful teeth that anyone ever had. And we show them some pictures of such teeth. We don't try to say that there is no such thing as fluorosis, even at 1.2 parts per million, which we are recommending, but you have got to have an answer. Maybe you have a better one. . . . We never use the term 'artificial fluoridation,'" he added. "There is something about that term that means a phony. . . . We call it controlled fluoridation."

Bull especially chided Katherine Bain of the Children's Bureau. Fluoride toxicity was probably the toughest issue facing promoters, he noted. "I noticed that Dr. Bain used the term 'adding sodium fluoride.' We never do that. That is rat poison," Bull said. "You add fluorides. Never mind that sodium fluoride business . . . all of those things give the opposition something to pick at, and they have got enough to pick at without our giving them any more. But this toxicity question is a difficult one. I can't give you the answer on it."

17. Some of the best information on this spying comes from the archive of author Donald McNeil. While writing influential articles on fluoride for *The Nation*, among others, McNeil had compiled an extensive list of antifluoride opponents. He was helped in his list-keeping by the American Water Works Association. He had written to its executive secretary Raymond J. Faust, on May 5, 1954, asking for the names of fluoride opponents: "I have a dossier on every anti I have ever heard of in the country. This includes even names and what background I know about of a person who might only have written a letter to the editor. By collecting the background on EVERYONE I hope to find a pattern which will eventually lead to an intelligent labeling of the opposition" (emphasis in original).

    Faust wrote to McNeil in reply, "We have from time to time developed some background information on some opponents of fluoridation. I will send you a copy of the material we have produced, however, it is sent to you with the understanding that it must be kept confidential or if used the source of the material must not be made public. We have been very careful to keep this material under lock and key as you may well understand." A document entitled "List of Rabid Opponents of Fluoridation" is found alongside. Faust to McNeil, May 13, 1954, File ADA 53–56, and ADA Misc, Box 1, McNeil Collection, Wisconsin State Historical Society.
18. *Science*, vol. 153 (September 23, 1966), p. 1498. See also letter from Donald McNeil to Peter Goulding, Director of Public Information, ADA, February 24, 1961. "Dear Pete . . . I see your powerful hand at work. . . . Today, along with your letter, I received one from Dr. Van Rensselaer Sill, Information Officer, Division of Dental Public Health and Resources, Department of Health, Education, and Welfare. He wrote to me care of *The Nation*, which was strange, but because of what he said, I can only conclude that he has been in touch with you or someone from the ADA. He said that he was interested in the 'paperback on the opponents' [a book McNeil was then shopping] and that he had 'some material on these learned gentlemen.' He hoped that I would call on him for any of the information they had in their offices." File 15, ADA Correspondence 60–63, Box 1, McNeil Collection, Wisconsin State Historical Society.
19. Several examples of professional censure for opposing fluoridation are cited in G. L. Waldbott et al., *Fluoridation: The Great Dilemma* (Lawrence, KS:

Coronado Press, 1978), p. 324. "In 1961 Dr. Max Ginns of Worcester, Massachusetts was dropped from his state dental society after he refused to discontinue use of a petition, circulated in 1953, which listed 119 dentists and 59 physicians in Worcester who opposed fluoridation. . . . [In 1962] the ADA House of Delegates voted to uphold the expulsion."

20. F. B. Exner and G. L. Waldbott, *The American Fluoridation Experiment* (New York: Devin-Adair, 1957), p. 232, letter exchange between John W. Knutson, assistant surgeon general, chief dental officer PHS, and Mr. James Rorty. Feltman was charged in his PHS grant, according to Knutson, with determining "the efficiency (in preventing dental caries) of the addition of measured doses of fluoride salts to pregnant women and children." His funding was cut off because he had "not reached his objective and was not likely to do so." Letter to Rorty from Knutson, August 9, 1956.
21. Waldbott's books on fluoride include *Fluoridation: The Great Dilemma*, with Albert W. Burgstahler and H. Lewis McKinney; *The American Fluoridation Experiment*, with F. B. Exner and James Rorty; and *A Struggle with Titans*.
22. B. Hileman, "Fluoridation of Water," *Chemical and Engineering News*, vol. 66 (August 1, 1988), pp. 26–42.
23. Dr. Exner had served six terms as secretary of the Association of American Physicians and Surgeons. For McNeil's subterfuge, see File ADA 53-56, McNeil Collection, Wisconsin State Historical Society.
24. Irene R. Campbell, *The Role of Fluoride in Public Health: The Soundness of Fluoridation of Communal Water Supplies, A Selected Bibliography*, Supported by Research Grant DE-01493 (Formerly D-1493) from the National Institute of Dental Research, Public Health Service, U.S. Department of Health, Education, and Welfare. As an example of the censorship in the bibliography, there is not a solitary citation for the published works of Dr. George Waldbott discussing fluoride's toxic effects in low doses, nor of the published studies of French epidemiologist Lonel Rapport, who linked fluoride in water to mongolism, also known as Down syndrome.

    Rapport's work is discussed in Waldbott et al., *Fluoridation*. The following citation of Rapport's work included: "Rapaport, I.: 'Les opacifications du crystalline mongolisme et cataracte senile.' *Rev. Anthropol. (Paris)*, Ser. 2,3: 133–135, 1957. 'Contribution a l'etude du mongolisme. Role pathogenique du fluor.' *Bull. Acad. Natl. Med. (Paris)*, 140: 529–531, 1956. 'Contribution a l'etude etiologique du mongolisme. Role des inhibiteurs enzymatiques. *Encephale*, 46: 468–481, 1957. 'Nouvelles recherches sur le mongolism. A propos du role pathogenique du fluor.' *Bull. Acad. Nat. Med. (Paris)*, 143:367–370, 1959. 'Oligophrenie mongolienne et ectodermoses congenitatles. *Ann. Dermatol. Syphiligr.*, 87: 263–278, 1960. 'A propos du mongolisme infantile. Une deviation du metabolisme de tryptophane ches es enfants mongoliens. C. R. Hebd. *Acad. Sci.* 251: 474–476, 1960. 'Oligophrenie mongolienne et caries dentaires.' *Rev. Stomatol.* 64: 207–218, 1963."
25. From 1957 to 1973 the ADA received $6,453,816 from the federal government, according to Waldbott et al., *Fluoridation*, p. 294, citing "Directory

of Dental Consultants and Executive Personnel and Representatives of the American Dental Association to National Agencies and Societies," Bureau of Public Information, Am. Dent. Assoc., October 19, 1955. Direct funding for fluoridation from the PHS is harder to ascertain. According to S. J. Kreshover, director of the National Institute of Dental Research, the Office of Management and Budget "advises that a breakdown of budgeted funds spent specifically on such programs or portions of projects dealing with fluorides is not available." Cited in *National Fluoridation News*, vol. 21. no. 1 (October–December 1975), p. 4.

26. American Dental Association Radio Script, National Children's Health Day, "Fluoridation Fights Tooth Decay," ADA Duplicates, Box 1.
27. This note was found in Donald McNeil's papers. It is marked "ADA Files." The newspapers identified as carrying the identical story are listed as: Hot Springs, AR, *Sentinel Record*, August 20, 1952; Lead, SD, *Daily Call*, August 19, 1952; *Idaho Evening Statesman*, Boise, Idaho, August 18, 1952; Poplar Bluff, MO, *American Republic*, August 21, 1952; *Newton Daily News* (Iowa) reprinted in Boone, Iowa *News-Republican* on August 22, 1952. The note is in a file marked "ADA duplicates," Box 1, McNeil Collection, Wisconsin State Historical Society.

Chapter 13

1. See *Reynolds Metals Company* vs. *Paul Martin*. Appellant's Brief, Appeal from Final Judgements of the District Court for the District of Oregon, Honorable William G. East, Judge. May 14, 1956, p. 3. U.S. Court of Appeals, 9th Circuit Ct. of Appeals, San Francisco, Court Case Papers and Printed Matter, Case #14990, transcript of Record in six volumes, Folder 14990–14992, Box 5888–5890, RG 276.
2. The attorneys were Frank Seamans, for Alcoa; Gordon Martin, for Kaiser Aluminum and Chemical Corporation; E. J. Epielman, Louis C. Viereck, and Lawrence A. Harvey, for Harvey Aluminum; B. W. Davis, for West Vaco Chemical Division of Food Machinery and Chemical Corporation; Lon P. MacFarland, for Monsanto Chemical Company; and R. E. McCormick and Francis R. Kirkham, for Olin Mathieson Chemical Corportaion. Brief Amicus Curiae, In the US Court of Appeals for the Ninth Circuit, Rehearing en banc on Appeal from Final Judgments of the District Court for the District of Oregon, File 18, Box 63, RAK Collection.
3. At Harvard, Hunter had studied with Dr. Philip Drinker. *Reynolds Metals Company* vs. *Paul Martin*, plaintiffs direct examination, p. 471, US Court of Appeals, 9th Circuit Ct. of Appeals, San Francisco, Court Case Papers and Printed Matter, Case #14990, transcript of Record in six volumes, Folders 14990–14992, Boxes 5888–5890, RG 276.
4. Testimony of Dr. Donald Hunter, p. 492.
5. Ibid., p. 473.
6. Ibid., p. 475.

7. Ibid., p. 476. Hunter was an examiner at Cambridge University. Cambridge had been a fluoride poison gas research center during the war. Sir Rudolph Peters also did his enzyme studies at Cambridge. R. E. Banks, ed., *Fluorine Chemistry at the Millenium* (Amsterdam and New York Elsevier, 2000), R. E. Banks (ed) p. 500.]
8. University of Rochester, Progress Report for October, 1944—Abstracts, Dr. Harold Hodge, p. 478. "The results indicated that the inhibition of esterase activity produced by T [code for uranium] was small compared with that by C-216 [code for fluorine]. Thus 0.025 ppm C-216 caused the same percentage inhibition of esterase activity as 100 ppm T (33 percent). From these results it is concluded that in a mixture of T and C-216 in which the amount (by weight) of T is not more than 50-fold that of C-216 the effect of the T upon the activity of liver esterase can be neglected." Also, "The useful range of this curve for determining C-216 concentrations was from 0–0.5 ppm, C-216." Document #SO9FO1B227, p. 19, ACHRE, RG 220. For a discussion of the role of fluoride on enzyme inhibition, and for comprehensive citations, see Waldbott et al., *Fluoridation: The Great Dilemma.*
9. Court of Appeals for 9th Circuit, , Brief Answer to Petition for Rehearing, Appeals from the Final Judgements of the District Court for the District of Oregon, p. 5, Folders 14990–14992.
10. *Reynolds Metals Company* vs. *Paul Martin*. Plaintiffs direct examination, p. 500, U.S. Court of Appeals, 9th Circuit Ct. of Appeals, San Francisco, Court Case Papers and Printed Matter, Case #14990, transcript of record in six volumes, Folders 14990–14992,Boxes 5888–5890, RG 276.
11. Ibid., p. 492.
12. Ibid., p. 1913, deposition of Paula Martin.
13. Ibid., pp. 259 and 213, direct and cross examination of Richard Capps.
14. Ibid., p. 245.
15. Ibid., p. 197.
16. See chapters 3 and 9.
17. Direct examination of Robert Kehoe, *Reynolds Metal Company* vs. *Paul Martin*, pp. 995 and 997.
18. Robert Kehoe to Edward Largent, February 13, 1956, File 5, Box 76, RAK Collection.
19. Manufacturing Chemists Association, Inc. Minutes of the Air Pollution Abatement Committee, November 2, 1955. Via Environmental Working Group searchable database.
20. Appellant's Brief, Appeal from Final Judgments of the District Court for the District of Oregon, Honorable William G. East, Judge. May 14, 1956, p. 7.
21. Following the Martin trial, the company put Largent directly on its payroll as a health and environment consultant. In the years to come Reynolds and its health consultants would be preoccupied with another citizen protest, this time from Mohawk Indians on the Akwesasne reservation on the New York–Canada border, who lived downwind of a newly built Reynolds alu-

minum plant and who claimed that their health and economy were being destroyed by fluoride (see chapter 15).

Chapter 14

1. *Reynolds Metals Company* vs. *Paul Martin*. Petition for Rehearing en banc, p. 6, and Appellant's Brief, p. 32 Appeal from Final Judgments of the District Court for the District of Oregon, Honorable William G. East, Judge. May 14, 1956. P. 3. RG 276, US Court of Appeals, 9th Circuit Ct. of Appeals, San Francisco, Court Case Papers and Printed Matter, Box 5888–5890, Folder 14990 to 14992, case #14990, transcript of record in six volumes.
2. The judges were told that *Our Children's Teeth* included "the statements of one medical and scientific expert after another, all to the effect that fluorides in low concentrations (such as are present around aluminum and other industrial plants) present no hazard to man." Brief Amicus Curiae, In the U.S. Court of Appeals for the Ninth Circuit, Rehearing en banc on Appeal from Final Judgements of the District Court for the District of Oregon, p. 8, File 18, Box 63, RAK Collection.
3. Statement of Robert Kehoe, *Our Children's Teeth, A Digest of Current Scientific Opinion Based on Studies of Fluorides in Public Water Supplies*, prepared by the Committee to Protect Our Children's Teeth, Inc., submitted to the Mayor and the Board of Estimate of the City of New York (1957), p. 31.
4. R. Kehoe, "Memorandum on the Present Status and the Future Needs, with Respect to Information Deriving from Observation and Investigation of the Behavior of Inorganic Compounds of Fluorine in the Animal Organism," February 1, 1956, File 5, Box 76, RAK Collection.
5. Robert Kehoe to James M. McMillan, September 20, 1961, cc: Mr. Frank Seamans, Box 63, RAK Collection.
6. R. Kehoe, "Memorandum on the Present Status and the Future Needs, with Respect to Information Deriving from Observation and Investigation of the Behavior of Inorganic Compounds of Fluorine in the Animal Organism," February 1, 1956, File 5, Box 76, Kettering Files.
7. The corporations "which are concerned mainly with the results of exposure to fluorides in various occupations" included "The Pennsylvania Salt Manufacturing Company, Aluminum Company of America, Reynolds Metals Company, Universal Oil Products Company, American Petroleum Institute, Kaiser Aluminum and Chemical Corporation, Tennessee Valley Authority, The Harshaw Chemical Company, [and] Minnesota Mining and Manufacturing Corporation," Kehoe told the Medical Director of the Division of Industrial Hygiene, Dr. Seward Miller.
8. "In a meeting a little while ago," Kehoe wrote to Dr. Miller, "the question was raised, naturally, as to the long-term influence of small quantities of fluorides, such as those which might be taken in with drinking water, both in areas in which fluorides occur in somewhat unusual concentrations in the drinking water as well as those areas in which fluorides are being added to community water supplies . . . I feel that I should transmit to you the

opinions expressed by this group and by the industries for whom they speak, not as a matter of their right to request any activity on the part of the Public Health Service, but rather as evidence of their interest in a broad problem of public health," Kehoe wrote. "That this interest has been aroused by their concern for the employees of their own companies, is a phenomenon which seems to me to be of some public consequence." Kehoe to Miller, May 20, 1952, RAK Collection.

9. The Bartlett Cameron study examined the health of 116 people in Bartlett and 113 from Cameron. George Waldbott noted that there was no information about how much fluoride the Cameron residents might have consumed in food, which had perhaps been grown in the nearby Bartlett area, or elsewhere in Western Texas, known as a high fluoride region. Although the study reported no "significant differences" in the health status of the two populations, there was a high incidence of cataracts, bone changes, arthritis and deafness in *both communities*, compared to the national average. Also, mortality in Bartlett was 265 percent higher in Bartlett than in Cameron. Furthermore, using data on just 116 individuals to justify adding fluoride to the drinking water of 50 million people meant that, according to George Waldbott, "if 1 in 117 were to suffer ill effect from fluoride in water, the number of those so afflicted among the fifty million citizens would be 427,350—a sizable incidence. Thus the sampling in the Bartlett survey was far too small to assure the safety of millions of people drinking fluoridated water." G. L. Waldbott, *A Struggle with Titans*, p. 296. (Leone's Bartlett research was published as "Medical Aspects of Excessive Fluoride in a Water Supply," *Public Health Report*, vol. 69, no. 10 (October 1954). Submitted as part of the amicus curiae brief in the Martin trial, it was additionally published as N. C. Leone, et al. 1955. "Review of the Bartlett-Cameron survey: A Ten-Year Fluoride Study," *J. Amer. Dent. Assoc.*, vol. 50, pp. 277–281. And Leone et al., *Am J. Roentgen*, vol. 74 (1955), p. 874.
10. "This undoubtedly was the paper Dr. Leone referred to in our long distance telephone conversation while I was engaged in the trial of the Martin personal injury case." W. T. Lennon to Robert Kehoe, March 15, 1957, cc: R. W. Anderson, Alcoa, File 5/6, Box 76, RAK Collection.
11. Leone had given Lennon the reference to a version of the Bartlett Cameron study, published as "A Roentgenologic Study of a Human Population Exposed to High-Fluoride Domestic Water" in *Am. J. Roentgenology, Radium Therapy and Nuclear Medicine,* vol. 74, no. 5, November 1955. The paper included reference to an autopsy, Lennon wrote Kehoe. "Evidently the autopsy was only complete to the extent of bone analysis as the paper contained no comment on soft tissue. I was wondering whether or not you had any talks with Dr. Leone regarding this autopsy and whether or not any examination was made of soft tissue." W. T. Lennon to Robert Kehoe, March 15, 1956, cc: R. W. Anderson, Alcoa, File 5/6, Box 76, RAK Collection.
12. Dr. Leone to Dr. Irwin, letter sent on March 5, 1957, File 5/6, Box 76, RAK Collection.

13. D. A. Greenwood, "Background for Studies in Utah County," paper given at the 1957 Kettering Fluoride Symposium. Greenwood was Professor of Biochemistry and Pharmacology, Utah State University. Another figure claims that the legal claims against U.S. Steel in Utah were for $30 million. Butler C., Proceedings: National Conference On Air Pollution, November 18–20, 1958, p. 268.
14. Leone was an unapologetic propagandist for fluoride. For example, in 1983 he helped organize a conference at Utah State University. In the proceedings he writes, "Further publicizing the importance [of fluoride] in the treatment of selected cases of osteoporosis can help us achieve control of another facet of the fluoride problem. By emphasizing and appraising the older members of our aging population as to the beneficial aspects of fluoride at levels in the neighborhood of 5 mg per day, we can make known the obvious safety of fluoride levels at higher than the advocated (1 ppm) in the prevention of dental caries in children. The process would thus give supportive evidence as to the safety and desirability of fluorides in human diets." J. L. Shupe, H. B. Peterson, and N. C. Leone, eds., "*Fluorides: Effects on Vegetation, Animals, and Humans*" (Salt Lake City, UT: Paragon Press, 1983), p. 361.
15. Dudley Irwin to Frank Seamans, March 13, 1957, 42.17, RAK Collection.
16. *Reynolds Metals Comp* vs. *Yturbide*, 258 F. 2d 321 (9th Cir.) cert. den. 358 U.S. 840 (1958), p. 25.
17. Motion for Leave to File Brief Amicus Curiae, p. 2, and Brief, p. 5, File 18, Box 63, RAK Collection.
18. Kehoe notes of meeting, Folders 18, 19, and 23, Box 63, RAK Collection. For the relationship of the Medical Advisory Committee to the Fluorine Lawyers, see Seamans to Medical Advisory Committee, April 16, 1957: "The legal representatives of the several companies interested in the Kettering Research project have agreed that it would be advantageous if the principal liaison with Kettering were undertaken by persons of competent technical background . . . [to] conduct the necessary liaison between the Kettering Institute and the lawyers' group by a system of regularly scheduled visitations to Kettering and regular reports to the lawyer's group," File 17, Box 42, RAK Collection.
19. Leone's Bartlett Cameron study, comparing two Texas communities with low and high natural fluoride in water, was cited. So was his work in Provo, Utah, where U.S. Steel's giant plant was being blamed for widespread injury to crops and livestock, and where Leone was serving as a consultant to R. A. Call, who was studying fluoride deposition in soft tissues. Leone was also working with Harold Hodge and Frank Smith at Rochester, studying the soft tissues of people who had died in areas with varying levels of fluoride in water. Much of this work, including a summary of Call's work, was brought together in the 1957 Symposium at the Kettering Laboratory and published by editor Philip Drinker in the *Archives of Industrial Health*, vol. 21 (1960). See also *Public Health Report.*, no. 80 (1965), pp. 529–538, for an expanded version of Call's report. In regards to the Call study (Leone was not

publicly listed as an author or 'consultant' on the work), George Waldbott noted, "Their grants were not renewed, according to Dr. Call's letter to the author, June 22, 1964. Therefore, the study of ill-effect of airborne fluoride on kidney disease which their research had disclosed was abandoned." G. L. Waldbott, *A Struggle with Titans: Forces Behind Water Fluoridation* (New York: Carleton Press, 1965), p. 251.

20. The final ruling of the Appeals Court on absolute liability was a victory for industry. "This case can no longer be cited for the proposition that in a case of this kind absolute liability exists. Thus, the companies filing amicus curiae briefs at least succeeded in winning the major point which they argued. This may be of doubtful value because of the view taken on proof of negligence but at least we succeeded on this point." Legal memo from Frank Seamans, sent to Robert Kehoe, June 13, 1958, File 18, Box 64, RAK Collection.
21. Ibid.
22. Kehoe to Willard Machle, May 29, 1956, Box 42; and Drinker to Kehoe, July 8, 1958, File 17, Box 42, RAK Collection.

Chapter 15

1. Boscak, 1978. EPA report No. EPA-450/3-78-109. Cited in EPA, "Summary Review of Health Effects . . . ," EPA/600/8-89/002F (December 1988), pp. 3–5. It states, regarding HF manufacturing plants and additional sources of industrial air exposure, "The figure is naturally higher when other fluoride or HF sources are considered."
2. In 1975, 350,000 men and women in 92 occupations were exposed to fluorides. National Institute for Occupational Safety and Health, 1975, DHEW/NIOSH-76-103. Cited in EPA, "Summary Review of Health Effects," pp. 3–5. Also, 22,000 workers were potentially exposed to hydrogen fluoride gas alone, in 57 occupations. Criteria For a Recommended Standard . . . Occupational Exposure to Hydrogen Fluoride, NIOSH DHEW/PUB/NIOSH-76-143, cited pp. 3–5, EPA, "Summary Review of Health Effects."
3. Kehoe to Derryberry, January 9, 1956, File 18, Box 63, RAK Collection.
4. "Memorandum Concerning the Objectives of the Investigative Program on the Behavior of Fluoride in the Human Body and Concerning the Purposes and Policies of the Kettering Laboratory and the University of Cincinnati, in the Prosecution of This Investigative Program," Prepared by Robert Kehoe, November 10, 1956. Box 42, RAK Collection.
5. Some 2,845 pounds a day of Reynolds's fluoride had spilled over the Martin ranch as hydrofluoric acid gas and also as tiny particles of fluoride dust. Memorandum from Frank Seamans summarizing the finding of the Appeals Court en banc, June 12, 1958, File 18, Box 63, RAK Collection.
6. R. Kehoe, "Memorandum on the Present Status and the Future Needs, with Respect to Information Deriving from Observation and Investigation of the Behavior of Inorganic Compounds of Fluorine in the Animal Organism," February 1, 1956, File 5, Box 76, RAK Collection.
7. Ibid.

8. Minutes of the meeting of the Fluoride Committee on October 10, 1956, at 10.00 AM in Room 207, College of Medicine, Folders 18, 19, and 23, Box 63, RAK Collection.
9. Hosted by the Kettering Laboratory, the symposium had been planned earlier in the year at the May 20 meeting at the Kettering Laboratory, following the Martin Appeals Court verdict. It was arranged by Alcoa's Dudley Irwin, Robert Kehoe, and the government's Dr. Nicholas Leone as part of their "strategic" information plan.
10. R. A. Kehoe, handwritten notes, "A World of Welcome on Behalf of the Kettering Laboratory," 1957 Fluoride Symposium, File 42, Box 17, RAK Collection.
11. For aluminum employees, see amicus curiae brief, *Reynolds* vs. *Martin*, p. 2.
12. Other papers on fluoride safety were given by NIDR officials Nicholas Leone, Isadore Zipkin, and Harold McCann. Another study by Richard A. Call on the effects of fluoride air pollution on humans was being conducted in Utah. That project had been explained by the NIDR's Dr. Leone, who described himself as a "consultant" on the project, to Alcoa's Dudley Irwin in a letter of March 5, 1957: "As you know, it has been proven beyond a question of a doubt that similar conditions have an effect upon animals," wrote Leone. He explained that the Public Health Service was financing the human studies "with funds supplied by another Bureau." They were being conducted in the laboratories of the Mormon Latter Day Saints Hospital in Provo, Utah. Urine levels were being recorded. The bones and tissues of individuals who died suddenly were examined. "Inmates of a mental institution close by comprise the study material," Leone noted. The study of forty-eight autopsied bodies that had experienced sudden death concluded that "no histologic abnormities attributable to fluorides were recognized." Nevertheless, 29.3 percent of the "major causes of death" in the study area were listed as "respiratory tract" in origin, compared with just 5.9 percent in the control group. Nicholas D. Leone, MD, Chief Medical Investigations NIDR, to Dudley A. Irwin, MD, Alcoa Medical Director, March 5, 1957, RAK Collection.
13. Kehoe to Dudley Irwin, Alcoa, December 4, 1959. "Dear Dudley: The Symposium has been accepted for publication by Phil Drinker in the [AMA] Archives [of Industrial Health] and it will appear in the April or May number. It will be made available in one volume in reprints and, therefore, it is now time to decide how and in what numbers we wish to have it assembled. . . . I would suggest that the sponsors be polled for the numbers of copies they desire, that this information, together with the addresses to which the reprints and the bill for them are to be sent, be forwarded to me, so that I can hand all of this, in a complete and orderly manner, to Phil Drinker. The sooner this is done the better it will be, I believe, since I would like to be sure that the sponsors get just what they want." File 17, Box 42, RAK Collection. (The editorial board of the Archives of Industrial Health included Du Pont's John Foulger, the Mellon Institute's Helmuth Schrenk, the Kettering

Institute's Frank Princi, and Herbert Stokinger, formerly of the University of Rochester's Atomic Energy Project.) The publication of the papers was part of the post-Martin strategy drawn up by Kehoe, Alcoa, and Dr. Leone from the NIDR at their planing meeting that spring. The collected papers appeared in the *Archives of Industrial Health*, vol. 21 (1960).

14. Frank Seamans to Robert Kehoe, April 16, 1957, File 17, Box 42, Kettering Files.
15. William Jolley died of colon cancer in the 1970s, the result of what Bingham and Jolley's family believe was radiation poisoning from his earlier work at the AEC's Mound Laboratory in Miamisburg, Ohio. According to a July 15, 2002, author interview with Eula Bingham, Davis "had worked up at Mound laboratory, the radiation laboratory up in Miamisburg, Ohio, Bill Jolley worked up there also. Bill died of colon cancer in the seventies and his family tried to file a lawsuit and they didn't get anything out of it. He got out of that job—they came here to the university because they really were worried about all the radiation up there." Bingham also believed radiation killed Jolley: "I feel so, too," she said.
16. An earlier draft of the report found in Kehoe's files records the investigators' shock at the results, and the discussion section notes that even the control animals had been hurt by a small amount of fluoride, to which they had somehow been exposed. "The principal findings in the lung were of peribronchial fibrosis and scattered granulomatous (inflammatory) lesions. . . . The striking enlargement of the tracheal lymph nodes was caused by a hyperplastic lymphadenitis. In the lungs there was a strikingly large amount of cholesterol, which, at present, has no clear explanation. . . . Some degree of 'reaction to injury' was encountered even in the lungs of the control dogs which sustained only a modest degree of incidental exposure to air borne calcium fluoride." Folders 18–20, Box 63, RAK Collection.
17. Albert A. Brust, Director, Toxicology Division, to Dudley Irwin, February 10, 1960, cc: R. A. Kehoe and R. K. Davis, File 17, Box 42, RAK Collection.
18. Industry's fear of lawsuits for emphysema damage can be seen in a 1966 symposium at the Mellon Institute in which the managing director of the Industrial Hygiene Foundation, Robert T. P. deTreville, MD, announced: "The Foundation's interest in emphysema stems partly from the concern reflected by its membership that the potential for abuse in the awarding of claims for compensation could easily dwarf that for silicosis at its worst." *Emphysema in Industry*, Medical Series Bulletin No. 10, Mellon Institute Library. (See also Epilogue for discussion of emphysema in industry in 2003.)
19. Charles McCarthy to Robert Kehoe, July 9, 1962, RAK Collection.
20. Dr. Arden Pope at the University of Utah recommended Phalen. Pope described Phalen as "honest and candid."
21. Two early influential members of ACGIH were Harold Hodge and Jim Sterner. (Both had attended the Conference on Fluoride Metabolism at the Hotel Pennsylvania.) Hodge and Sterner were bread-and-butter pragmatists, in Phalen's opinion, forging compromise in the real world of industry

smokestacks and worker paychecks. "If I wanted to harm someone's health, I would put their breadwinner out of a job," Phalen said. "It has a greater health effect dropping someone below the poverty level than becoming a heavy smoker. These people realized the critical nature of someone earning a living. They had seen the Depression. ACGIH decided it would establish limits that workers could be exposed to and most workers, the vast majority, not get ill . . . Harold was of that sort," Phalen added.

22. "No studies were located regarding respiratory effects in animals following inhalation of fluoride." Draft Toxicological Profile for Fluorides 2001 (Department of Health and Social Services, Public Health Service, ATSDR). p. 50.
23. ACGIH's current 2.5 mg/m standard is based on a 1963 paper by Dr. O. M. Derryberry of the Tennessee Valley Authority, a member of the Medical Advisory Committee that had shaped the original Kettering research program.
24. "If it is a study that you are saying is very important and clearly has some relevance, I think it is unlikely that we would ignore it, so I think that you might assume [that we didn't see it] . . . but I can't say that for sure," Brossard added. Author interview, July 22, 2002.
25. Of significance is the report by Laura Trupin, an epidemiologist at the University of California, San Francisco, in the *European Respiratory Journal* vol. 22, no. 3 (September 2003), that on-the-job exposure to dust or toxic fumes may cause as many as five million cases of a group of deadly lung diseases called chronic obstructive pulmonary disease [COPD.] According to *USA Today*, "this study suggests that workplace exposure to pollutants may be a more important cause of the disease than previously suspected. The new study found that workplace exposure may cause as much as 31 percent of all cases of COPD, which kills more than 100,000 Americans each year." *USA Today*, August 26, 2003, Section D, p. 7.
26. "Compensation for Illnesses Realized by Department of Energy Workers Due to Exposure to Hazardous Materials"—Hearings before Subcommittee on Immigration and Claims, September 21, 2000, Serial No. 132, p. 147.
27. Ibid., p. 142.
28. Harding recalled, "At any time, you could see that haze of smoke and smell a strong acrid odor, and you could taste it in your mouth. So you were literally breathing and eating uranium-containing gases and dusts and powder all the time."
29. For "Buchenwald," see J. G. Hamilton, University of California, to Shields Warren, DBM, AEC, November 28, 1950 ("Unfortunately, it will not be possible for me to be at the meeting on December 8"), Document #DOE-072694-B-45, p. 1, ACHRE, RG 220.
30. Congressional testimony of Rep. Ed Whitefield of the State of Kentucky. "Compensation for Illnesses Realized by Department of Energy Workers Due to Exposure to Hazardous Materials," Hearings before Subcommittee on Immigration and Claims, September 21, 2000. Serial No. 132, p. 123.
31. Ibid., pp. 234–235.

32. However, as of August 2003, according to an Internet posting from the worker advocacy group, the Alliance for Nuclear Accountability, "DOE has received over 17,000 claims requesting assistance with state workers' compensation for occupational illnesses, but as of June 25 had processed only 45 claims through its Physicians' Panels for a determination, and none of these claims had yet been paid. DOE has advised Congress that it expects it will take another 5 years to work through its backlog of claims."
33. "The Link Between Exposure to Occupational Hazards and Illness in the Department of Energy Contractor Workforce" (The National Economic Council, 2000), p. 18. This study provided the scientific foundation upon which the legislation was based. At Oak Ridge, a K-25 worker, Sam Vest, watched his father sicken with chronic fatigue syndrome. He watched an uncle get cancer. Both had worked in the K-25 uranium production plant at Oak Ridge, and both died in their fifties and sixties. Today they are buried alongside each other in an Oak Ridge cemetery. Vest continued to work at the plant during the 1990s. He now has bladder cancer, arthritis, and memory loss, he asserts. He was placed on disability in 1998. He describes Oak Ridge as "a tragedy," where sickness stalks former workers. "They all have joint and muscular problems, skeletal problems, a lot of them have memory problems similar to mine," says Vest. "A lot of people have respiratory problems." "Nobody wanted to work in the gaseous diffusion buildings," Vest added. "Deep down they knew they were being exposed to very hazardous chemicals, the HF and the hexafluoride and all the other things."
34. Although the legislation and compensation process did not create a special category for fluoride injury, fluoride had played a leading role in hurting atomic workers, Congress heard. One government-funded study found that 20 percent of former gaseous-diffusion employees have chronic bronchitis and/or emphysema. Exposure to "hydrofluoric acid and other powerful lung irritants in the gaseous diffusion process played a significant contributing role," in causing that illness, scientists said. Congressional testimony of Steven B. Markowitz, Director of the Center for the Biology of Natural Systems, Queens College, Flushing, NY, September 21, 2000, "Compensation for Illnesses Realized by Department of Energy Workers Due to Exposure to Hazardous Materials"—Hearings before Subcommittee on Immigration and Claims, September 21, 2000, Serial No. 132, p. 163.
35. Other investigators, while seemingly aware that hazards exist, are simply unwilling to evaluate the risk communities and workers face from fluoride. Arjun Makhijani, director of the Institute for Energy and Environment Research and one of the nation's most quoted nuclear-health experts, told me that he had "made a decision not to go there" in examining fluoride's health effects, choosing instead to focus on the risks from radiation. He confirmed that little accounting has yet been made of the health damage fluoride has inflicted on nuclear workers. "I don't know how to begin thinking about this question. It is a sleeper," Makhijani said.

36. Richard Wilson and John Spengler, eds., *Particles in Our Air: Concentrations and Health Effects* (Cambridge, MA: Harvard University Press, 1996), p. 212.
37. C. Schneider, *Death, Disease, and Dirty Power: Mortality and Health Damage Due to Air Pollution from Power Plants* (The Clean Air Task Force, October 2000). This report is a summary of a fuller report done by Abt Associates for the Clean Air campaign. On p. 5 it states, "The Abt Associates report further shows that hundreds of thousands of Americans suffer from asthma attacks, cardiac problems and upper and lower respiratory ailments associated with fine particles from power plants." Lung cancer study cited in the *New York Times*, March 6, 2002, Section A, p. 14, from *JAMA* study of same date.
38. Children breathe 50 percent more air per pound of body weight than adults. Children make up 40 percent of all asthma cases, while only 25 percent of the total population. "Asthma: A Public Health Response" (U.S. CDC), cited in *Death, Disease, and Dirty Power*, p. 9.

    One study found infants in high-pollution areas were 40 percent more likely to die of respiratory causes. Another found a 26 percent increase in the risk for sudden infant death syndrome. T. J. Woodruff et al., "The Relationship Between Selected Causes of Postneonatal Infant Mortality and Particulate Air Pollution in the United States," *Environmental Health Perspectives*, vol. 105, no. 6 (June 1997); cited in *Asthma: A Public Health Response*. U.S. CDC; cited in *Death, Disease, and Dirty Power*, p. 9.
39. *New York Times*, May 12, 2000, p. 32.
40. EPA Toxic Release Inventory data—1999 data, updated as of August 1, 2001.
41. For fluoride synergy, see A. S. Rozhkov and T. A. Mikhailova, *The Effect of Fluorine-Containing Emissions on Conifers*, trans. L. Kashhenko, Siberian Institute of Plant Physiology and Biochemistry, Siberian Branch of the Russian Academy of Sciences (Frankfurt: Springer-Verlag, 1993). Part of this text was excerpted on the Fluoride Action Network website. See also Stokinger et al., "The Enhancing Effect of the Inhalation of Hydrogen Fluoride Vapor on Beryllium Sulfate Poisoning in Animals," UR-68, University of Rochester, unclassified.
42. *Florida: from the late 1940s through the 1960s, multiple lawsuits were launched against several fertilizer manufacturers mining the state's rich natural phosphate beds.* U.S. Senate hearings were prompted when 25,000 acres of citrus land in Polk County were damaged and 150,000 acres of pasture abandoned as "fluorides gushed into the orange-blossom-scented air. . . . As cattle ate the grass, they absorbed fluorine into the bloodstream. Teeth decayed, joints stiffened, and bones became brittle," the Associated Press reported. Local citizens were also injured, according to news reports. According to one news account in Florida's Polk and Hillsborough County, "17 plants are clustered abound rich deposits of phosphate rock. Fumes from these plants have destroyed 25,000 acres of citrus trees and damaged vegetation for 50 miles in all directions. Cattle in Polk county have suffered from fluoro-

sis and died and people have been afflicted with sore throats and burning eyes and nosebleeds and respiratory problems. Millions of dollars in damage suits have been filed against phosphate plants." Ned Groth, *Pennisula Observer*, January 27–February 8, 1969. In 1966 the Chemical Manufacturers Association mandated a decrease in airborne fluoride emissions from the phosphate industry in Polk County, from 17 tons to 9 tons per day, according to Dr. D. R. Hendrickson, professor of Sanitary Engineering, University of Florida. Cited in Manufacturing Chemists Association minutes, January 26–28, 1966, CMA Archives, Document 085439, Environmental Working Group searchable database.

Human effects were also claimed by an attorney in Lakeland, Florida, A. R. Carver from the firm of Carver and Langston, whose letter to Dr. Robert Kehoe refers to "air pollution litigation" on behalf of "A man, his wife and two teenage children. They have been living for several years in close proximity; that is to say, within a circle, the radius of which extends five miles, would include eight producing super triple phosphate plants." May 8, 1956, A. R. Carver to Kehoe, RAK Collection.

A Lakeland resident, Mrs. Harriet Lightfoot, told the AP about the human effects of fluoride pollution: "It seemed as she came suddenly awake that a strangler's hands were at her throat. Madly she gasped for breath. Her head was pierced by a splitting pain. Her throat and eyes burned." *Pensacola News-Journal*, December 18, 1966.

See also the account "Death in Our Air," reported in *The Saturday Evening Post* by Ben H. Bagdikian, "Donald McLean, of Polk County, Fla, told a Senate subcommittee that since phosphate plants began putting seven tons of fluoride a day into the air he has had to sell his cattle and his citrus crops because cattle died, crops that used to mature in 80 days now take 200, barbed wire that used to last 20 years rots in 4, and he doesn't dare grow vegetables for his family for fear they will pick up the same chemicals that fall onto his pastures and groves. 'It eats up the paint and etches glass, it kills trees, it kills cattle. It is an irritant to mucous membrane, and we have sore throats, tears run out of our eyes, we sneeze, we have nosebleeds. Gentlemen, am I a fool to assume that that stuff [is] injurious to humans?'" (Date missing on article.)

- *1961, The Dalles, OR: Fairview Farms Inc. received $300,000 from the Harvey Aluminum Company's reduction plant because of damage to farmlands and animals.* Orchardist W. J. Meyer and his wife Mary Ann also received $485,000 for "willful damage" to cherry, apricot, and peach crops, according to news accounts. ("Harvey Loses Fluoride Case," *Hood River (OR) News*, October 29, 1970. Cited in G. L. Waldbott et al., *Fluoridation: The Great Dilemma*, p. 298. The company argued that pollution reduction equipment would cost $15 million and require 100 extra employees. *National Fluoridation News* (March–April 1965), p. 3.
- *1962, Vancouver, WA: Alcoa paid William Fraser $60,000 and, in the same year, $20,000 to Earl Reeder because of fluoride injury to their cattle on Sau-*

vies Island. Sauvies Island. *Portland (OR) Reporter*, June 26, 1962. Cited in Waldbott et al., *Fluoridation*, p. 298.

- *1962, Contra Costa County, CA: Cattle ranchers in California sued four chemical plants for damages to their herds*. Ned Groth, *Peninsula Observer*, January 27–February 8, 1969.
- *Garrison, MT: Human harm from fluoride pollution was alleged after the Rocky Mountain phosphate plant opened in 1963, with residents complaining of, among other symptoms, heart problems and asthma*. Lawsuits for $740,000 were filed. "Smog Battle Ends in Montana Town," *New York Times*, September 17, 1967. Cited in Waldbott et al., *Fluoridation*, p. 299. See also *New York Times*, December 1966: "It is charged among other things that fluoride-laden smoke from the phosphate plant has caused malformations and deteriorating teeth in cattle and horses, that trees have been afflicted by cancerous growths and that people have developed symptoms akin to bronchitis, sinus trouble and heart attacks." See also, B. Merson, "The Town That Refused to Die," *Good Housekeeping*, January 1969, lawsuits cited in *National Fluoridation News*, March–April 1965, p. 3. "People were made so ill that many were literally driven out of their homes," according to Ned Groth in *Peninsula Observer*.
- *1968: Cominco American Phosphate Company in Douglas Creek was successfully sued for $250,000*. L. Greenall, "Industrial Fluoride Pollution in British Columbia," Canadian Scientific Pollution and Environmental Control Society, Vancouver, mimeo, January 1971. Cited in M. Prival and F. Fisher, "Fluorides in the Air," *Environment*, vol. 15, no. 3 (April 1973), pp. 25–32.
- *Columbia Falls, MT, 1970: Six damage suits for $625,402 were filed on Sept 24 by residents for alleged fluoride damage caused by the Anaconda Aluminum Company and the Anaconda Wire and Cable Co, according to news accounts*. A week earlier a $21.5 million dollar action was filed against the two companies by Dr. and Mrs. Loren Kreck of Columbia Falls, and a suit filed by Mr. and Mrs. Harold Dehibom asked $1,650,000 from the same defendants. *National Fluoridation News* (September–October 1970), p. 4.
- *Tennesse, 1970: Reports of $3 million in fluoride claims against Alcoa*. "Alcoa Sued for Nearly $3 Million," *Knoxville (TN) Journal*, October 29, 1970, Cited in Waldbott et al., *Fluoridation*, p. 298.
- *1971: $9 million lawsuit in by the Sierra Club against the Harshaw Chemical Company for fluoride pollution, which, the Club charged, had corroded a main bridge over the Cuyahoga River.*
- *Ferndale, WA, 1972: $83,060 judgment by farmer against Intalco Aluminum Company in Ferndale, WA*. R. Park, "The Italco Trial," *Bellingham (WA), Northwest Passage* March 20–April 2, 1972, cited in Prival and Fisher.
- *1980: $150 million lawsuit against Reynolds Metals and Alcoa, alleging fluoride injury to cattle on the New York–Canadian St. Regis Reservation, during the period of 1960–1975, settled for $650,000*. Karen St. Hilaire, "St.

Regis Indians to Settle Fluoride Dispute," *Syracuse Post Standard*, January 8, 1985, cited in Griffiths "Fluoride: Commie Plot or Capitalist Ploy?" *Covert Action Information Bulletin*, no. 42 (fall 1992), p. 26.

43. Ned Groth, "Capitalist Plot? Air Is Fluoridated," *Peninsula Observer*, January 27–February 3, 1969. Also, Public Law 84-159 of the 84th Congress (1955) established the PHS's first air-pollution program. "At the time Public Law 84-159 was implemented, fluorides constituted the major industrial pollutant of immediate concern to agriculture," in "Six Years of Research in Air Pollution: A review of Grants in aid, Contracts, and Direct Operations Sponsored by the Division of Air Pollution, Bureau of State Services. July 1, 1955, to June 30, 1961." U.S. Department of Health, Education and Welfare.

44. *Agriculture Handbook*, No. 380, published by the Agriculture Research Service of the U.S. Department of Agriculture (1970), cited in E. Jerard and J. B. Patrick, "The Summing of Fluoride Exposures" *Intern. J. Environmental Studies*, vol. 4 (1973), pp. 141–155. "Whenever domestic animals exhibited fluorosis, several cases of human fluorosis were reported, the symptoms of which were one of more of the following; dental mottling, respiratory distress, stiffness in knees or elbows or both, skin lesion, or high level of F in teeth and urine. Man is much more sensitive than domestic animals to F intoxication," that report added.

45. Weinstein: "Whereas threshold concentrations for ozone or sulfur dioxide that will produce an irreversible effect [upon plants] were found to be generally above 0.05 ppm for exposure periods of about 7 days, and more than double that concentration and time for nitrogen dioxide, gaseous hydrogen fluoride could cause a metabolic or physiologic change and produce lesions on leaves of the most sensitive species at 0.001 ppm (1 ppb v/v, or 0.8 Mg HF m^3) or less for similar durations of exposure. Only peroxyacetylnitrate, a constituent of photochemical smog, can rival this extreme phytotoxicity."

Does fluoride have a role in acid rain? Weinstein, in this report in 1982, wrote, "Even less is known of effects of fluoride on soil structure and chemistry, micro- and macro flora, and on fluoride availability to the plant. Increased acidity in precipitation has heightened interest in these subjects" (p. 53). Also, "There are huge gaps in our knowledge with respect to effects on insects and other anthropods, soil microorganisms and aquatic flora and fauna" (p. 56). Weinstein's comments were made at an industry-funded conference of fluoride lawyers, government dentists, and former bomb-program scientists, held at Utah State University in 1982. L. H. Weinstein, "Effects of Fluorides on Plants and Plant Communities: An Overview," in J. L. Shupe, H. B. Peterson, and N. C. Leone, eds., *Fluorides: Effects on Vegetation, Animals and Humans* (Salt Lake City, UT: Paragon Press, 1983), p. 54. Attending this conference were Frank Seamans, Nicholas Leone, Harold Hodge, Frank Smith, David Scott (a former director of the National Institute of Dental Research), and B. D. Dinman, the vice president of health and safety for Alcoa. (Harold Hodge, Nicholas Leone, and fluoride lawyers

Frank Seamans and Keith Taylor organized that industry-funded conference, the book states.)

46. In 1966 Morris Katz, professor of atmospheric sanitation, explained at a Canadian National Conference on Pollution and Our Environment why atmospheric fluoride levels are measured in parts per billion, although maximum permissible levels for most atmospheric contaminants are calculated in parts per million. "Prolonged exposure to ambient air with concentrations of less than 1 part per thousand million part of air by volume may create a hazard. . . . In this respect fluorides are more than one-hundred times more toxic than sulfur dioxide." Elise Jerard and J. B. Patrick, "The Summing of Fluoride Exposures," *Intern. J. Environmental Studies*, vol. 4 (1973), pp. 141–155; citation from p. 143. Also see, cited in Jerard, a report in *Environmental Science and Technology* (August 1970) that states fluoride "compared to other pollutants is toxic at much lower concentration (0.5 ppb) and also acts as a cumulative poison. . . . Aside from the injury to vegetation there is a potential danger to animals and even human beings feeding on plants high in fluoride content."
47. According to historian Lynn Snyder, the U.S. military had designed the National Air Sampling Network. The network had, for example, measured protein in air as a marker for the presence of biological weapons. L. P. Snyder, *The Death-Dealing Smog*, p. 58, n. 50. According to Groth, fluoride had been one of the chemicals initially reported. After pressure from New York Congressman Richard L. Ottinger, national monitoring of fluoride pollution was reinitiated in 1968. See Groth, "Capitalist Plot? Air Is Fluoridated."
48. "Summary Review of Health Effects Associated with Hydrogen Fluoride and Related Compounds," U.S. Environmental Protection Agency, December 1988, pp. 2–9.
49. Globally that figure was an estimated 3.6 million tons in 1972. Ibid., Section 3, p. 2.
50. "Despite the fact that the further litigation which was anticipated with apprehension some years ago has failed to appear, the industries are vulnerable in the field of occupational disease hazard and in the field of community health relating to air pollution." Robert Kehoe to Reynolds's medical director, James MacMillan, September 20, 1961, cc: Frank Seamans, Box 63, RAK Collection.
51. See Taylor in *Fluorides: Effects on Vegetation, Animals, and Humans*, p. 359.
52. The six criteria pollutants were sulfur dioxide, carbon monoxide, hydrocarbons, nitrogen oxides, ozone, and particulate matter. (Lead was added in 1978.) Air pollutants were listed as hazardous by EPA according to whether those emissions, "can be expected to result in an increase in mortality or irreversible illness," according to an EPA official, D. F. Walters. In 1977 the U.S. Forestry Service asked EPA to fix a national ambient air quality standard (AAQS) for fluoride, to control fluoride damage in Montana. According to Walters, "EPA's reevaluation concluded that though there may be a number of local problems with fluoride damage to sensitive species around

industrial sources, the problem was not of a sufficiently national character to require a NAAQS. [National Ambient Air Quality Standard]." Also, for "permissive": "States may apply less stringent standards to sources when economic factors or physical limitations specific to those sources make less stringent standards significantly more reasonable," Walters added. D. F. Walters, "Regulatory, Economic, and Legal Aspects of Fluoride" in *Fluorides: Effects on Vegetation, Animals, and Humans*, pp. 351–358.

53. For D. F. Walters, see ibid. Perhaps a telling illustration of how fluoride has been "disappeared" or whitewashed as an air pollutant can be seen in the discussion surrounding the important study by Pope of the health improvement in local citizens following the temporary shuttering in the 1980s of the U.S. Steel mill in Provo, Utah. Although that plant was sued in the 1950s for some $30 million for fluoride pollution, by the time of the Pope study that history had so faded that there was little or no discussion of fluoride's role in the pollution-related health effects proved by the Pope study. C. A. Pope, "Respiratory Disease Associated with Community Air Pollution and a Steel Mill, Utah Valley," *Am. J. Public Health*, vol. 79 (May 1989), pp. 623–628.
54. The EPA concluded, "Fluoride pollutants were highly located in the vicinity of major point sources, in contrast to the other criteria pollutants which were more pervasive and widespread." See Walters, p. 351.
55. Reynolds had just concluded the Martin trial and was commissioning fresh studies at Kettering. But according to EPA official D. F. Walters, instead of instituting strict emission controls, it was not until the 1970s—a full decade after the plant was opened—that pressure from Canadian officials and lawsuits from farmers forced Reynolds to begin to install air-pollution control equipment. Ibid, Walters, p. 353.
56. J. Raloff, "The St. Regis Syndrome," *Science News*, vol. 118 (July 19, 1980), p. 42.
57. B. Carnow and S. A. Conibear, "Airborne Fluorides and Human Health, Report to the St. Regis Band on the Implications of Airborne Fluoride Contamination of Cornwall Island for the Health of its People," January 1978.
58. The transborder International Joint Commission, the U.S. Department of State, the Canadian Department of External Affairs, Canadian Department of the Environment, New York State, and the Ontario Ministry of the Environment were variously involved in addressing the dispute. See Walters, p. 353.
59. Curiously, in 1980 and 1981 the Mount Sinai School of Medicine and Selikoff received two awards totaling $446,975 from the National Institute of Dental Research to study "Long-term, low-level exposure to environmental agents (human)." See NIH CRISP awards, Project #5P30ES00928-08 and . . . 928-09.
60. "The increase noted in cardiovascular and respiratory morbidity/mortality rates in the older population (and females in particular) of the entire Band indicates a possible adverse effect from environmental exposure." Also: "The early infant mortality appears significant. Moreover, the higher number of hospital admissions . . . due to disease of the joints and connective tissue

could be related to fluorine effect." I. J. Selikoff, E. C. Hammond, and S. M. Levin, "Environmental Contaminants and the Health of the People of the St. Regis Reserve," *Fluoride: Medical Survey Findings* (Environmental Sciences Laboratory, Mount Sinai School of Medicine of the City University of New York), vol. 1, pp. 342–343.

61. "Should notable correlations between fluoride exposure and adverse health effects be found in Selikoff's epidemiological study of Cornwall Island residents, major changes in the way EPA looks at fluoride could result, including its reclassification as hazardous," *Science News*, vol. 118 (July 19, 1980), p. 43.

Chapter 16

1. The attorneys for the workers, Bruce McMath and Steve Napper of Little Rock, Arkansas, had signed up a hundred of these clients, known as the Beaty cases, for a claim against Reynolds to be mounted following the first Bareis trial, which is described in the following pages. The former group, which included Alan Williams and Jerry Jones (interviewed here), had been part of a team that developed a chemical process to dispose of the by-product waste of aluminum smelting. (The waste is called *treated spent potliner* and is described in the chapter.)
2. Author interview with EPA's Steve Silverman, June 18, 2002.
3. *Arkansas Business*, January 12, 1998, p. 23.
4. The old Reynolds Troutdale plant, which had injured the Martins, was designated a Superfund site, for example.
5. In the months after the Benton trial, Alan Williams would have open-heart and back surgery and lose most of his body hair.
6. Author interview, June 24, 2002.
7. A Reynolds memo as read in deposition states, "Alcoa expressed some concern that the actual soluble fluoride content in the kiln discharge is actually more than revealed—more than revealed by the TCLP. We are aware of this, but TCLP is the procedure used." Plaintiffs exhibit 173, in *George Bareis, et al.* vs. *Reynolds Metals*, Saline County Court, Case 97–703–2.
8. In December 1997 the EPA finally reversed course and reclassified the "treated" chemical waste as toxic. It was the first time the agency had taken back a delisting, said Peace. It was far too late, however, for Scotty and Dianne Peebles and the several hundred Hurricane Creek workers who had been breathing and handling the fluoride waste for years. And it was too late for the local environment, where thousands of tons of toxic waste had been buried in two mighty landfills. Eventually nearly 225,000 tons of treated potliner waste would be dumped in unlined pits at the Hurricane Creek site, according to the Associated Press, December, 2 1997.
9. Following the redesignation in December 1997 of the treated potliner as a hazardous material, new safety and disposal criteria were instituted.
10. The verdict, OSHRC Docket No. 98–0057, was voided on December 14, 2001 on jurisdiction grounds.

11. Kehoe to James MacMillan, medical director, Reynolds Metals, September 20, 1961, Box 63, RAK Collection.
12. Nevertheless, as Mullenix described the beagle study to the jury on October 20, 2000, McMath attempted to sneak in some history and context. Hadn't the Reynolds study been done in the 1950s, he asked Mullenix, "in connection with some litigation they had going at that time?" Johnson was ready. "Your Honor, objection," he exclaimed. "We ruled on this in chambers, didn't we?" McMath retreated. "I'll withdraw the question," he conceded.
13. In the end it seems that McMath's hunch about the jury was correct. Polled after the trial, a majority sided with Reynolds. The Benton claimants were simply looking for easy money, according to juror Marilyn Schick. "It was a situation where [workers] were exposed to a lot of dust, but as far as the ALROC [the name Reynolds had given to the treated spent potliner] being toxic to them, I just wasn't convinced that it was," she told me.

    But there were some jurors who did lean in favor of the workers against Reynolds. "It was a big company not caring about some low-class workers," said juror Sue Magness. "So what if it cost them some health problems—they had to get the job done." She blames the "excellent" Reynolds lawyers for portraying the plaintiffs as "sorry" drunks and drug addicts. "They weren't looking at them as people. They were looking at them as just bringing this lawsuit to get a buck. They didn't strike me that way," she added, about the workers. Magness had wanted a chance to talk with the other jurors and maybe influence them to rule in favor of the workers, she said. "I can be pretty persuasive. Sometimes people don't pick up on things, and when you bring it up in a jury room and they get to thinking about it, they change their minds," she said.

Chapter 17

1. "No deleterious systemic effects have occurred," he added. HR 2341 "A Bill to Protect the Public Health From the Dangers of Fluorination of Water," Hearings Before the Committee on Interstate and Foreign Commerce, House of Representatives, 83rd Congress, May 25–27, 1954, p. 470.
2. Philip R. N. Sutton, *Fluoridation: Errors and Omissions in Experimental Trials* (Melbourne: Melbourne University Press, 1959); "United Kingdom Mission Report (1953): The Fluoridation of Public Water Supplies in North America as a Means of Controlling Dental Caries" (London: Her Majesty's Stationary Office); World Health Organization (1958) Expert Committee on Water Fluoridation, First Report, Technical Report Series No. 146 (Geneva: World Health Organization); New Zealand Commission of Inquiry, "The Fluoridation of Public Water Supplies" (Wellington: Government Printer, 1957).
3. "I accept the whole of the evidence given by Professors Hodge [and others]," Justice Kenny wrote in his ruling verdict, which had the effect of imposing fluoridation on Ireland's entire population, a situation that remains to this day. M. Stanley, "Fluoridation of Public Water Supplies in Ireland," New

Jersey State Dental Soc., vol. 37 (1966), p. 242, cited in Frank McClure, *Water Fluoridation: The Search and the Victory* (NIDR, 1970), p. 275.

4. J. V. Kumar and P. A. Swango, *Community Dent. Oral Epidemiol.* vol. 27, no. 3 (June 1999), pp. 171–180, L. L. Lininger, G. S. Leske, E. L. Green, and V. B. Haley, "Changes in dental fluorosis and dental caries in Newburgh and Kingston, New York," *Am. J. Public Health*, vol. 88, no. 12 (December 1998), pp. 1866–1870.
5. *Boston Globe*, November 11, 1999; "Cincinnati's dental crisis," *Cincinnati Enquirer* October 6, 2002; *Washington Post*, March 5, 2002; and J. Kozol, *Savage Inequalities* (New York: HarperPerennial, 1991).
6. J. A. Lalumandier and R. G. Rozier, *Pediatric Dentistry* (January–February 1995), pp. 19–25, cited in *Medical Abstracts Newsletter*, July 1995, p. 28. Also, the University of York's fluoridation review found that up to 48 percent of children in fluoridated areas in the United Kingdom had some form of fluorosis. M. McDonagh, et al. "A Systemic Review of Public Water Fluoridation," NHS Center for Reviews and Dissemination, 2000, Executive Summary, p. 3.
7. M. Teotia, S. P. Teotia, and K. P. Singh, "Endemic chronic fluoride toxicity and dietary calcium deficiency interaction syndromes of metabolic bone disease and deformities in India: year 2000," *Indian J Pediatr.*, vol. 65, no. 3 (May–June 1998), pp. 371–381.
8. The Australian scientist Mark Diesendorf writes that "infants who are bottle fed with milk formula reconstituted with fluoridated water . . . receive 100 times the daily fluoride dose of breast-fed babies and at least 4–6 times that recommended by medial authorities for fluoride supplementation in unfluoridated areas." M. Diesendorf and A. Diesendorf, "Suppression by Medical Journals of a Warning About Overdosing Formula-Fed Infants with Fluoride," *Accountability in Research*, vol. 5 (1997), pp. 225–237. Also, the chicken in infant food can reach 8.38 micrograms per gram. J. R. Heilman et al., "Fluoride Concentrations in Infant Food," *JADA* (July 1997), p. 857. (Mechanically boned meat can include higher fluoride content. Fluoride concentrates in bone, therefore when some of that bone is found in the "boned" meat, the fluoride content can rise.)
9. The American fluoride researcher H. V. Smith, who codiscovered the fact that fluoride caused dental mottling, wrote, "Mottling, no matter how mild, is an external sign of internal distress," Letter from H. V. Smith to George Waldbott, June 1, 1964, cited in Waldbott, *A Struggle with Titans*, p. 65.
10. Christa Danielson, MD, Joseph L. Lyon, MD, et al., "Hip Fractures and Fluoridation in Utah's Elderly Population," *JAMA*, vol. 268, no. 6 (August 12, 1992), p. 746.
11. For hip fracture rate, see U.S. National Research Council, *Diet and Health* (Washington, DC: National Academy Press, 1989), p. 121. For arthritis data, see *Newsweek*, September 3, 2001, pp. 39–46.

12. See, for example, Y Li et al., "Effect of Long-Term Exposure to Fluoride in Drinking Water on Risks of Bone Fractures," *J. Bone and Mineral Research*, vol. 16 (2001), no. 5, pp. 932–939.
13. M. T. Alarcon-Herrera et al., "Well Water Fluoride, Dental Fluorosis, Bone Fractures in the Guadiana Valley of Mexico," *Fluoride*, vol. 34, no. 2 (2001), pp. 139–149.
14. One published account, quoting data from the U.S. National Center for Health Statistics, reported that bone fractures in male children and adolescents may be increasing. Joel Griffiths, "Fluoride: Commie Plot or Capitalist Ploy?" *Covert Action Information Bulletin*, no. 42 (fall 1992), p. 65.
15. *Fluoridation Facts* (published since 1956 by the American Dental Association). Paul R. Thomas, program officer at the Food and Nutrition Board of the National Academy of Sciences wrote in a March 18, 1991, letter to Darlene Sherrell, 'The statement you quote from the ADA pamphlet on water fluoridation—"The Academy found that the daily intake required to produce symptoms of chronic toxicity . . . is 20 to 80 milligrams or more . . . ' may be misleading." It was an easy lie to perpetuate, however. For example, even the "Recommended Daily Allowances for Fluoride" published in 1989 by the National Academy of Sciences, stated that "chronic toxicity . . . occurs after years of daily exposures of 20 to 80 mg of fluorine, far in excess of the average intake in the United States." That, too, was hugely disingenuous, conveying the impression that toxicity was found only at this elevated threshold.
16. H. C. Hodge, "The Safety of Fluoride Tablets or Drops," in *Continuing Evaluation of the Use of Fluorides*, eds. E. Johansen, D. R. Taves, and T. O. Olsen, AAAS Selected Symposium (Westview Press, 1979), p. 255.
17. *Review of Fluoride Benefits and Risks* (Public Health Service, Department of Health and Human Services, 1991), p. 45.
18. National Research Council, *Health Effects of Ingested Fluoride* (Washington, DC: National Academy Press, 1993), p. 59.
19. South Carolina was suing the EPA, objecting to the federal requirement to remove fluoride in water supplies that exceeded the threshold.
20. Presented in part as the David Murray-Cowie Memorial Lecture, University of Michigan, Ann Arbor, October 12, 1951. Published in full in S.Z. Levine, ed., *Advances in Pediatrics* (New York: Interscience Publishers, 1955), pp. 13–51.
21. Safe Drinking Water Committee, *Drinking Water and Health* (National Research Council, NAS, 1977), p. 389.
22. National Toxicology Program (NTP) (1990), *Toxicology and Carcinogenesis Studies of Sodium Fluoride in F344/N Rats and B6C3f1 Mice* (Technical report Series No. 393, NIH Publ. No 91-2848, National Institute of Environmental Health Studies, Research Triangle Park, NC).
23. W. Marcus, "Fluoride Conference to Review the NTP Draft Fluoride Report," Memorandum dated May 1, 1990, from Wm. L. Marcus, senior science adviser, Office of Drinking Water (ODW), U.S. EPA, to Alan B. Hais, acting direc-

tor, Criteria & Standards Division, ODW, U.S. EPA. See also: "Such a trend associated with the occurrence of a rare tumor in the tissue in which fluoride is known to accumulate cannot be causally dismissed," *Environmental Health Criteria*, no. 227 (WHO 2002), p. 169.

24. *The Lancet*, vol. 336, no. 8717 (September 22, 1990), U.S. Department of Labor, Case # 92-TSC-5, Recommended Decision and Order, p. 27.

25. There has been a great deal of information associating fluoride with cancer. Cancer has been experimentally linked to fluoride since the early 1950s, when Alfred Taylor at the University of Texas in Austin found that cancer-prone mice drinking water containing 1 ppm NaF, and eating food with a negligible fraction of fluoride, developed mammary tumors at an earlier age than similar mice fed nonfluoridated water. A. Taylor, "Sodium Fluoride in the Drinking Water of Mice," *Dental Digest*, vol. 60 (1954), pp. 170–172. Cited in Waldbott et al., *Fluoridation: The Great Dilemma* (Lawrence, KS: Coronado Press, 1978), p. 223.

For cancer in fluoride workers and around fluoride industrial plants, see A. J. deVilliers and J. P. Windish, "Lung Cancer in a Fluorspar Mining Community. Radiation, Dust, and Mortality Experience," *Br. J. Ind. Med.*, vol. 21 (1964), pp. 94–109; N. N. Litvinov, M. S. Goldberg, and S. N. Kimina, "Morbidity and Mortality in Man Caused by Pulmonary Cancer and Its Relation to the Pollution of the Atmosphere in the Areas of Aluminum Plants," *Acta Unio Int. Contra Cancrum*, vol. 19 (1963), pp. 742–645, V. A. Celilioni, "Lung Cancer in a Steel City [Hamilton, Ontario]: Its Possible Relation to Fluoride Emission," *Fluoride*, vol. 5 (1972), pp. 172–181, cited in Waldbott et al., *Fluoridation*, p, 236.

The late John Yiamouyiannis—a biochemist and antifluoride activist, and a retired National Cancer Institute biochemist, Dean Burke, reported more cancer in fluoridated communities in the United States. J. Yiamouyiannis and D. Burk, "Fluoridation and Cancer: Age-Dependence of Cancer Mortality Related to Artificial Fluoridation," *Fluoride*, vol. 10 (1977), pp. 102–123. And J. Yiamouyiannis, "Fluoridation and Cancer: The Biology and Epidemiology of Bone and Oral Cancer Related to Fluoridation," *Fluoride*, vol. 26 (1993), pp. 83–96.

For bone cancer and fluoridated water, see A. Takahashi, K. Akiniwa, and K. Narita, "Regression Analysis of Cancer Incidence Rates and Water Fluoride in the U.S.A. based on IACR/IARC (WHO) data (1978–1992)," *J. Epidemiol.*, vol. 11, no. 4 (July 2001), pp. 170–179, abstracted in *Fluoride*, vol. 34, no. 3 (May 2001). In this study the researchers found that "cancers of the oral cavity and pharynx, colon and rectum, hepato-biliary and urinary organs were positively associated with FD [fluoridation of drinking water]. This was also the case for bone cancers in males, in line with results of rat experiments."

In 1991 the National Cancer Institute found that the occurrence of osteosarcoma in young males was, in fact, significantly higher in fluoridated versus unfluoridated communities. However, the researchers concluded that the increased was *unrelated* to water fluoridation. According to the U.S. Public Health Service, "Although the increase in rates of osteosarcoma for males during this period was greater in fluoridated than nonfluoridated areas, extensive analyses revealed that these patterns were unrelated to either the introduction or duration of fluoridation." R. N. Hoover, S. Devesa, K. Cantor, and J. F. Fraumeni Jr., "Time Trends for Bone and Joint Cancers and Osteosarcomas in the Surveillance, Epidemiology and End Results" (SEER) Program, National Cancer Institute," in *Review of Fluoride: Benefits and Risks, Report of the Ad Hoc Committee on Fluoride of the Committee to Coordinate Environmental Health and Related Programs* (U.S. Public Health Service, 1991), pp. F 1–177. Despite those assurances, similar increases in bone cancer in young men were also found in New Jersey in a 1992 study. In that report, between the years 1970 and 1989 the rate of osteosarcoma (among ten- to nineteen-year-old males) was found to be 3.5 to 6.3 times greater in the fluoridated areas versus the unfluoridated ones. P. D. Cohn, *An Epidemiologic Report on Drinking Water and Fluoridation* (Trenton, NJ: New Jersey Department of Health, 1992). The latter two references are cited on the Fluoride Action Network webpage.

26. Interview with Paul Connett, May 1998. This taped interview can be obtained from GG Video, 82 Judson Street, Canton, NY 13617.

27. The researchers reported that fluoridated drinking water helped to carry aluminum to the brain in experimental rats, producing "irregular mincing steps characteristic of senile animals." Autopsies revealed brain damage. The data are "the latest of several studies hinting at some link between aluminum in the environment and Alzheimer's," according to the *Wall Street Journal*, October 28, 1992, section B, p. 6.

    Also, J. A. Varner, C. Huie, W. Horvath, K. F. Jensen, R. L. Issacson, "Chronic AlF_3 Administration: II. Selected Histological Observations," *Neuroscience Research Communications*, vol. 13, no. 2 (1993), pp. 99–104. R. L. Isaacson, J. A. Varner, and K. F. Jensen, "Toxin-Induced Blood Vessel Inclusions Caused by the Chronic Administration of Aluminum and Sodium Fluoride and Their Implications for Dementia," *Neuroprotective Agents. Annals of the New York Academy of Sciences*, no. 825 (1997), pp. 152–166, J. A. Varner, K. F. Jensen, W. Horvath, R. L. Isaacson, "Chronic Administration of Aluminum-Fluoride or Sodium-Fluoride to Rats in Drinking Water: Alterations in Neuronal and Cerebrovascular Integrity," *Brain Research*, no. 784 (1998), pp. 284–298.

28. *Fluoride, the Pineal Gland, and Melatonin: An Interview with and Presentation by Dr. Jennifer Luke.* Videotape, length: 40 minutes. Available from GGVideo, 82 Judson Street, Canton, NY. GGVideo [Grassroots and Global Video] (1999).

29. *The Newburgh Times*, January 27, 1954: "The 283 heart deaths in Newburgh in the year were equal to a rate of 882 deaths per 100,000 population. This was more than the rate for the nation as a whole, 507 per 100,000. It was also higher than the Middle Atlantic States, 590 heart deaths per 100,000."

30. For Michigan, see T. L. Hagen, M. Pasternack, and G. C. Scholz, "Waterborne Fluorides and Mortality," *Public Health Rep.*, vol. 69 (1954), pp. 450–454, cited in Waldbott et al., *Fluoridation*, p. 158; see also p. 160. For fluoride's effect on chicken embryo hearts, see also, J. D. Ebert, "The First Heartbeats," *Scientific American*, vol. 56 (1959), pp. 4–7: "At low concentrations [fluoride] primarily affects the heart. . . . At any given stage of development . . . the locations of the cells destroyed by fluoride coincide with the sites that have the greatest capacity to form heart muscle, and with the areas that have the greatest capacity for the synthesis of actin and myosin."
31. T. G. Reeves, *Water Fluoridation: A Manual for Engineers and Technicians* (U.S. Public Health Service, CDC Division of Oral Health, 1986) and *Water Fluoridation; A Manual for Water Plant Operators* (U.S. Public Health Service, CDC Division of Oral Health, April 1994), cited in M. Coplan and R. D. Masters, "Why Have U.S. Health Agencies Refused to Test Silicofluorides for Health Safety?" (unpublished, 2001), via authors.
32. For risk of cancer at the trace levels (up to 1.6 parts per billion) of arsenic found in water that is fluoridated by silicofluoride, see *Arsenic in Drinking Water: 2001 Update* (National Academies Press, 2001). See discussion on p. 7 in Summary, of linear nature of toxic effects at low doses. For lead, see R. D. Masters, M. J. Coplan, B. T. Hone, J. E. Dykes, "Association of Silicofluoride Treated Water with Elevated Blood Lead," *Neurotoxicology*, vol. 21, no. 6 (December 2000), pp. 1091–1100. See also Chairman, Subcommittee on Energy and Environment, Cong. Ken Calvert (R-CA), May 8, 2000, letter to Carol M. Browner, EPA administrator.
33. Almost no fluorspar is mined domestically. Of a reported 2001 consumption of 536,000 tons of fluorspar, 353,000 tons were imported from China. U.S. Geological Survey, *Minerals Yearbook* (2001).
34. Only a tiny fraction of the recovered silicofluoride waste is now converted for use as industrial fluoride—4,700 tons, for AlF_3, for aluminum smelting. U.S. Geological Survey, *Minerals Yearbook* (2001). (The fluosilicic acid recovered from the phosphate industry must first be converted into fluorspar, or aluminum fluoride, before being reused by industry.) But the potential of the Sunshine State as a source of industrial fluoride remains. In 2001 65,200 tons of fluosilicic acid were recovered from the phosphate industry. That's one-fifth of the nation's potential industrial fluoride needs, according the U.S. Geological Survey.
35. http://www.fluoride-journal.com/
36. University of Rochester, Progress Report for October, 1944—Abstracts, Dr. Harold Hodge, p. 478. "The results indicated that the inhibition of esterase activity produced by T [code for uranium] was small compared with that by C-216 [code for fluorine]. Thus 0.025 ppm C-216 [code for fluorine] caused the same percentage inhibition of esterase activity as 100 ppm T [code for uranium] (33 percent). From these results it is concluded that in a mixture of T and C-216 in which the amount (by weight) of T is not more than 50-fold that of C-216 the effect of the T upon the activity of liver esterase can

be neglected." Also: "The useful range of this curve for determining C-216 concentrations was from 0–0.5 ppm, C-216." Document #SO9FO1B227, ACHRE, RG 220.

37. Twenty-fifth ISFR Conference Abstracts, *Fluoride*, vol. 35, no. 4 (2002), p. 244.
38. *A Century of Public Health: From Fluoridation to Food Safety* (CDC, Division of Media Relations, April 2, 1999).

Epilogue

1. PFCs are "organic" chemicals, which means that they are based on carbon. In a PFC chemical, the fluorine atom is joined to the carbon molecule with a much stronger "covalent" bond, rather than the weak "ionic" bond in fluorides.
2. In September 2000 EPA officials met with a lobbying group known as the Fluoropolymer Manufacturers Group, composed of DuPont and Dow Chemical, plus the giant European and Japanese chemical manufacturers Elf Atofina and Asahi Glass Fluoropolymers. The industry representatives impressed upon the EPA the importance of PFOA chemicals in scores of vital commercial products, upon which industries worth an estimated $25 billion depended, from aerospace to automobiles to medical devices, according to records of that meeting. Despite "repeated attempts," industry declared, there had been "no success" in finding alternatives.
3. T. Midgley Jr. and A. L. Henne, *Ind. Eng. Chem.* vol. 22 (1930), p. 542. On December 31, 1928, General Motors' Frigidaire Division was issued the first patent for CFCs: US#1,886,339. A new company called Kinetic Chemicals, owned by DuPont and General Motors, was incorporated on August 1, 1930. By 1935, 8 million new refrigerators had been sold in the United States, filled with DuPont's patented "Freon" CFC gas. Global CFC production continued to soar; it increased from 150,000 tons in 1960 to 800,000 tons in 1974.
4. The secret PFC called "Joe's Stuff" that was delivered to Columbia University in December 1940 was named after Professor Joseph Simons from Penn State University. Simons invented a process known as "electro-chemical fluorination" which used electricity to replace the hydrogen with fluoride in hydrogen-carbon bonds, producing fluorocarbons. After the war the technology would be licensed to the 3M corporation, who would use it to make, among other things, the fabric protector Scotchgard. J. H. Simons, ed., *Fluorine Chemistry*, vol. 1 (New York: Academic Press, 1950), p. 423. T. Abe, "Electrochemical fluoridation as a locomotive for the development of fluorine chemistry at NIRIN, Nagoya," and John Colin Tatlow, "Fluorine Chemistry at the University of Birmingham: A Cradle of the Subject in the UK" in *Fascinated by Fluorine* (Amsterdam and New York: Elsevier, 2000), pp. 273 and 476. H. Goldwhite, *J. Fluorine Chem.*, vol. 33, p. 113. *Industrial and Engineering Chem.* vol. 39, no. 3 (March 1947), p. 292.
5. Colborn has since learned that some organofluorines are "really nasty" endocrine disrupters, she told me in an e-mail.

6. "It would be desirable," Col. Stafford Warren told Dr. John Foulger in a letter dated August 12, 1944, "to have the work on the toxicity of fluorocarbons being done in your laboratory parallel the investigations being made on similar compounds elsewhere. For that reason it would be appreciated if Dr. Harold Hodge of the University of Rochester could visit your laboratory in the near future and an exchange of ideas be effected. . . . The Medical Section has been charged with the responsibility of obtaining toxicological data which will insure the District's being in a favorable position in case litigation develops from exposure to the materials." Warren to Dr. John Foulger, Box 25, Accession #72C2386, Atlanta FRC, RG 326.
7. In a document titled "Research Plans for the Division of Pharmacology 1946–47," a subsection, "Industrial Hygiene," lists item "k" as "Investigation of the Nature of Fluoride in Blood." Fluoride exists in blood in "an organic and an inorganic state," while "organic fluorine compounds appear to be more toxic than the fluoride ion," the research summary noted. The Rochester team now planned "to investigate the nature of the compounds of fluorine existing in the blood, devoting special attention to the so-called organic fraction." Additional questions the bomb program researchers wanted answered were as follows:

    - An investigation of the possible relations between fluorides, iodide and calcium levels and the thyroid gland.
    - The effect of fluorine upon enzyme systems of the blood, particularly by means of an in vivo experiment.
    - The relation between fluorine and non-diffusable (protein bound) blood calcium.
    - How high can the blood fluoride level be raised before ill effects are raised in animals.

    The document concluded: "These experiments are intended to give fundamental information regarding the mode of action and metabolism of fluorine in the system. The information would appear to be of value for the following reasons. . . . Exposure to fluoride is of industrial significance, particularly since the advent of atomic energy programs," and that, "the determination of base levels is of immediate practical value in the impending litigation between the DuPont Co. and residents of New Jersey areas." DOE's HREX search engine, found at 0712317, document numbers 1075992, 1076012, 1076013. Where are the results of these experiments?
7. DuPont bulletin No. X-59a.
8. "Two types of reaction have been noted in humans as the result of accidental inhalation of the products of heated polymer. 1) a condition similar to metal fever; and 2) a condition in which there may be an irritation of the lungs leading to pulmonary edema." DuPont bulletin No. X-59a. Du Pont conducted human experiments giving volunteers Teflon-laced cigarettes to investigate fume fever. J. W. Clayton, "Fluorocarbon Toxicity and Biological Action," *Fluorine Chem. Reviews*, vol. 1, no. 2 (1967), pp. 197–252.

9. Harold D. Field to the Kettering Laboratory, January 23, 1958. Albert Henne to Robert Kehoe, October 15, 1958. "Teflon Coated Cooking Utensils," File 12, Box 15, RAK Collection. In the early 1930s Henne, a Belgian immigrant, had invented a manufacturing process for the first CFC Freon gas. He had also done fluoride work for the Manhattan Project.
10. *Nature*, vol. 217 (March 16, 1968), pp. 1050–1051.
11. "Little has been published about the metabolic handling and toxicology of perfluorinated fatty acid derivatives. Computer assisted literature searches using Medline, Toxline and Chemcon developed no information on these subjects." W. S. Guy, D. R. Taves, and W. S. Brey, "Organic Fluorocompounds in Human Plasma," *Biochemistry Involving Carbon-Fluorine Bonds* (American Chemical Society, 1976), p. 132.

    On the subject of collaboration, "3M got concerned apparently," Taves told me. "They would come check with me periodically—they wouldn't tell me what they were doing," he said, "but they wanted to know what I knew."
12. Taves's 1976 observation that "little has been published" on the toxicity of PFCs deserves scrutiny. During the cold war Taves was a leading arbiter of fluoride safety for the National Academy of Sciences. (Taves is listed on p. 396 of the 1977 document "Drinking Water and Health" by his initials as an author. This research was conducted by the National Research Council for the National Academy of Sciences and the EPA.) Donald Taves may also have buried evidence of fluoride's harm to humans on behalf of his Rochester colleagues, such as Harold Hodge, who worked for the nuclear program.

    In 1963 another colleague of Dr. Taves at Rochester, Dr. Christine Waterhouse, reported a case in which a patient at the Strong Memorial Hospital, a female nurse, "convulsed, aspirated and died suddenly" following kidney dialysis. Waterhouse and a team of scientists watched as the forty-one-year-old nurse suffered a collapse of her central nervous system. "A bizarre neuromuscular irritability characterized by a twitching of the right arm with occasional generalized convulsive seizures developed five days after the third dialysis," Waterhouse reported. Kidney dialysis can greatly concentrate the amount of fluoride in blood, scientists suspected. But the Waterhouse team never mentioned fluoride as a possible cause of the woman's symptoms or death. L. H. Kretchmar, W. M. Greene, C. W. Waterhouse, and W. L. Parry, "Repeated Hemodialysis in Chronic Uremia," *J. Am. Med. Assoc.*, vol. 184, no. 41 (1962), pp. 1037–1044.

    Two years later Dr. Donald Taves reported the same case in the medical literature. He discussed the high levels of fluoride found in the patient's bones and blood. He speculated as to a possible "beneficial" effect from the fluoride. *But Taves failed to report that the patient had died an hour after dialysis, that she had died in agony, and that the fatality had been reported by his Rochester colleague a year earlier.* (He claimed that he was unaware of Dr. Waterhouse's *JAMA* paper in which she reported the patient death. However, in the acknowledgments in his own work he thanked none other than his colleague, Dr. Christine Waterhouse.)

"Did they tell you how the patient had fared?" I asked Taves. "No, I don't think I ever heard," he said. "You were interested in fluoride and dialysis but you didn't follow up or ask what had happened to the patient?" I asked. "Right," Taves replied. (D. R. Taves, R. Terry, F. A. Smith, and D. E. Garnder, "Use of Fluoridated Water in Long-Term Hemodialysis," *Chronic Uremia., J. Am. Med. Assoc.*, vol. 184 [1963], pp. 1030–1031.) Both Rochester papers were funded by the U.S. Public Health Service. Neither mentioned the secret AEC kidney studies on human patients performed at Strong Memorial Hospital nor the government's interest in fluoride.

Did Taves censor his paper at the behest of Drs. Waterhouse and Hodge? In the 1960s Dr. Waterhouse was at the center of cold-war human experimentation, monitoring Harold Hodge's Rochester patients who had been given plutonium injections. (See Eileen Welsome, *The Plutonium Files* [New York: Dial Press, 1999].) "Waterhouse was uncomfortable with me publishing [the 1965 kidney paper]," Taves told me. "She didn't want me to do anything that sounded antifluoridation. Just like Hodge didn't. They were all biased that way. Hodge had gotten on the bandwagon of being in favor of fluoridation so his blinders were up," Taves added.

Similarly, the effects of fluoride on kidneys were another critical concern of the scientists overseeing health conditions inside the nuclear factories, and Rochester and Kettering researchers each performed multiple human experiments. Hodge's researchers performed secret human experiments in the 1940s at Rochester, giving fluoride to "patients having kidney diseases" to determine how much fluoride their damaged kidneys could excrete, according to declassified papers. Extra fluoride was stored in the bones of those injured patients, the government scientists found. *Quarterly Technical Report*, AEC No. UR-38, 1948. Also cited in Kettering Laboratory unpublished report, "Annual Report of Observations on Fluorides—October 25, 1954." Kettering did similar experiments on patients with damaged kidneys, according to the unpublished report.

13. Again, there is not a solitary reference to organofluorines in the book.
14. There may also be a link between accounts of birds dying, injured humans, and carpets impregnated with fluorochemicals, such as Scotchgard. In the early 1990s CNN and other media reported on families who claimed that they had been poisoned by newly installed carpets. One family told the BBC (in an interview conducted by the author) that their caged birds had died soon after the new carpet arrived. See also U.S. Court of Appeals for the Fourth Circuit No. 94-1882 Sandra Ruffin; Catherine Ruffin, by and through her Guardian Ad Litem, C. Timothy Williford, *Plaintiffs-Appellants,* vs. *Shaw Industries, Incorporated*; Sherwin-Williams Company, Decided: July 16, 1998. "With their motion for summary judgment, defendants submitted the affidavit of Larry D. Winter, an analytical chemist for Minnesota Mining and Manufacturing Company (3M). Mr. Winter specializes in the analysis of fluorochemicals such as those used in the manufacturing of 3M's Scotchguard carpets, the type involved in the present case." The case was dismissed.

15. *Scientific American*, March 1, 2001, pp. 16–17. Also, when 3M announced that the company was phasing out Scotchgard, the EPA praised 3M's openness in sharing data about the toxicity of PFCs. But Purdy is not so sure. As soon as the ecotoxicologist arrived at 3M in 1981, he says he grew concerned about the impact of PFCs on the environment, proposing new testing. "I could see that this could be a potential problematic class of chemicals, and so did everybody else in the ecological group," says Purdy. "We were very suspicious that we were seeing the tip of an iceberg. There was a proposal to do a lot of different testing—and it wasn't done."

    Former Michigan State scientist Kurunthachalam Kannan is not sure either about the 3M announcement in 2000 to phase out PFOS chemicals. "I work closely with 3M so I know what is really going on. But in terms of the words 'phase out,' when we try to talk to them, their people are not sure what it really means [laughs]. It is only a fraction of what they really manufacture in terms of organofluorines." Author interview, 2002.

    Also on 3M's internal studies, see the collection of documents in possession of the Environmental Working Group. In 1976, 3M company medical tests showed that some employees had levels of fluorocarbons in their blood as high as 30 parts per million. Although those exposure levels fell for a while, in 1984 blood contamination "remained constant or increased," according to 3M documents. That situation prompted concern about "employee health" and "corporate liability," according to the documents (thirteen tests showed values of over 10 ppm). Subsequently 3M workers showed abnormal liver function tests and "high kidney function tests," while other workers had lung abnormalities, described as "cases of pleural thickening." (Internal memo from 3M doctor Larry Zobel to D. W. Dworak dated March 20, 1987, entitled "Medical Examinations.") Also, in the late 1970s, 3M ran toxicity tests for the fluorocarbon PFOS on rhesus monkeys. All the animals died. (J. Morris, "Did 3M and DuPont Ignore Evidence of Health Risks?" *Mother Jones*, September–October 2001, online edition.)
16. *Scientific American*, March 1, 2001, pp. 16–17.
17. "3M's Big Clean Up," *Business Week*, June 5, 2000 via online edition.
18. "3M's Big Clean Up," *Business Week*, June 5, 2000; *Scientific American*, March 1, 2001, pp. 16–17.
19. Kannan et al., "Perfluorooctane Sulphonate in Fish Eating Water Birds Including Bald Eagles and Albatrosses," *Environmental Science and Technology*, vol. 35, pp. 3065–3070.
20. *Scientific American*, March 1, 2001, pp. 16–17.
21. http://www.ewg.org/issues/pfcs/
22. It is not the first time DuPont chemicals have been linked to eye defects in children. In the early 1990s a DuPont fungicide marketed as Benlate was discovered to contain a fluorine chemical called flusalizole, which was not licensed for use in the United States. Benlate provided one of the most disastrous and expensive episodes in U.S. corporate history. Some of the lawsuits blamed Benlate for causing children to be born without eyes.

DuPont has since paid $1.3 billion in costs and settlements with farmers who used Benlate and whose crops were damaged. In July 2003 the Florida Supreme Court also reinstated a $4 million jury award to the family of a boy born without eyes, in what the Associated Press described as "a birth defect linked to the agricultural pesticide Benlate." (Associated Press, July 3, 2002.) And although another judge threw out a ruling that DuPont had engaged in "racketeering," by allegedly concealing evidence in the Benlate saga, a similar case in Atlanta was settled when DuPont agreed to pay $2.5 million dollars to each of Georgia's four law schools.

Judge Hugh Lawson explained that settlement made a statement about the importance of legal ethics, according to the *New York Times*, January 2, 1999, section A, p. 12. How much was learned about legal ethics is not clear. DuPont was also accused of destroying evidence in the West Virginia PFC litigation. "In April 2003 a Judge in West Virginia found that in 2002, Du Pont had destroyed evidence relevant to ongoing litigation on PFOA brought by 3000 citizens of West Virginia and Ohio." Press Release, Environmental Working Group, June 6, 2003.

The billion-dollar DuPont/Benlate debacle may be an example of one of fluoride's best-known chemical properties gone tragically awry. As early as 1949 the Atomic Energy Commission reported that fluoride had a synergistic ability to boost the toxicity of beryllium. When fluoride was added, twice as many rats were killed, according to experiments performed at the University of Rochester. (H. Stokinger et al., "The Enhancing Effect of the Inhalation of Hydrogen Fluoride Vapor on Beryllium Sulfate Poisoning in Animals," UR-68, University of Rochester, unclassified.) Similarly, during World War II, Hitler's chemists discovered that fluoride could dramatically boost the toxicity of nerve gases. Sarin—the same gas used by Saddam Hussein on the Kurds of Halabja and used in the deadly subway attack in Tokyo—is a fluorinated chemical, named after the German scientists who invented it. (*Fascinated by Fluorine*, p. 515). Today drug companies know that adding even a single fluorine atom to a drug molecule can boost chemical potency. Numerous modern drugs now contain small amounts of fluoride, including the antidepressant Prozac and the powerful antianthrax antibiotic Cipro. "Just one fluorine placed at a strategic site in an organic molecule can hot up its activity," says the English scientist Eric Banks. "The opportunities for finding something useful for society are truly mind blowing." Unfortunately, adding fluorine to drugs may also make them quite literally "mind blowing." Cipro, for example has numerous reported side effects, including central-nervous-system problems such as acute anxiety. And recently several fluorine-containing drugs have been withdrawn because of their side effects, including:

- Baycol, a cholesterol-lowering drug taken by 700,000 Americans, and linked to 31 deaths in the United States, with at least nine other fatalities worldwide;

- Cisapride ("Propulsid"), withdrawn in 2000 because it caused severe cardiac side effects;
- Mibefradil ("Posicor"), withdrawn in 1998 after it was shown that in patients with congestive heart failure the drug produced a trend to higher mortality;
- Flosequinan, withdrawn in 1993 after it was shown that the beneficial effects on the symptoms of heart failure did not last beyond the first three months of therapy. After the first three months of therapy, patients on the drug had a higher rate of hospitalization than patients taking a placebo;
- Astemizole (allergy drug), withdrawn in 1999 because it also became associated with life-threatening cardiac adverse events;
- The "weight loss" drugs fenfluramine and dexfenfluramine, withdrawn in 1997 because of serious adverse cardiac health effects, generating almost a billion dollars in lawsuits;
- Tolrestat (antidiabetic), withdrawn in 1997 after the appearance of severe liver toxicity and deaths;
- Temafloxacin ("Omniflox"), withdrawn in 1992. The antibiotic had caused deaths and liver dysfunction;
- Grepafloxacin, removed from the market in 1999 because of serious cardiac events.

(List courtesy of Andreas Schuld and Wendy Small, Parents of Fluoride Poisoned Children [PFPC], Vancouver, BC, Canada.)

Fluoride's potential role in drug toxicity has not been well studied. An expert on the withdrawn diet drug dexfenfluramine, Dr. Kenneth Weir at the University of Minneapolis, said that he had no information on whether fluoride played a role in that drug's toxic action on the human heart. Central-nervous-system problems, such as depression, were also reported among the drug's unwanted effects. "It seems an intriguing question," notes Dr. Weir, "if you broke it down into its constituent parts, whether they would have a toxic effect." A mighty paradox exists. Just as fluoride performs some of the heaviest lifting in modern industry—but gets a glancing scrutiny from regulators and health officials—it is also routinely added to drugs to boost their chemical effect but mostly overlooked for its potential role in toxicity. Dr. Phyllis Mullenix points her finger at the not-too-distant past. She believes the sweeping cold-war-era assurances on fluoride safety from such scientists as Robert Kehoe and Harold Hodge have left a "black hole" in our understanding of fluoride's biological effects, and a failure by regulators to consider the toxicity of fluoride compounds. "Any drug that has a fluoride component should be automatically red-flagged," Mullenix says. "It simply is not done."

23. "PFOS caused postnatal deaths (and other developmental effects) in offspring in a two-generation reproductive-effects rat study," EPA official Charles Auer noted in a May 16, 2000, e-mail, referring to the PFC used in Scotchgard, "At higher doses in this study," the summary continued, "*all* progeny in the first

generation died while [at the lower level] many of the progeny from the *second* generation died. It is very unusual to see such second generation effects" (emphasis in the original). The e-mail concluded, "PFOS accumulates to a high degree in humans and animals. It has an estimated half-life of 4 years in humans. It thus appears to combine Persistence, Bioaccumulation, and Toxicity to an extraordinary degree. . . . EPA's preliminary risk assessment indicated potentially unacceptable margins of exposure (MOE's) for workers and possibly the general population."

DuPont has concerns about PFC toxicity, too. In the 1990s, for example, the company worried about the cancer risk from PFCs. "We may have a product stewardship issue if we have a [Teflon] finish that contains a suspect carcinogen," a 1994 Dupont document noted. "The worst-case scenario is that [PFOA] could be classified as a large 'C' carcinogen," a 1996 company memo added. *Mother Jones*, September-October 2001, online edition.

That "scenario" may be scientific reality. Working on a grant from the U.S. Air Force, Michigan State's Brad Upham collected evidence that the PFOS and PFOA fluorocarbons disrupt intercell communication, allowing potentially tumor-producing cells to multiply. "We have very good reasons to think that they could contribute to cancer," the scientist told me. (Author interview).

24. Richard Hefter, chief, High Production Volume Chemicals Branch, USEPA, to A. Michael Kaplan, director, Regulatory Affairs and Occupational Health, DuPont Haskell Laboratory, May 22, 2003. Andrea V. Malinowski to Richard Hefter, chief, High Production Volume Chemicals Branch, USEPA, June 20, 2003. Ken Cook, president, EWG, to EPA Administrator Christine Todd Whitman, April 11, 2003.

25. DuPont worries about a public-relations catastrophe and has shied from media attention regarding its blood-seeking fluorochemicals. When farmers Wilbur and Sandra Tennant of Parkersburg, West Virginia blamed PFC pollution from the DuPont factory for killing their cattle and harming their health, DuPont asked U.S. District Judge Joseph Goodwin to prevent the Tennants from testifying at an EPA hearing in March 2000, according to court documents cited by investigative reporter Jim Morris at *Mother Jones* magazine. (J. Morris, "Did 3M and DuPont Ignore Evidence of Health Risks?" *Mother Jones*, September-October 2001, online edition.)

DuPont's attorney, John Tinney, blamed Hollywood for the company's woes and for the necessity of a restraining order against the farmers. "The court need look no further than the movies for practical application," the lawyer told Judge Goodwin, citing "the enormous success at the box office of *Erin Brockovich* and *A Civil Action*." The company, however, need not have worried. Although no restraining order was issued, media attention was limited, according to *Mother Jones*.

DuPont also claims that there is no risk to Teflon workers. The company's recent employee monitoring has found no elevation of PFOA-class chemicals in employees directly involved in production, according to comments by spokesperson Dave Korzeniowski in the journal *Environmental Science and Technology*. DuPont seems reassured by that data. It was 3M's discovery of high PFOS levels in its employees, for example, that helped to lead to the promised phase-out of Scotchgard. "PFOS appears to behave differently from our products," Korzeniowski states. (R. Renner, *Environmental Science and Technology*, vol. 35, no. 7 [April 1, 2001], pp. 154A–160A.)

26. Cited in letter from Kenneth Cook, president of Environmental Working Group to Mr. Richard H. Hefter, chief of High Production Volume Chemicals Branch, United States EPA, August 15, 2003. At web location www.ewg.org/issues/pfcs/20030813/.

    The company also told workers that "a female who has an organic fluorine level above background level should consult with her personal physician prior to contemplating pregnancy." Washington Works Proposed Communication to Females Who Had Worked in Fluoropolymers Area, embedded as link in above letter. Cook to EPA, August 15, 2003.

27. Xiang et al., "Effect of Fluoride in Drinking Water on Children's Intelligence," *Fluoride*, May 2003, J. A. Varner, K. F. Jensen, W. Horvath, and R. L. Isaacson, "Chronic Administration of Aluminum-Fluoride or Sodium-Fluoride to Rats in Drinking Water: Alterations in Neuronal and Cerebrovascular Integrity," *Brain Research*, vol. 784 (1998), pp. 284–298.
28. *Sunday Telegraph*, November 24, 1996.
29. L. Trupin et al., "The Occupational Burden of Chronic Obstructive Pulmonary Disease," *European Respiratory Journal*, vol. 22, no. 3 (September 1, 2003), pp. 462–469.
30. March 18, 2002, comments submitted to the EPA, on DowAgroSciences petition to establish fluoride and sulfuryl fluoride tolerances for a large number (40) of raw and processed foods. *Federal Register*, February 15, 2002, U.S. EPA Docket control number PF-1068, submitted by Paul Connett, professor of Chemistry, St. Lawrence U., Canton, NY, and Ellen Connett, editor, *Waste Not*, Canton, NY.

Index